METHODS IN MOLECULAR BIOLOGY

For further volumes:
http://www.springer.com/series/7651

For over 35 years, biological scientists have come to rely on the research protocols and methodologies in the critically acclaimed *Methods in Molecular Biology* series. The series was the first to introduce the step-by-step protocols approach that has become the standard in all biomedical protocol publishing. Each protocol is provided in readily-reproducible step-by step fashion, opening with an introductory overview, a list of the materials and reagents needed to complete the experiment, and followed by a detailed procedure that is supported with a helpful notes section offering tips and tricks of the trade as well as troubleshooting advice. These hallmark features were introduced by series editor Dr. John Walker and constitute the key ingredient in each and every volume of the *Methods in Molecular Biology* series. Tested and trusted, comprehensive and reliable, all protocols from the series are indexed in PubMed.

Optogenetics

Methods and Protocols

Edited by

Armin Baumschlager

Bern, Switzerland

Humana Press

Editor
Armin Baumschlager
Bern, Switzerland

ISSN 1064-3745 ISSN 1940-6029 (electronic)
Methods in Molecular Biology
ISBN 978-1-0716-4046-3 ISBN 978-1-0716-4047-0 (eBook)
https://doi.org/10.1007/978-1-0716-4047-0

This Humana imprint is published by the registered company Springer Science+Business Media, LLC, part of Springer Nature.
The registered company address is: 1 New York Plaza, New York, NY 10004, U.S.A.

Preface

Light: A Source of Life and Human Enlightenment

Light has been engrained in the evolutionary journey of all life domains: Light is an energy source in photosynthetic processes, light is used as an information carrier in the context of vision, light synchronizes life's rhythm via circadian clocks, and cells sense and respond to light to protect themselves against the damaging effects caused by radiation… to highlight a few.

Controlling Life with Light

Since the dawn of biological research light has been essential for observation and measurement. In recent biology, light has also become a bioengineering tool to steer and control biological functions.

This up-and-coming field of *optogenetics* uses cellular light-sensing components and employs genetic engineering for bioengineering purposes to control proteins and biological processes. What makes optogenetics an attractive research tool is its high spatio-temporal precision and its non-invasiveness. Because of the ability for precise, remote and dynamic control, optogenetics enables novel applications and insights and has reached most fields of biological research and engineering.

This book shows highly interesting examples and applications for the control of biological cell functions. Rich literature covers neuronal optogenetics due to the great impact of light-gated ion channels in this field, thus the focus of this book is set on other optogenetic methods. In general, as such a book can only highlight a few examples in the field, this edition is structured to aid researchers in two main ways: first, help with the understanding and application of the concepts of optogenetics and some of the underlying bioengineering principles with concrete examples (Parts 1, 2, and 3); second, aid in establishing the required technical light-illumination setups for administering light inputs and analysis of experimental outcomes (Part 4). Naturally, as chapters describe complete specific methods, their content can be associated with multiple parts. Hopefully, this book will assist researchers in establishing optogenetics for their research questions.

Part 1 focuses on *intracellular optogenetic components for the control and investigation of specific cell functions*. The first four chapters describe methods for *light-regulated gene expression*. Due to the wide applicability of gene expression, multiple approaches have been developed. The following two chapters explain the use of optogenetic regulators for the activation of three major types of regulated cell death as well as YAP, a central regulator of the Hippo-YAP signaling axis that modulates organ growth and regeneration. Chapter 7 highlights the advantages and use of transient light-controlled target binding for in vivo imaging, specifically actin binding and imaging. The part closes with two methods that describe the optogenetic release of (therapeutic) proteins either via secretion and cell injection or via release from engineered living materials.

Part 2 focuses on *externally supplied light regulators*, which do not require genetic manipulation of target cells. These function either through photosensitive chemicals in Chapter 10 or with externally supplied optogenetic proteins in Chapters 11 and 12.

Part 3 highlights methods for *optogenetic control of organelles*. This includes light-controlled association and morphology control of cell organelles, as well as light-induced formation of synthetic organelles.

Part 4 introduces *technical tools* required for light induction in optogenetic experiments, as well as a method for performing and analyzing optogenetic cell-cell adhesion.

Bern, Switzerland ***Armin Baumschlager***

Contents

Contributors

ANJA ARMBRUSTER • *INM—Leibniz Institute for New Materials, Saarbrücken, Germany; Signalling Research Centres BIOSS and CIBSS, University of Freiburg, Freiburg, Germany; Faculty of Biology, University of Freiburg, Freiburg, Germany*

PETR BROZ • *Department of Immunobiology, University of Lausanne, Lausanne, Switzerland*

SEBASTIAN M. CASTILLO-HAIR • *Department of Bioengineering, Rice University, Houston, TX, USA; Department of Electrical and Computer Engineering, University of Washington, Seattle, WA, USA*

JIAJIE DIAO • *Department of Cancer Biology, University of Cincinnati College of Medicine, Cincinnati, OH, USA*

ANDREAS DIEPOLD • *Department of Ecophysiology, Max Planck Institute for Terrestrial Microbiology, Marburg, Germany; Center for Synthetic Microbiology (SYNMIKRO), Marburg, Germany*

THOMAS DREPPER • *Institute of Molecular Enzyme Technology, Heinrich Heine University Düsseldorf, Forschungszentrum Jülich GmbH, Jülich, Germany*

LITING DUAN • *Department of Biomedical Engineering, The Chinese University of Hong Kong, Hong Kong, China*

CHRISTINA G. GANGEMI • *Flinders Health and Medical Research Institute (FHMRI), College of Medicine and Public Health, Flinders University, Bedford Park, SA, Australia; Australian Regenerative Medicine Institute (ARMI), Faculty of Medicine, Nursing and Health Sciences, Monash University, Clayton, VIC, Australia; European Molecular Biology Laboratory Australia (EMBL Australia), Monash University, Clayton, VIC, Australia*

ELLIOT J. GERRARD • *Monash Institute of Pharmaceutical Sciences, Monash University, Parkville, VIC, Australia; Australian Regenerative Medicine Institute (ARMI), Faculty of Medicine, Nursing and Health Sciences, Monash University, Clayton, VIC, Australia; European Molecular Biology Laboratory Australia (EMBL Australia), Monash University, Clayton, VIC, Australia*

SEBASTIAN GROSSMANN • *Department of Ecophysiology, Max Planck Institute for Terrestrial Microbiology, Marburg, Germany*

DANIEL J. HALLER • *Systems, Synthetic, and Physical Biology Program, Rice University, Houston, TX, USA*

ZACHARY P. HARMER • *Department of Biomedical Engineering, University of Wisconsin-Madison, Madison, WI, USA*

FABIAN HOGENKAMP • *Institute of Bioorganic Chemistry, Heinrich Heine University Düsseldorf, Forschungszentrum Jülich GmbH, Jülich, Germany*

MAXIMILIAN HÖRNER • *Signalling Research Centres BIOSS and CIBSS, University of Freiburg, Freiburg, Germany; Faculty of Biology, University of Freiburg, Freiburg, Germany; Prolific Machines, CAEmeryville, USA*

PEIYUAN HUANG • *Department of Biomedical Engineering, The Chinese University of Hong Kong, Hong Kong, China*

WEIQI HUANG • *College of Computer and Information Sciences, Fujian Agriculture and Forestry University, Fuzhou, Fujian, China*

KARL-ERICH JAEGER • *Institute of Molecular Enzyme Technology, Heinrich Heine University Düsseldorf, Forschungszentrum Jülich GmbH, Jülich, Germany; Institute of Bio- and Geosciences IBG-1: Biotechnology, Forschungszentrum Jülich GmbH, Jülich, Germany*

MORGANE JAEGER • *Aix Marseille Université, Centre National de la Recherche Scientifique, Institut National de la Santé et de la Recherche Médicale, CIML, Turing Center for Living Systems, Marseille, France*

HARALD JANOVJAK • *Australian Regenerative Medicine Institute (ARMI), Faculty of Medicine, Nursing and Health Sciences, Monash University, Clayton, VIC, Australia; European Molecular Biology Laboratory Australia (EMBL Australia), Monash University, Clayton, VIC, Australia; Flinders Health and Medical Research Institute, Flinders University, Bedford Park, SA, Australia*

HARALD JANOVJAK • *Flinders Health and Medical Research Institute (FHMRI), College of Medicine and Public Health, Flinders University, Bedford Park, SA, Australia; Australian Regenerative Medicine Institute (ARMI), Faculty of Medicine, Nursing and Health Sciences, Monash University, Clayton, VIC, Australia; European Molecular Biology Laboratory Australia (EMBL Australia), Monash University, Clayton, VIC, Australia*

FABIENNE KNAPP • *Institute of Molecular Enzyme Technology, Heinrich Heine University Düsseldorf, Forschungszentrum Jülich GmbH, Jülich, Germany*

KOURTNEY L. KROLL • *Department of Biochemistry and Molecular Biology & The Institute for Biophysical Dynamics, The University of Chicago, Chicago, IL, USA*

YUTO KUWASAKI • *Graduate School of Arts and Sciences, The University of Tokyo, Tokyo, Japan*

RÉMI LASSERRE • *Aix Marseille Université, Centre National de la Recherche Scientifique, Institut National de la Santé et de la Recherche Médicale, Institut Paoli-Calmettes, CRCM, Turing Center for Living Systems, Marseille, France*

MANJIA LI • *Department of Chemical and Biological Engineering, The Hong Kong University of Science and Technology, Hong Kong, China*

ZHAOXIA LI • *College of Marine and Biological Engineering, Yancheng Institute of Technology, Yancheng, Jiangsu, China*

FLORIAN LINDNER • *Department of Ecophysiology, Max Planck Institute for Terrestrial Microbiology, Marburg, Germany*

GÜNTER MAYER • *LIMES, University of Bonn, Bonn, Germany; Center of Aptamer Research & Development, University of Bonn, Bonn, Germany*

MEGAN N. MCCLEAN • *Department of Biomedical Engineering, University of Wisconsin-Madison, Madison, WI, USA; University of Wisconsin Carbone Cancer Center, University of Wisconsin School of Medicine and Public Health, Madison, WI, USA*

TAKAHIRO NAKAJIMA • *Kanagawa Institute of Industrial Science and Technology, Kawasaki-shi, Kanagawa, Japan; Graduate School of Arts and Sciences, The University of Tokyo, Tokyo, Japan*

TAKAHIRO OTABE • *Kanagawa Institute of Industrial Science and Technology, Kawasaki-shi, Kanagawa, Japan; Graduate School of Arts and Sciences, The University of Tokyo, Tokyo, Japan*

SEUNG-HYUN PAIK • *Institute of Molecular Enzyme Technology, Heinrich Heine University Düsseldorf, Forschungszentrum Jülich GmbH, Jülich, Germany*

BYUNG MIN PARK • *Department of Chemical and Biological Engineering, The Hong Kong University of Science and Technology, Hong Kong, China; Department of Imaging Physics, Faculty of Applied Sciences, Delft University of Technology, Delft, the Netherlands*

Jörg Pietruszka • *Institute of Bioorganic Chemistry, Heinrich Heine University Düsseldorf, Forschungszentrum Jülich GmbH, Jülich, Germany; Institute of Bio- and Geosciences IBG-1: Biotechnology, Forschungszentrum Jülich GmbH, Jülich, Germany*

Kangqiang Qiu • *Department of Cancer Biology, University of Cincinnati College of Medicine, Cincinnati, OH, USA*

Christopher A. Raab • *University of Münster Institute of Physiological Chemistry and Pathobiochemistry, Münster, Germany*

Christian Renzl • *LIMES, University of Bonn, Bonn, Germany*

Ronald S. Rock • *Department of Biochemistry and Molecular Biology & The Institute for Biophysical Dynamics, The University of Chicago, Chicago, IL, USA*

Shrikrishnan Sankaran • *INM—Leibniz Institute for New Materials, Saarbrücken, Germany*

Moritoshi Sato • *Kanagawa Institute of Industrial Science and Technology, Kawasaki-shi, Kanagawa, Japan; Graduate School of Arts and Sciences, The University of Tokyo, Tokyo, Japan*

Kateryna Shkarina • *Institute of Innate Immunity, University Hospital Bonn, Bonn, Germany*

Yutong Song • *Department of Biomedical Engineering, The Chinese University of Hong Kong, Hong Kong, China*

Tobin R. Sosnick • *Department of Biochemistry and Molecular Biology & The Institute for Biophysical Dynamics, The University of Chicago, Chicago, IL, USA*

Fei Sun • *Department of Chemical and Biological Engineering, The Hong Kong University of Science and Technology, Hong Kong, China; Greater Bay Biomedical InnoCenter, Shenzhen Bay Laboratory, Shenzhen, China; Research Institute of Tsinghua, Pearl River Delta, Guangzhou, China*

Jeffrey J. Tabor • *Systems, Synthetic, and Physical Biology Program, Rice University, Houston, TX, USA; Department of Bioengineering, Rice University, Houston, TX, USA; Department of Biosciences, Rice University, Houston, TX, USA; Rice Synthetic Biology Institute, Rice University, Houston, TX, USA*

Varun Sai Tadimarri • *INM—Leibniz Institute for New Materials, Saarbrücken, Germany*

Alexandra-Madelaine Tichy • *Australian Regenerative Medicine Institute (ARMI), Faculty of Medicine, Nursing and Health Sciences, Monash University, Clayton, VIC, Australia; European Molecular Biology Laboratory Australia (EMBL Australia), Monash University, Clayton, VIC, Australia*

Pearlyn Jia Ying Toh • *Mechanobiology Institute, National University of Singapore, Singapore, Singapore; Novo Nordisk Foundation Center for Basic Metabolic Research, Faculty of Health and Medical Sciences, University of Copenhagen, Copenhagen, Denmark*

Renaud Vincentelli • *Aix Marseille Université, Centre National de la Recherche Scientifique, AFMB, Marseille, France*

Wilfried Weber • *INM—Leibniz Institute for New Materials, Saarbrücken, Germany; Signalling Research Centres BIOSS and CIBSS, University of Freiburg, Freiburg, Germany; Department of Materials Science and Engineering, Saarland University, Saarbrücken, Germany*

Seraphine V. Wegner • *University of Münster Institute of Physiological Chemistry and Pathobiochemistry, Münster, Germany*

XIUQIONG XU • *Department of Cancer Biology, University of Cincinnati College of Medicine, Cincinnati, OH, USA*
SHOTA YAMAMOTO • *Graduate School of Arts and Sciences, The University of Tokyo, Tokyo, Japan*
KAI ZHANG • *Department of Biochemistry, School of Molecular and Cellular Biology, University of Illinois at Urbana-Champaign, Urbana, IL, USA*

Chapter 1

Optogenetic Control of *B. subtilis* Gene Expression Using the CcaSR System

Daniel J. Haller, Sebastian M. Castillo-Hair, and Jeffrey J. Tabor

Abstract

Optogenetics enables precise control of gene expression in a variety of organisms. We recently developed the first system for optogenetic control of transcription in *Bacillus subtilis*. This system is based on CcaSR, a light-responsive two-component regulatory system originally derived from *Synechocystis* PCC 6803. The so-called *B. subtilis* CcaSR v1.0 enables activation of gene expression with green light and deactivation with red. As a result, *B. subtilis* CcaSR v1.0 can be used to program gene expression with high quantitative, spatial, and temporal resolution. The expression levels of the CcaS light sensing histidine kinase and the CcaR response regulator are set by the addition of chemical inducers in *B. subtilis* CcaSR v1.0, enabling adjustment of the basal expression level and optimization of the magnitude of gene expression induction. In principle, *B. subtilis* CcaSR v1.0 should be compatible with expression of any target gene of interest. Here, we provide growth, strain engineering, and light treatment protocols for working with *B. subtilis* CcaSR.

Key words Optogenetics, *Bacillus subtilis*, Two-component system, CcaSR, Synthetic biology

1 Introduction

CcaSR is a light-sensing two-component system (TCS) that originally evolved in the Gram-negative cyanobacterium *Synechocystis* PCC 6803 [1, 2]. CcaSR consists of CcaS, a sensor histidine kinase belonging to the cyanobacteriochrome family, and CcaR, an OmpR/PhoB-family response regulator. In order to sense light, CcaS requires phycocyanobilin (PCB), a linear tetrapyrrole that binds covalently to the GAF domain in CcaS to form a holo-CcaS complex. PCB is synthesized in two metabolic steps from cellular heme pools. Synthesis of PCB requires heme oxygenase 1 (encoded by *ho1*) and phycocyanobilin:ferredoxin oxidoreductase (encoded by *pcyA*) [3]. CcaS is produced in the ground state with low autokinase activity, and upon absorbing a green light photon (λ_{max}= 535 nm), it transitions to an active state with high autokinase activity. Active CcaS phosphorylates itself and then transfers

Armin Baumschlager (ed.), *Optogenetics: Methods and Protocols*, Methods in Molecular Biology, vol. 2840, https://doi.org/10.1007/978-1-0716-4047-0_1,

this phosphoryl group to CcaR. Phosphorylated (active) CcaR then binds to and induces transcription from the P_{cpcG2} promoter. Under red light, CcaS (λ_{max}= 672 nm) is thought to dephosphorylate CcaR, leading to reduced transcription [2].

The ability of the CcaSR system to activate transcription in response to green light and deactivate it in response to red light has made it an attractive target for engineering optogenetic control of gene expression in bacteria. Previously, we ported the system to *Escherichia coli* and improved its performance by optimizing the expression of each of the four component proteins, truncating the output promoter to decrease leakiness and increase dynamic range, and miniaturizing CcaS to further improve dynamic range [4–7].

Numerous important physiological pathways such as sporulation, biofilm formation, and competence are regulated by stochastic or time-varying patterns of gene regulator activity in *B. subtilis*. Such processes could be rigorously investigated using artificial, light-driven patterns of gene expression in batch culture or single cells. Thus, optogenetic tools could substantially advance understanding of *B. subtilis* biology.

To control gene expression with light in *B. subtilis*, we ported the so-called CcaSR v3.0 system from *E. coli* into *B. subtilis* and reoptimized its function for this new bacterium [8]. These optimizations included codon optimization of the first 15 codons of *ho1*, *sfgfp*, and *CcaS* as well as the entire sequence of *pcyA* for expression in *B. subtilis*, engineering a translational fusion of Ho1 and PcyA to increase PCB production, and fusing $P_{cpcG2\text{–}172}$ to the strong constitutive *B. subtilis* promoter P_{veg} (resulting in $P_{cpcG2\text{-}veg}$) to increase target gene expression levels. The resulting system was named *B. subtilis* CcaSR v1.0. It produces a 70-fold increase in transcription rate of a target gene in response to green versus red light and the change in target protein levels can be observed within minutes [8].

B. subtilis CcaSR v1.0 comprises three genetic modules: a PCB production module (PPM), a light-sensing module (LSM), and a transcriptional output module (TOM) (Fig. 1). The PPM encodes an expression cassette for the *ho1-pcyA* fusion. The LSM encodes the so-called *ccaSmini#10* variant [6] under the control of the xylose-inducible promoter $P_{xylA(+47)}$, which we engineered for superior xylose response during the construction of *B. subtilis* CcaR v1.0, as well as its cognate repressor *xylR* (*see* **Note 1**) [8, 9]. The TOM encodes *ccaR* under the control of IPTG-inducible promoter $P_{hy\text{-}spank}$, its cognate repressor gene *lacI*, and the target gene under control of $P_{cpcG2\text{-}veg}$.

Construction of a strain for optical control of a novel target gene requires first building a custom TOM that places the target gene under the control of $P_{cpcG2\text{-}veg}$, replacing *sfgfp*. We recommend replacing *sfgfp* via Golden Gate assembly, a molecular cloning strategy that uses restriction enzymes and a ligase to scarlessly assemble

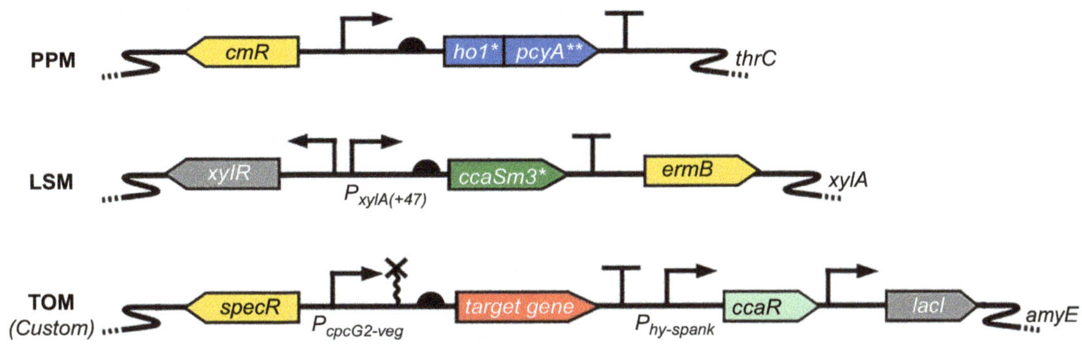

Fig. 1 CcaSRv1.0 strain overview. The complete system includes three linear integration modules in separate genomic loci

several DNA fragments [10, 11] (Subheading 3.1). The resulting product constitutes a linear integration module (LIM), which is a double-stranded DNA molecule with homology to the genome, enabling integration of DNA into the chromosome through homologous recombination. LIMs can be produced through assembly techniques (e.g., Gibson Assembly, Golden Gate) and/or amplified via PCR. The LIM containing the custom TOM is transformed to wild-type *B. subtilis* (Subheadings 3.2 and 3.3). After confirming successful integration, the genome of the TOM strain is purified to use as a genome integration module (GIM) (Subheading 3.4). A GIM is a purified genome that contains a single LIM integrated somewhere in the chromosome. It can be transformed to a *B. subtilis* strain in order to integrate the LIM with high efficiency due to the high degree of homology between the GIM and the recipient chromosome at the LIM locus. Both the published LSM (in the *xylA* locus of the *B. subtilis* genome) and the customized TOM (in the *amyE* locus) must be transformed as GIMs to the published base strain containing the PPM in the *thrC* locus (Subheadings 3.5 and 3.6).

To run an optogenetic experiment with this system, we recommend exposing engineered cells growing at exponential phase to light signals in a Light Plate Apparatus (LPA), a commercially available open-source device that can deliver two independent light signals to each well of a 24-well plate [12] (Subheadings 3.7, 3.8, and 3.9). The cells are exposed to custom light programs, enabling quantitative control of gene expression.

2 Materials

Strains from the original report of *B. subtilis* CcaSR v1.0 [8] can be obtained from the Bacillus Genetic Stock Center (BGSC). For those bearing a single LIM, the annotated sequence of the LIM is available through GenBank. Strain genotypes, BGSC numbers, and

Table 1
BGSC strains

Name	Description	BGSC code	GenBank accession number
S028	*thrC*::PPM v0.3	1A1606	MK995048
S051	*amyE*::TOM v0.2	1A1608	MK995068
S061	*xylA*::LSM v0.3a	1A1607	MK995072
S063	CcaSR v1.0 (*thrC*::PPMv0.3; *amyE*::TOMv0.2; *xylA*::LSMv0.3a)	1A1609	N/A

GenBank accession numbers can be found in Table 1. When preparing media, filters sterilize all components and combine aseptically. Standard PCR equipment and reagents are required.

2.1 Strains

1. PY79 (wild-type *B. subtilis* can also be obtained from BGSC).
2. S028.
3. S051.
4. S061.

2.2 Media and Antibiotics

1. Iron chloride solution with citric acid: 50 mM $FeCl_3$ and 100 mM $C_6H_8O_7$ in water. Filter sterilize. Store at 4 °C protected from light.
2. T-base: Add 9.91 g ammonium sulfate ($(NH_4)_2SO_4$), 69.68 g potassium phosphate dibasic (K_2HPO_4), 29.94 g potassium phosphate monobasic (KH_2PO_4), and 5.15 g sodium citrate dihydrate ($Na_3C_6H_5O_7{\cdot}2H_2O$) to a sterile bottle. Water up to 500 mL. Mix thoroughly and heat to dissolve. Filter sterilize.
3. M9 media formulated for *B. subtilis* optogenetics: For 1 L, add the following to 700 mL filtered deionized water in sterile water: 200 mL 5X M9 salts, 20 mL 10% casamino acids, 20 mL iron chloride solution with citric acid, 8 mL 50% glycerol, 2 mL 50 mM $MnSO_4$, 2 mL 1 M $MgSO_4$., and 1 mL 100 mM $CaCl_2$. Add more water up to 1 L. *See* **Notes 2** and **3**.
4. SpC media: 20 mL 10x T-base, 4 mL 10% yeast extract, 3 mL 1.2% w/v $MgSO_4$, 2 mL 50% glucose, 0.5 mL 10% casamino acids, water up to 200 mL.
5. SpII media: 20 mL 10x T-base, 14 mL 1.2% w/v $MgSO_4$, 2 mL 50% glucose, 2 mL 10% yeast extract, 1 mL 100 mM $CaCl_2$, 0.2 mL 10% casamino acids, water up to 200 mL.
6. SpII-EGTA: 1000 μL 10x T-base, 700 μL 1.2% w/v $MgSO_4$, 200 μL 0.1 M EGTA, 100 μL 50% glucose, 100 μL 10% yeast extract, 100 μL 1% casamino acids, 50 μL 100 mM $CaCl_2$, water to 10 mL.

7. Concentrated SpII-EGTA: 2500 μL 10x T-base, 1750 μL 1.2% w/v $MgSO_4$, 500 μL 0.1 M EGTA, 250 μL 50% glucose, 250 μL 10% yeast extract, 250 μL 1% casamino acids, 125 μL 100 mM $CaCl_2$, water up to 10 mL.
8. D-xylose: 50% w/v in water. *See* **Note 4**.
9. Isopropyl β-d-1-thiogalactopyranoside (IPTG): 1 M in water. *See* **Note 4**.
10. Spectinomycin, 100 mg/mL in water.
11. Erythromycin, 0.5 mg/mL in ethanol.
12. Chloramphenicol, 35 mg/mL in ethanol (optional, *see* **Notes 5** and **6**).
13. Kanamycin, 50 mg/mL in water (optional, *see* **Notes 6** and 7).
14. Tetracycline, 10 mg/mL in water (optional, *see* **Note** 7).
15. Starch plates: Combine the following in a sterile 1 L bottle: 25 g LB Miller, 15 g bacteriologial agar, 10 g starch, water up to 1 L. Autoclave the bottle containing the suspension. After autoclaving, shake the bottle to mix the preparation and pour into plates as normal, 10–15 mL per plate. Allow plates to dry under flame, then store in a bag at 4 °C. Prewarm before use.
16. Gram's iodine, stabilized.
17. 50% glycerol.

2.3 PCR and Golden Gate Supplies

1. BsaI restriction enzyme (NEB).
2. 10x NEB T4 buffer.
3. NEB T4 ligase.
4. DMSO is recommended for all PCRs (*see* **Note 8**).

2.4 Equipment and Kits

1. Wizard Genomic DNA Purification Kit from Promega.
2. Wizard Gel Purification Kit from Promega.
3. Light Plate Apparatus (LPA). The LPA is a device that can deliver two independent light signals to each well of a 24-well plate [12]. A guide to building, programming, and using the LPA is available in Gerhardt et al. (2019) [13]. Alternatively, LPAs can be purchased through Flash Genetics (flashgenetics.com).
4. ArcticWhite opaque 24-well clear-bottom plates (arcticwhiteusa.com).
5. Reflective aluminum foil seals (VWR, Radnor, PA, USA).

3 Methods

3.1 Design and Construction of the Transcriptional Output Module

1. Design and order primers for Golden Gate assembly of the custom TOM, which will place the target gene downstream of $P_{cpcG2\text{-}veg}$ (*see* **Notes 9–11**). One pair of primers must amplify the gene of interest and introduce recognition sites for the Type II restriction enzyme BsaI, as well as sticky ends with homology to adjacent fragments. Two other pairs will amplify the 5′ and 3′ fragments of the TOM (flanking the gene of interest). We have provided the primer pairs necessary for PR amplification of the 5′ and 3′ fragments, as well as primer templates for amplifying your gene of interest—see Table 2. These primers introduce the sticky end 5′-TATG-3′ upstream of the gene of interest, and 5′-TAAC-3′ downstream (*see* **Note 12**). Figure 2 provides a reference for the cloning scheme.
2. Perform PCR amplification of the 5′ and 3′ fragments of the TOM (using S051 as a template), as well as the PCR amplification of your gene of interest. Table 3 provides recommended PCR reaction conditions (*see* **Notes 13–15**).
3. Perform agarose gel electrophoresis (recommended conditions: 1% agarose gel, 90 V, ~30–50 min depending on the total size of PCR product) on your PCR reactions and identify the appropriately sized band for your product.
4. Perform gel extraction and purification to obtain purified DNA fragments. This avoids off-target recombination in later transformations due to residual genomic DNA.

Table 2
Recommended primers for PCR

Primer name	Sequence	Reaction partner
prTOM_5_FWD	GTTCAGCTCAGTGATACCTGC	prTOM_5_REV
prTOM_5_REV	TTAGGTCTCG***CATA***GTAGTTCCTCCTTATGTAAGCTT	prTOM_5_FWD
prTOM_3_FWD	TTAGGTCTCG***TAAC***CAGGCATCAAATAAAACGAAAGG	prTOM_3_REV
prTOM_3_REV	GCGAAGCCGGTAAAGAAGAAG	prTOM_3_FWD
prTOM_GOI_FWD (custom)	TTAGGTCTCG***TATG***__________________ [insert sequence to amplify gene of interest (GOI), 5′ end]	prTOM_GOI_REV
prTOM_GOI_REV (custom)	TTAGGTCTCG***GTTA***__________________ [insert sequence to amplify gene of interest, 3′ end reverse complement]	prTOM_GOI_FWD

Bold text indicates Golden Gate sticky ends. Underlined text indicates the BsaI recognition sequence

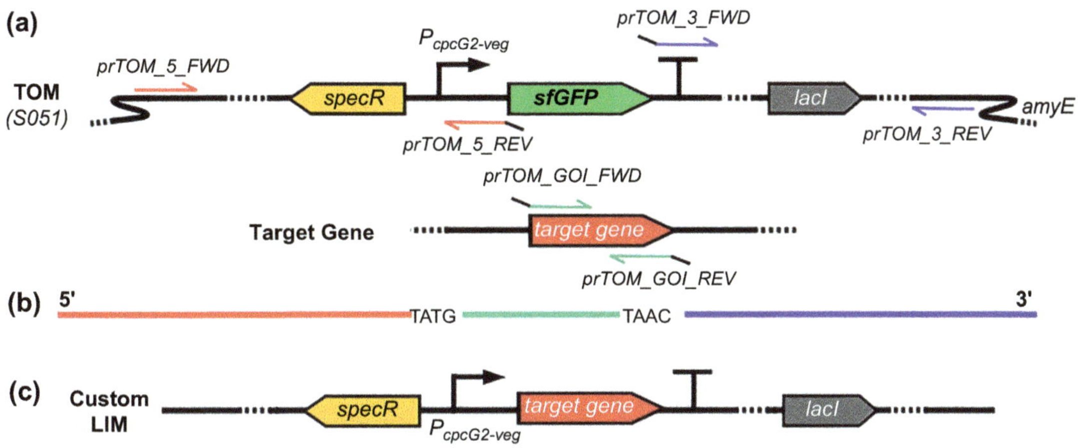

Fig. 2 Cloning scheme for CcaSRv1.0 with a new target gene. (**a**) Three PCR reactions are required to generate the fragments necessary for the Golden Gate reaction. (**b**) A Golden Gate reaction is employed to ligate three fragments according to sticky ends introduced by the primers. (**c**) The final custom LIM is generated by the Golden Gate reaction

Table 3
Recommended PCR conditions

Fragment	Forward primer	Reverse primer	Annealing temperature [°C]	Length [kb]	Extension time	Template
5′	prTOM_5_FWD	pr_TOM_5_REV	63	3.8	170 s	S051
3′	prTOM_3_FWD	prTOM_3_REV	63	4.6	210 s	S051
GOI	prTOM_GOI_FWD	prTOM_GOI_REV	–	–		–

5. Perform a Golden Gate reaction to link the 5′ and 3′ portions of the TOM to your target gene. On ice, mix together equimolar quantities of each of the three purified DNA fragments in a PCR tube. Total DNA mass should be around 200 ng, with a volume of less than 11.5 μL. Add water up to 11.5 μL. Then add 1.5 μL 10x NEB T4 Buffer, 1 μL BsaI enzyme, and 1 μL NEB T4 Ligase. Flick or pipette gently to mix. Then run the reaction using the thermocycler conditions provided in Table 4. *See* **Notes 16–17**.

3.2 Preparation of Competent B. subtilis

1. Streak out the recipient strain (PY79) on an LB agar plate and grow overnight at 37 °C.
2. Inoculate a single colony of the recipient strain into 2 mL of SpC media and grow at 37 °C, shaking.

Table 4
Golden gate thermocycler program

Step	Name	Temperature [°C]	Time [min]
1	Restriction	37	3
2	Ligation	16	4
3	Go to 1, 24 times		
4	Final restriction	50	5
5	Enzyme inactivation	80	5
6	Hold	4	Forever

3. Grow until early stationary phase (*see* **Note 18**). This usually takes about 4.5 h. Early stationary phase can be detected when the OD_{600} value of the culture stops changing significantly over time.
4. Dilute the liquid culture 1:10 into prewarmed SpII media. 1 mL of diluted SpII culture will yield enough cells for two transformations; scale the volume of SpII media accordingly.
5. Grow for 90 min at 37 °C, shaking.
6. Pellet the cells (3 min at 4000× *g*, room temperature) and decant the supernatant—save the supernatant. It contains secreted competence factors. *See* **Note 19**.
7. Resuspend pellets in a volume of supernatant equal to 10% of the culture volume that was spun down. For example, if 1 mL of culture was spun down, resuspend in 100 μL of supernatant.
8. At this stage, cells are ready for transformation. Either transform immediately or add sterile glycerol to a final concentration of 15% and freeze at −80 °C for future use. *See* **Note 20**.

3.3 Chromosomal Integration of TOM in Wild-type PY79

1. Add the following to a 1.5 mL microcentrifuge tube: 50 μL competent cells (PY79), 20 μL *concentrated* SpII-EGTA, and 30 μL Golden Gate reaction (*see* **Notes 21–23**).
2. Incubate the mixture at 37 °C, shaking, for 1 h (*see* **Note 24**).
3. Plate the cells on LB with spectinomycin (100 μg/mL) for selection and incubate overnight at 37 °C (*see* **Note 25**).

3.4 Confirmation of TOM Integration and Sequence

1. Grow colonies from the transformation plate to the early exponential phase in LB media.
2. Make temporary freezer stocks.
3. Spot plate cultures on prewarmed starch plates to perform a starch test (*see* **Notes 26** and **27**).
4. Grow starch plates overnight at 37 °C.

5. The following day, pour Gram's iodine solution (1–2 mL) over the plates such that it forms a thin coat. Allow to incubate at room temperature for 5 min, then pour off the iodine solution. Cells with an intact *amyE* locus are able to degrade starch and will form a red/yellow halo after incubation with iodine; cells with a disrupted locus (implying successful integration) will be unable to form the halo.
6. Once integration is confirmed, grow the cells overnight at 37 ° C in liquid LB media and amplify the locus via PCR. The following primers amplify the TOM for sequencing (not including the spectinomycin resistance marker): `CCATAAAAC-TAAAGTAAGTGTAAACCTATTCATTG` (FWD) and `CAT-AAAGGTCATTGTTGACGCG` (REV). If using NEB Phusion polymerase, use an annealing temperature of 63 °C. Total length is 4.1 kb, plus the length of your gene of interest.
7. Separate the PCR product using gel electrophoresis as before.
8. Extract and purify the band corresponding to the appropriate size.
9. Submit the purified locus for sequencing (Sanger or amplicon).
10. Following sequence confirmation of the locus, perform a genome extraction of the new strain and store the purified genome at −20 °C. We recommend the Wizard Genomic DNA Purification kit from Promega.

3.5 Integration of the LSM on PPM Base Strain to Produce S065

1. Purify the genome of *B. subtilis* strain S061, containing the LSM integrated into the *xylA* locus (*see* **Notes 28–31**).
2. Prepare competent cells of *B. subtilis* strain S028, containing the PPM in the *thrC* locus.
3. Add the following to a 1.5 mL microcentrifuge tube: 50 μL competent cells, 50 μL (nonconcentrated) SpII-EGTA, and 500 ng purified S061 genome.
4. Incubate the mixture at 37 °C, shaking, for 2.5 h (*see* **Note 24**).
5. Plate the cells on LB + erythromycin (0.5 μg/mL) and incubate overnight at 37 °C.
6. Inoculate candidate colonies into LB media and make temporary glycerol stocks.
7. We recommend checking colonies for the presence of both the PPM and the LSM in their respective loci by PCR. Resequencing is typically not necessary.

3.6 Integration of the TOM on S065

1. Prepare competent cells of the chassis strain S065.
2. Add the following to a 1.5 mL microcentrifuge tube: 50 μL competent cells, 50 μL (nonconcentrated) SpII-EGTA, and 500 ng purified genome containing custom TOM.

3. Incubate the mixture at 37 °C, shaking, for 1 h (*see* **Note 24**).
4. Plate the cells on LB + spectinomycin (0.5 μg/mL) and incubate overnight at 37 °C.
5. Inoculate candidate colonies into LB media and make temporary glycerol stocks.
6. Confirm the successful integration of the TOM using the starch test as described in Subheading 3.4. This strain will be referred to as the "experimental strain."

3.7 Preparation of Glycerol Aliquots

1. Streak out the experimental strain on LB agar plates and grow overnight at 37 °C for preparation of glycerol aliquots (*see* **Notes 32** and **33**).
2. Prepare M9 media and add inducers (IPTG and xylose) in the same concentration that you plan to use for the experiment. See Subheading 3.9 for notes on choosing these concentrations.
3. Pick a colony, inoculate in 2 mL M9 media, and grow until early exponential phase (OD_{600}~0.05). This typically requires 45–90 min.
4. Combine 1.4 mL culture with 600 μL of 50% glycerol as if making a freezer stock. Mix thoroughly. This will provide enough volume for 20 glycerol aliquots (50 μL each).
5. Measure the OD_{600} of the mixture. Use as a blank a mixture of 700 μL M9 media and 300 μL glycerol (50%). Record this OD.
6. Dispense 50 μL of the mixture into each of 20 PCR tubes. Place the tubes in a secondary container such as a 50 mL Falcon tube and label with the strain, date, and OD.
7. Place the tubes at −80 °C. Alternatively, aliquots can be flash-frozen using liquid nitrogen.

3.8 Designing a Light Program

1. Use Iris to design your desired light program for the LPA [13] (*see* **Notes 34–40**).
2. Download the light program file onto an SD card containing the necessary dc and gcal files (*see* **Note 41**).

3.9 Running an Experiment

1. Prepare the necessary quantity of M9 media. You will need 500 μL per well, along with a pilot tube of media (2 mL is usually sufficient) for OD_{600} measurements at the end of the experiment if applicable.
2. Add inducers (IPTG and xylose) to the media based on the required dynamic range and basal expression level (*see* **Notes 42–45**). Mix thoroughly. We recommend final inducer concentrations of 17.0 μM IPTG, 0.025% w/v xylose.
3. Prewarm media to 37 °C.

4. Separate the media into tubes for each strain being used for your experiment, as well as one pilot tube (uninoculated).
5. Thaw a single-use frozen aliquot for each strain and subinoculate into the prewarmed media. We recommend employing a starting OD_{600} between 10^{-5} and 10^{-4}. Discard any remaining unused volume from the single-use aliquot.
6. Distribute media into wells of the optogenetic 24-well plate (ArticWhite opaque 24-well clear-bottom plates). We recommend 500 μL per well.
7. Seal the plate with a reflective seal aluminum foil seal and secure to the LPA device.
8. Activate the light program and allow cells to incubate at 37 °C, shaking, while exposed to light signals. We recommend an experimental duration of 8–10 h.
9. Remove plates from the incubator and remove the seal (*see* **Note 46**).
10. Perform endpoint assays on the cultures. Possible assays include measurement of optical density, microscopy, flow cytometry or FACS, Western blotting, or RNA-seq (*see* **Note 47**).

4 Notes

1. $P_{xylA(+47)}$ is an engineered form of P_{xylA} from plasmid pAX01, modified to eliminate disruptive elements [8, 9].
2. Glycerol is used as the carbon source because glucose interferes with $P_{xylA(+47)}$.
3. A large batch of M9 media can be prepared ahead of time, separated into aliquots for different experiments, and stored at room temperature protected from light. However, if using this approach, the iron chloride solution and the calcium chloride solution must be withheld from the batch media preparation and added fresh before the experiment.
4. We recommend making large batches of 1 M IPTG and 50% w/v xylose, dividing them into small aliquots (5–200 μL), and storing at −20 °C.
5. The PPM bears a chloramphenicol resistance marker. A chloramphenicol selection is not necessary for this protocol, but you may wish to confirm chloramphenicol resistance in your strains. Carry out chloramphenicol selections at 5 μg/mL.
6. Chloramphenicol and kanamycin tend to be weak/unreliable selective markers in *B. subtilis.* We recommend using others when possible, or reconfirming antibiotic resistance after the original selection if using these two markers.

7. Additional modules, such as fluorescent reporters or genomic knockouts, may also be integrated into this strain. We recommend integrating additional modules into the *lacA* locus if necessary. Tetracycline and/or kanamycin can be used as selective markers at concentrations of 10 μg/mL and 5 μg/mL, respectively.
8. DMSO can be added to a final concentration of 3% v/v to disrupt the secondary structure of template DNA during PCRs.
9. To enable strong expression of almost any gene of interest, we also include the transcriptional insulator RiboJ and the strong RBS MF001 upstream of the target gene.
10. The published TOM, available in the genome of strain S051, uses *sfGFP* as the target gene. To replace it, we will construct a new linear integration module (LIM) with a new target gene, but maintaining homology for the *amyE* locus.
11. LIMs are double-stranded DNA molecules containing the DNA payload to be integrated, flanked by homology arms that enable integration into genomic DNA via homologous recombination [14]. Transformation of in vitro-assembled linear DNA fragments with sufficient homology to the genome (~2 kb on either flank) will result in genomic integration by homologous recombination.
12. Note that these sticky ends assume that your gene of interest begins with an ATG start codon and ends with a TAA stop codon; they can be modified if either of these is not true.
13. Boiled cell culture can be used as a template (2 μL/50 μL reaction).
14. Table 3 provides the recommended annealing temperature and extension time if using NEB Phusion polymerase. Adjust these parameters if using a different polymerase.
15. We recommend the use of GC buffer for PCR reactions using *B. subtilis* genomic DNA as a template.
16. Note that the temperature of the restriction digestion step (**Step 1**) may need to be adjusted if using an enzyme other than BsaI.
17. Keep completed Golden Gate reactions at 4 °C until transformation.
18. We will leverage the natural competence of *B. subtilis*, which arises under conditions of nutrient deprivation.
19. Avoid pelleting cells at low temperatures (4 °C or less), as *B. subtilis* cells will lyse.
20. We recommend using fresh competent cells when possible, and only using fresh competent cells when transforming Golden Gate reactions.

21. The custom TOM is first integrated into the genome of wild-type cells in order to enable identification of constructs with the correct sequence, without the presence of the full system, which may exert a mutagenic pressure on the TOM construct. Additionally, the genomes of the strain containing only the TOM may be purified and used for easy integration of the TOM on other strains.
22. If you have less than 30 μL Golden Gate reaction, complete with water.
23. 400 ng of DNA in 30 μL of Golden Gate reaction is usually recommended; 200 ng is sufficient if the resistance marker is spectinomycin (as it is for the TOM).
24. When the resistance marker is spectinomycin, 1 h is sufficient. Otherwise, 2.5 h is recommended.
25. When transforming DNA to *B. subtilis* recipient strains already bearing selective markers in other loci, only the antibiotic corresponding to the selective marker in the donor DNA should be used for selection. Including additional selective markers is unnecessary and places excessive stress on cells, sometimes reducing transformation efficiency.
26. Alternatively, PCR amplification of the *amyE* locus can be used to confirm integration; make sure to design the PCR reaction such that at least one primer is outside the linear fragment that was transformed.
27. We recommend including wild-type cells (PY79) as a negative control and the published TOM strain S051 as a positive control for this test. These strains can be struck out on LB agar and grown overnight along with the transformation plate, then grown to exponential phase and spot-platted onto the starch plates along with the new strains being tested.
28. A strain containing both the PPM and LSM forms the chassis for the engineered system. We refer to this strain as S065. S065 can be constructed by transforming the purified genome of S061 to competent cells of S028.
29. Note that S065 only needs to be constructed once, and can be used as a recipient strain for any custom TOM.
30. When transforming a GIM, always ensure both the GIM strain and recipient strain are the same base strain (e.g., PY79) of *B. subtilis* to avoid an uncertain genotype.
31. Also, we recommend transforming one GIM containing a single new integration module at a time.
32. Generating precise and reproducible experimental results from optogenetic experiments requires consistent patterns of cell growth on different days. To achieve this goal, we begin

experiments from single-use aliquots of cells frozen during exponential growth at a specified optical density at 600 nm (OD_{600}). We have found that this approach enables reproducible growth on different days, as well as minimizes lag phase.

33. Alternatively, exponential phase precultures can be diluted to a starting OD_{600} between 10^{-5} and 10^{-4} and used to start the experiment.
34. LPA devices generate custom light signals according to a Light Program File (.lpf) placed on an SD card. Gerhardt et al. (2019) describe how to use the Iris software to program basic to advanced light signals [13].
35. We recommend employing a total experiment duration of 8–10 h, with 3–5 h of red light preconditioning at 20 μmol m^{-2} s^{-1}. This preconditioning step allows the cells to grow and adjust to experimental conditions before the CcaSR system is activated by green light.
36. When determining how many wells to use for each experimental condition, consider the number of replicates necessary to achieve the desired statistical power. Also consider the required amount of material for downstream analysis, remembering that the experimental time course should conclude before the cells enter stationary phase, as the function of the CcaSR system is likely to change during stationary phase growth.
37. Be sure to set aside wells for appropriate controls (e.g., cells lacking fluorescent reporters to be used as measures of autofluorescence).
38. If light exposures of various durations are required, consider using a staggered-start approach to ensure all samples will finish simultaneously, enabling easier downstream assays [15].
39. The maximum recommended green light intensity is 50 μmol m^{-2} s^{-1}.
40. When programming light signals, consider that patterns of target gene expression will not resemble the dynamical light signals. Green and red light intensity control the fraction of active sensor kinase in the cell, which in turn influences the amount of target gene transcription according to a Hill function. If simple behaviors (e.g., "turn on expression at $t = 6$ hours" or "slowly increase expression over the course of 2 hours") are required, light programs can be easily designed to achieve these behaviors. However, if more complex and/or precise patterns of gene expression are required, a mathematical model of how light signals determine gene expression is necessary. A simple ODE model that replicates light response dynamics is provided by Castillo-Hair et al. (2019) [8, 16].

41. We recommend inserting the card into the LPA device and starting the device to ensure the file exhibits the anticipated behavior.
42. The dynamic range of the CcaSR system will be determined by the level of steady-state expression of the CcaS and CcaR proteins within the cell. For example, a higher steady-state level of CcaR is predicted to increase the rate of basal expression of the target gene due to increased auto-dimerization and binding to $P_{cpcG2\text{-}veg}$. Chemical inducer levels should be selected to achieve the desired expression range and basal expression level (i.e., leaky expression of the target gene in red light).
43. If using our M9 media formulation, the following inducer concentrations can be used: 17.0 μM IPTG, 0.025% w/v xylose. These inducer concentrations enable a 13.5x fold change in steady-state expression of *sfGFP* from saturating red light to saturating green light, with a basal expression of 1.9×10^3 MEFL under saturating red light—see calibration data in Fig. 3.
44. If a larger fold change is desired, inducer concentrations of 10.0 μM IPTG and 0.010% xylose (w/v) can be employed with reduced iron (III) citrate concentration (reduced from 50 to 1 mM) to achieve a 70x fold change with similar basal expression. This is the inducer scheme used in Castillo-Hair et al. (2019) [8], and inducer calibration data is available in the

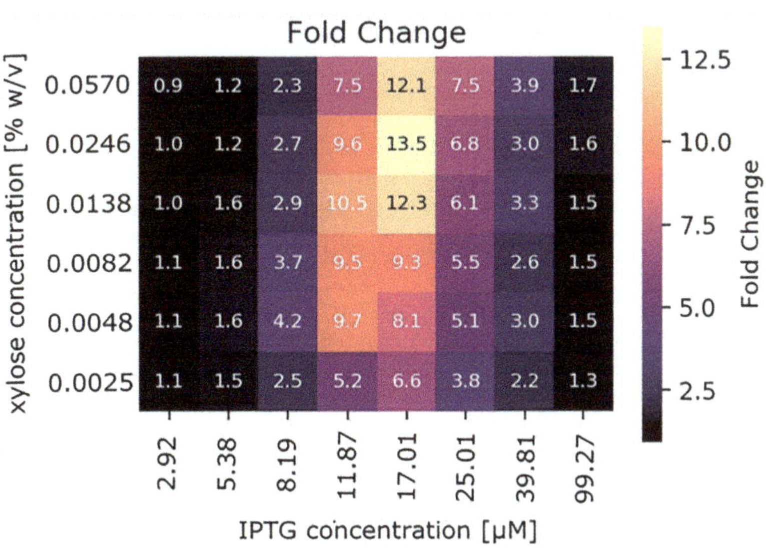

Fig. 3 Inducer concentration calibration heatmap. Data represent fold change in sfGFP expression levels grown under red and green light, as measured by the ratio of fluorescence levels obtained under the two conditions. Data were obtained from single replicates

supplementary material. Note that reducing iron (III) citrate concentration increases the rate of mutational inactivation of the PPM, preventing the use of glycerol aliquots. Subinoculation of an exponentially growing culture must be used instead.

45. The basal expression level and dynamic range generated by the provided inducer combinations will not be suitable for all applications. For example, if a toxic gene is being expressed, basal expression levels will likely be too high, even under saturating red light, and inducer concentrations will need to be adjusted. If the target gene product is easy to measure (e.g., a fluorescent protein), various combinations of inducers can be tried, with each combination being exposed to saturating green light and saturating red light, as in Fig. 3. This approach enables determination of inducer levels that offer desirable dynamic range and basal expression levels. Alternatively or in addition, genetic design tools, such as alteration of the ribosome binding site preceding the target gene, as well as the addition of protein degradation tags, can be employed [17].
46. Because the CcaSR light response is fast (i.e., seconds to minutes), cell phenotype could change rapidly upon exposure to ambient light after removal from the LPA device. Therefore, we recommend keeping samples covered as long as possible and performing assays immediately upon removal.
47. Perturbations to *B. subtilis* gene regulatory networks may produce altered phenotypes, such as increased rates of sporulation, chaining, or biofilm matrix production. Ensure that assays employed are robust to such variation.

References

1. Hirose Y, Shimada T, Narikawa R et al (2008) Cyanobacteriochrome CcaS is the green light receptor that induces the expression of phycobilisome linker protein. Proc Natl Acad Sci 105:9528–9533. https://doi.org/10.1073/pnas.0801826105
2. Hirose Y, Narikawa R, Katayama M et al (2010) Cyanobacteriochrome CcaS regulates phycoerythrin accumulation in Nostoc punctiforme, a group II chromatic adapter. Proc Natl Acad Sci 107:8854–8859. https://doi.org/10.1073/pnas.1000177107
3. Frankenberg N, Mukougawa K, Kohchi T et al (2001) Functional genomic analysis of the HY2 family of ferredoxin-dependent bilin reductases from oxygenic photosynthetic organisms. Plant Cell Online 13:965–978. https://doi.org/10.1105/tpc.13.4.965
4. Tabor JJ, Levskaya A, Voigt CA (2011) Multichromatic control of gene expression in *Escherichia coli*. J Mol Biol 405:315–324. https://doi.org/10.1016/j.jmb.2010.10.038
5. Schmidl SR, Sheth RU, Wu A et al (2014) Refactoring and optimization of light-switchable *Escherichia coli* two-component systems. ACS Synth Biol 3:820–831. https://doi.org/10.1021/sb500273n
6. Nakajima M, Ferri S, Rögner M et al (2016) Construction of a miniaturized chromatic acclimation sensor from Cyanobacteria with reversed response to a light signal. Sci Rep 6:37595. https://doi.org/10.1038/srep37595
7. Ong NT, Tabor JJ (2018) A miniaturized *Escherichia coli* green light sensor with high dynamic range. Chembiochem 19:1255–1258. https://doi.org/10.1002/cbic.201800007
8. Castillo-Hair SM, Baerman EA, Fujita M et al (2019) Optogenetic control of *Bacillus subtilis* gene expression. Nat Commun 10:3099.

https://doi.org/10.1038/s41467-019-10906-6

9. Härtl B, Wehrl W, Wiegert T et al (2001) Development of a new integration site within the *Bacillus subtilis* chromosome and construction of compatible expression cassettes. J Bacteriol 183:2696–3699. https://doi.org/10.1128/JB.183.8.2696-2699.2001
10. Bird JE, Marles-Wright J, Giachino A (2022) A user's guide to Golden Gate cloning methods and standards. ACS Synth Biol 11:3551–3563. https://doi.org/10.1021/acssynbio.2c00355
11. Engler C, Kandzia R, Marillonnet S (2008) A one pot, one step, precision cloning method with high throughput capability. PLoS One. https://doi.org/10.1371/journal.pone.0003647
12. Gerhardt KP, Olson E, Castillo-Hair SM et al (2016) An open-hardware platform for optogenetics and photobiology. Sci Rep 6:35363. https://doi.org/10.1038/srep35363
13. Gerhardt KP, Castillo-Hair SM, Tabor JJ (2019) DIY optogenetics: building, programming, and using the Light Plate Apparatus. Methods Enzymol 624:197–226. https://doi.org/10.1016/bs.mie.2019.04.018
14. Landry BP, Palanki R, Dyulgyarov N et al (2018) Phosphatase activity tunes two-component system sensor detection threshold. Nat Commun 9:1433. https://doi.org/10.1038/s41467-018-03929-y
15. Olson E, Hartsough L, Landry B et al (2014) Characterizing bacterial gene circuit dynamics with optically programmed gene expression signals. Nat Methods 11:449–455. https://doi.org/10.1038/nmeth.2884
16. Olson EJ, Tzouanas CN, Tabor JJ (2017) A photoconversion model for full spectral programming and multiplexing of optogenetic systems. Mol Syst Biol 13:926. https://doi.org/10.15252/msb.20167456
17. Guiziou S, Sauveplane V, Chang HJ et al (2016) A part toolbox to tune genetic expression in *Bacillus subtilis*. Nucleic Acids Res 44:7495–7508. https://doi.org/10.1093/nar/gkw624

Chapter 2

The Yeast Optogenetic Toolkit (yOTK) for Spatiotemporal Control of Gene Expression in Budding Yeast

Zachary P. Harmer and Megan N. McClean

Abstract

Optogenetic systems utilize genetically encoded light-sensitive proteins to control cellular processes such as gene expression and protein localization. Like most synthetic systems, generation of an optogenetic system with desirable properties requires multiple design-test-build cycles. A yeast optogenetic toolkit (yOTK) allows rapid assembly of optogenetic constructs using Modular Cloning, or MoClo. In this protocol, we describe how to assemble, integrate, and test optogenetic systems in the budding yeast *Saccharomyces cerevisiae*. Generating an optogenetic system requires the user to first define the structure of the final construct and identify all basic parts and vectors required for the construction strategy, including light-sensitive proteins that need to be domesticated into the toolkit. The assembly is then defined following a set of standard rules. Multigene constructs are assembled using a series of one-pot assembly steps with the identified parts and vectors and transformed into yeast. Screening of the transformants allows optogenetic systems with optimal properties to be selected.

Key words Optogenetics, Yeast, Modular cloning, Synthetic biology, Multigene constructs, Biological parts

1 Introduction

Optogenetic systems utilize genetically encoded light-sensitive proteins to control cellular processes such as gene expression and protein localization. In the important model and industrial organism *Saccharomyces cerevisiae*, optogenetic systems have been used to control activation and repression of gene expression [1, 2], protein localization [3–5], translation [6], and intracellular and intercellular signaling activity and metabolism [7–11]. This in turn has allowed for light-based control of yeast physiology both to investigate basic biological questions [12, 13] as well as achieve optimized control of engineered metabolic pathways [14] and synthetic consortia [15].

Armin Baumschlager (ed.), *Optogenetics: Methods and Protocols*, Methods in Molecular Biology, vol. 2840,
https://doi.org/10.1007/978-1-0716-4047-0_2,

Optogenetic systems consist of light-sensitive proteins combined with appropriate effector domains to drive the cellular process of interest in a light-induced manner. For example, fusion of an activation domain (AD) to one-half of a light-sensitive protein-protein interaction and a DNA-binding domain (DBD) to the other half generates a split optogenetic chimeric transcription factor. Functional optogenetic systems require appropriate choice of light-sensitive proteins, functional arrangements of the light-sensitive components and effector domains, as well as appropriate expression levels. Thus, optogenetic systems go through many design-test-build cycles before desirable or optimal function is achieved. To facilitate rapid generation of optogenetic constructs, this protocol describes the use of a yeast optogenetic toolkit, which allows for Modular Cloning (MoClo) of constructs to combine and express light-sensitive proteins fused to appropriate effector domains. Modular cloning systems exist in many different organisms [16–22] and take advantage of Type IIS-based restriction enzymes and Golden Gate cloning. MoClo is a hierarchical assembly system in which transcription units are assembled from a first level containing basic "parts" (e.g., promoters, terminators, and coding sequences for light-sensitive proteins). These transcriptional units are then assembled into multigene constructs which can be transformed and integrated into the yeast genome to express all necessary elements for the optogenetic system. In this chapter, we describe how to use a yeast optogenetic toolkit [1] to make functional optogenetic systems in *Saccharomyces cerevisiae*.

2 Materials

2.1 General Coning

1. Luria-Bertani (LB) medium: 800 mL deionized water, 10 g Tryptone, 5 g NaCl, 5 g yeast extract, 15 g bacto agar (only for solid media), adjust final volume to 1 L with deionized water, sterilize by autoclaving.
2. Antibiotics: chloramphenicol, carbenicillin (used here instead of ampicillin), and kanamycin. Filter-sterilized stocks of Chloramphenicol are prepared at 34 mg/mL in 10 mL EtOH. Filter-sterilized stocks of kanamycin or carbenicillin are prepared at 50 mg/mL in 10 mL EtOH. Stocks are stored in 1 mL aliquots at −20 °C. Dilute each stock solution 1:1000 in medium after the medium has been autoclaved and cooled to 65 °C.
3. 1.5 mL tubes.
4. 50 μL aliquots DH5α chemically competent *Escherichia coli* cells (New England Biolabs) (or another strain of cloning competent *E. coli* cells, such as TOP10 from ThermoFisher Scientific).

5. Plasmid Miniprep Kit (e.g., QIAwave Plasmid Miniprep Kit, Qiagen; Monarch Miniprep Kit, and New England Biolabs).
6. Water bath, 42 °C.
7. 37 °C incubator.
8. 4–4.5 mm glass plating beads (e.g., Zymo Research, Thermo-Fisher Scientific, and New England Biolabs).
9. Nanodrop.
10. UV transilluminator (optional).

2.2 Generation of New Parts

1. DNA insert. DNA inserts can be ordered from a DNA synthesis company (e.g., Integrated DNA Technologies) as gene blocks or generated by PCR that amplifies the target sequence with desired flanking sequences. DNA to be integrated into a part vector must be flanked by part type-specific overhangs, followed by a BsaI site, followed by a BsmBI site. The BsmBI sites should generate TCGG (at the 5′ end) and ACTC (at the 3′ end) overhangs for assembly into the entry vector.
2. Part entry vector plasmid (pYTK001 [19]).

2.3 Golden Gate Assembly

1. BsmBI-v2 Golden Gate Assembly Kit (New England Biolabs).
2. BsaI Golden Gate Assembly Kit (New England Biolabs).
3. 200 μL PCR tubes.

2.4 Transformation and Screening of Yeast

1. Appropriate yeast strain, for example MATα HAP1+ ura3Δ0 leu2Δ0 HIS3 LYS2 TRP1 could be used for transforming constructs with the URA3 marker and URA3 homology for integration.
2. 20 U/μL NotI-HF, CutSmart Buffer (for linearizing constructs for genomic integration) (New England Biolabs).
3. Synthetic Complete (SC) medium: 900 mL deionized water, 6.7 g yeast nitrogen base without amino acids, and 20 g bacto agar (only for solid media). Autoclave media and cool to 65 °C. Add 50 mL 20x amino acid mix (with appropriate nutrient dropped out, such as lacking uracil for SC-URA media) and 50 mL 20x glucose solution (40% w/v, autoclaved separately). Amino acid mix: Dissolve1.44 g dry powder stocks (5 g adenine hemisulfate, 2 g uracil, 5 g tryptophan, 2 g histidine-HCl, 2 g arginine-HCl, 2 g methionine, 3 g tyrosine, 10 g leucine, 3 g isoleucine, 10 g lysine-HCl, 5 g phenylalanine, 10 g aspartic acid, 15 g valine, 20 g threonine, and 40 g serine) into 50 mL filter-sterilized deionized water at 60 °C.
4. YPD medium: Add 950 mL deionized water, 20 g bacto peptone, 10 g yeast extract (add 24 g bacto agar for solid media), and autoclave. Add 50 mL filter-sterilized 40% w/v glucose solution.

5. 1.0 M LiAc solution: Prepare as a 1.0 M stock in deionized water and filter-sterilize.
6. 50% w/v Polyethylene glycol (PEG): Add 50 g PEG to 50 mL water and filter-sterilize with a 0.45 μm filter. Store at room temperature.
7. TE buffer: 10 mM Tris-HCl (pH 8.0), 1.0 mM EDTA.
8. Single-stranded carrier DNA (ssDNA): Add 200 mg of the DNA into 100 mL of TE buffer. Mix by pipetting and then on a magnetic stirrer for 2–3 h or until fully dissolved. Aliquot DNA into 100 μL volumes and store at −20 °C. Prior to use, heat ssDNA to 95 °C for 10–30 min and cool on ice.
9. LiAc SDS buffer: mix 50 μL 1 M LiAc, 25 μL filter-sterilized 10% SDS, 175 μL sterile, deionized water.
10. DNA oligos for PCR verification of construct integration.
11. 30 °C incubator.

2.5 Optogenetic Screening of Yeast

1. 96-well plate (black-walled, glass bottom; for example, Cellvis 96 Well glass bottom plate with #1.5 cover glass).
2. Illumination device (e.g., optoPlate, LITOS, optoWELL-24).
3. Plate reader or flow cytometer (e.g., Tecan Spark Multimode Microplate Reader or ThermoFisher Attune NxT Flow Cytometer).

3 Methods

Perhaps the most important step in the generation of a functional yeast optogenetic system is planning the design of the system including which light-sensitive proteins to employ, the structure of the final construct, and the assembly strategy. This process is described below.

3.1 Selection of Light-Sensitive Proteins

The choice of which light-sensitive proteins to use in the design of the optogenetic system is constrained by several considerations including the inherent photocycle kinetics of the light-sensitive proteins as well as which wavelengths of light can be used to stimulate the optogenetic system. For example, if the optogenetic system is to be used in combination with fluorescence microscopy this limits which wavelengths can be used to excite the light-sensitive proteins. Tolerance of the light-sensitive protein to effector fusion can also be a consideration. Optobase is a valuable resource that provides a summary of light-sensitive proteins excited by different wavelengths of light as well as associated reference publications [23]. This database can be consulted as a started point to choose several potential light-sensitive proteins. In addition, Figueroa et al. [24] recently compiled a list of light-sensitive proteins that have already proven functional in yeast. It is often advisable to try several different designs, and even several different light-sensitive proteins and empirically test their function, which is why the ability to rapidly clone different optogenetic systems into yeast using the yeast optogenetic toolkit is so powerful.

3.2 Selection of Standard Parts and Vectors

The yeast optogenetic toolkit, like many MoClo toolkits, is implemented as a hierarchical assembly process. Assembly of a multigene cassette, capable of integration into the yeast genome to express all aspects of the optogenetic system, requires identification of all of the basic parts needed. "Parts" are the most basic DNA sequence elements that can be assigned a function. For example, promoters, terminators, and coding sequences are all parts as are yeast marker genes and assembly connectors (Fig. 1a). An important decision is what the most basic light-sensitive part should be in the design scheme. For example, one approach is to encode the light-sensitive protein and effector domain as separate coding sequence parts (as in Ref. [2]). Alternatively, the light-sensitive protein fused to the appropriate effector domain can be encoded as a single coding sequence part (as in Ref. [1]). If a large number of light-sensitive proteins and effectors are to be screened for function in the

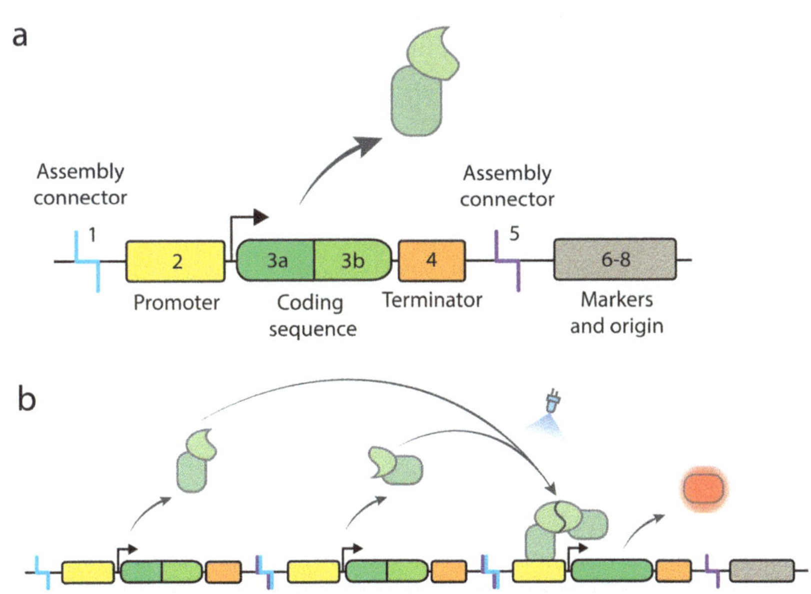

Fig. 1 (**a**) Representative cassette plasmid formed from BsaI assembly of individual part plasmids (numbered). (**b**) Multigene plasmid formed from BsmBI assembly of multiple cassette plasmids. The first two cassettes express the first and second components of a two-component split optogenetic transcription factor. Light-induced dimerization of the split transcription factor induces expression of the fluorescent reporter on the third cassette

optogenetic system, it would make sense to encode light-sensitive proteins and effectors as separate 3a and 3b parts so that different combinations can be easily assembled. These parts are then assembled into cassettes, which contain promoter and terminator elements for expression. These cassettes can be stand-alone and contain a yeast marker and integration homology. Alternatively, if multiple cassettes are needed to express multiple components in the optogenetic system (e.g., for a split light-sensitive transcription factor, which would need at least two cassettes), cassettes can be further combined into multigene cassettes using assembly connectors (Fig. 1b). Figure 1 should be consulted to design an assembly strategy. Many parts are already available [1, 2, 19] (see Table 1) and can be obtained from Addgene (https://www.addgene.org/Megan_McClean/).

3.3 Assembly and Integration of Multigene Cassettes

3.3.1 Generation of Parts

Parts plasmids consist of a DNA functional unit flanked by Type IIS restriction enzyme recognition sites that leave type-specific overhang sequences that allow appropriate assembly order during Golden Gate. While many parts exist (Table 1), if new parts are needed, they can be added to the toolkit as follows:

1. Set up a BsmBI Golden Gate reaction by pipetting into a 200 μL PCR tube 75 ng of the entry vector plasmid (pYTK001), the insert DNA containing the new part sequence at a 2:1 (insert:vector) molar ratio, 2 μL 10x T4 DNA Ligase Buffer, 1 μL BsmBI-v2 NEB Golden Gate Enzyme Mix, and adjusting the final volume to 20 μL with sterile, deionized water.
2. Thermal cycler incubation: (42 °C, 1 min → 16 °C, 1 min) × 30 → 60 °C, 5 min (*see* **Note 1**).
3. Transform Golden Gate reaction product into DH5α chemically competent *E. coli* cells (*see* **Note 2**). Thaw a tube of DH5α competent *E.coli* cells (50 μL, stored at −80 °C) on ice for 5 min.
   (a) Add 8 μL assembly reaction product and gently mix by flicking the tube several times. Incubate the tube on ice for 30 min.
   (b) Perform a heat shock for 30 s at 42 °C. Return the tube to ice for 2 min.
   (c) Add 950 μL of LB medium (or a higher nutrient medium, such as super optimal broth). Incubate at 37 °C for 60 min with constant mixing, such as on a roller drum.
   (d) Pellet cells by centrifugation and decant the supernatant. Resuspend cells in 200 μL LB medium and plate on LB + chloramphenicol (if more even colony distribution is desired, plate 20 μL on one plate and the remaining 180 μL on a separate plate), incubating overnight at 37 °C.

Table 1
Plasmids containing available toolkit parts

ID	Type	Description	*E. coli* marker	References
pMM0923	3	Gal4AD-CIB1	CamR	Harmer et al. 2023 *ACS Synth Biol*
pMM1235	3	mScarlet-I	CamR	Harmer et al. 2023 *ACS Synth Biol*
pMM0918	3a	Gal4DBD	CamR	Harmer et al. 2023 *ACS Synth Biol*
pMM1245	3a	eMagB	CamR	Harmer et al. 2023 *ACS Synth Biol*
pMM1246	3b	eMagA	CamR	Harmer et al. 2023 *ACS Synth Biol*
pMM1258	3a	eMagBF	CamR	Harmer et al. 2023 *ACS Synth Biol*
pMM1259	3b	eMagAF	CamR	Harmer et al. 2023 *ACS Synth Biol*
pMM1260	3a	eMagBM	CamR	Harmer et al. 2023 *ACS Synth Biol*
pMM1247	3b	Gal4AD	CamR	Harmer et al. 2023 *ACS Synth Biol*
pMM0920	3b	CRY2(535)	CamR	Harmer et al. 2023 *ACS Synth Biol*
pMM0921	3b	CRY2PHR	CamR	Harmer et al. 2023 *ACS Synth Biol*
pMM0922	3b	CRY2FL	CamR	Harmer et al. 2023 *ACS Synth Biol*
pMM0531	3	VP16-CIB1	CamR	An-adirekkun et al. 2020 *Biotechnol Bioeng*
pMM0530	3	Zif268DBD-CRY2PHR	CamR	An-adirekkun et al. 2020 *Biotechnol Bioeng*
pMM0529	3	Zif268DBD-CRY2	CamR	An-adirekkun et al. 2020 *Biotechnol Bioeng*
pMM769	3a	dCas9-NLS	CamR	Geller et al. 2019 *Cell Mol Bioeng*
pMM768	3b	dCas9-NLS	CamR	Geller et al. 2019 *Cell Mol Bioeng*
pMM767	3	dCas9-NLS	CamR	Geller et al. 2019 *Cell Mol Bioeng*
pMM766	3b	dCas9-NLS-Stop	CamR	Geller et al. 2019 *Cell Mol Bioeng*
pMM765	3	dCas9-NLS-Stop	CamR	Geller et al. 2019 *Cell Mol Bioeng*
pMM729	3	Mxi1-dCas9	CamR	Geller et al. 2019 *Cell Mol Bioeng*
pMM728	3b	Mxi1	CamR	Geller et al. 2019 *Cell Mol Bioeng*
pMM727	3	Mxi1	CamR	Geller et al. 2019 *Cell Mol Bioeng*
pMM726	3a	Mxi1	CamR	Geller et al. 2019 *Cell Mol Bioeng*
pMM725	4a	Linus-Mxi	CamR	Geller et al. 2019 *Cell Mol Bioeng*
pMM587	3b	dCas9	CamR	Geller et al. 2019 *Cell Mol Bioeng*
pMM586	3	dCas9	CamR	Geller et al. 2019 *Cell Mol Bioeng*
pMM585	3a	dCas9	CamR	Geller et al. 2019 *Cell Mol Bioeng*
pMM582	4a	Linus	CamR	Geller et al. 2019 *Cell Mol Bioeng*
pMM1088	3b	CRY2PHR L348F	CamR	Harmer et al 2024 *ACS Synthetic Biology*
pMM1089	3b	CRY2PHR W349R	CamR	Harmer et al 2024 *ACS Synthetic Biology*
pMM1134	3a	LexA	CamR	Harmer et al 2024 *ACS Synthetic Biology*
pMM1137	2	LexA	CamR	Harmer et al 2024 *ACS Synthetic Biology*
pMM1431	3	NLS-VP16-EL222(A79Q)	CamR	Harmer et al 2024 *ACS Synthetic Biology*

(continued)

Table 1 (continued)

ID	Type	Description	*E. coli* marker	References
pMM1432	3	NLS-VP16-EL222AQTrip	CamR	Harmer et al 2024 *ACS Synthetic Biology*
pMM1435	3b	p65AD	CamR	Harmer et al 2024 *ACS Synthetic Biology*
pMM1446	3b	Msn2AD	CamR	Harmer et al 2024 *ACS Synthetic Biology*
pMM1447	3b	VP16	CamR	Harmer et al 2024 *ACS Synthetic Biology*

4. Transformant screening. BsmBI assembly of a part plasmid will remove sfGFP from the entry vector plasmid and replace it with the desired functional unit part insertion. Thus, *E. coli* that take up the modified plasmid will contain the gene for chloramphenicol resistance (allowing them to grow on LB + chloramphenicol plates), but will not contain sfGFP (meaning they won't fluoresce green). GFP-negative colonies should be selected for culturing (*see* **Note 3**).
5. GFP-negative colonies are cultured and plasmids are purified using a plasmid miniprep kit according to manufacturer instructions. Construct assembly efficiency is high for most inserts, so preparing two different colonies will typically ensure a correct plasmid is obtained.
6. Concentrations of purified plasmids are measured by nanodrop, and the sequence of the insert and flanking restriction sites is verified by sequencing. Sanger sequencing (e.g., Functional Biosciences, GENEWIZ) from just outside the insert is a reliable method; however, due to lower costs of whole plasmid sequencing (e.g., Plasmidsaurus), this is also an option.

3.3.2 Cloning of Cassette Plasmids

Cassette plasmids contain promoter and terminator elements for expressing a desired coding sequence (and should contain part types 1–8; see Fig. 1a). They also contain machinery for replication in *E. coli* (origin and marker) and may also contain yeast markers, homology to the yeast genome, or CEN/ARS4 or 2micron elements (*see* **Note 4**). These plasmids are generated by BsaI Golden Gate assembly of the desired part plasmids into a cassette vector, such as pYTK095. If multiple cassettes are to be assembled into a multigene plasmid, they should contain the appropriate connector sequences (part types 1 and 5). For example, a cassette being integrated into the first position of a multigene plasmid should contain the connectors ConLS and ConR1 (or for the second position, ConL1 and ConR2, and so on).

1. Set up a BsaI Golden Gate reaction by pipetting into a 200 μL PCR tube 75 ng of the vector plasmid (such as pYTK095 when downstream assembly of multiple cassettes into a multigene

plasmid is desired), the insert plasmids at a 2:1 (insert:vector) molar ratio, 2 μL 10x T4 DNA Ligase Buffer, 1 μL BsaI NEB Golden Gate Enzyme Mix, and adjusting the final volume to 20 μL with sterile, deionized water.

2. Thermal cycler incubation: (37 °C, 5 min → 16 °C, 5 min) × 30 → 60 °C, 5 min (*see* **Note 1**).
3. Transform Golden Gate reaction product into DH5α chemically competent *E. coli* cells (*see* **Note 2**) as described in Subheading 3.3.1, **Step 3**. Plate cells on LB + selection marker of vector (e.g., carbenicillin for pYTK095).
4. Transformant screening. BsaI assembly of a cassette plasmid will remove sfGFP from the vector plasmid and replace it with the desired functional unit part sequences. Thus, *E. coli* that take up the modified plasmid with all inserts will contain the resistance gene (allowing them to grow on, e.g., LB + carbenicillin plates), but will not contain sfGFP (meaning they won't fluoresce green). GFP-positive colonies can be readily identified by placing the plate on a UV transilluminator (see Subheading 3.3.1, **Step 4**), allowing for selection of GFP-negative colonies for culturing.
5. GFP-negative colonies are cultured and plasmids are purified using a plasmid miniprep kit according to manufacturer instructions. Construct assembly efficiency is high for most sequences, so preparing two different colonies will typically ensure a correct plasmid is obtained.
6. Concentrations of purified plasmids are measured by nanodrop and the sequence of the insert and flanking restriction sites are verified by sequencing. Sanger sequencing (e.g., Functional Biosciences, GENEWIZ) from just outside the insert or whole plasmid sequencing (e.g., Plasmidsaurus) can be performed.

3.3.3 Multigene Assembly

Cassettes can be transformed into yeast individually, or if additional transcriptional elements are required can be assembled into multigene plasmids by BsmBI assembly into a vector plasmid, such as pYTK096 (see Fig. 1b). For two-component split transcription factors, each of the two components will typically be encoded in a separate cassette of the multigene plasmid, with an optional reporter cassette also included in the multigene plasmid (*see* **Note 5**).

1. Set up a BsmBI Golden Gate reaction by pipetting into a 200 μL PCR tube 75 ng of the vector plasmid (such as pYTK096 for integration at URA3), the insert plasmids at a 2:1 (insert:vector) molar ratio, 2 μL 10x T4 DNA Ligase Buffer, 1 μL BsmBI NEB Golden Gate Enzyme Mix, and adjusting the final volume to 20 μL with sterile, deionized water.

2. Thermal cycler incubation: (42 °C, 5 min → 16 °C, 5 min) × 30 → 60 °C, 5 min (*see* **Note 1**).
3. Transform Golden Gate reaction product into DH5α chemically competent *E. coli* cells (*see* **Note 2**). As described in Subheading 3.3.1, **Step 3**. Plate cells on LB + selection marker of vector (e.g., kanamycin for pYTK096).
4. Transformant screening. BsmBI assembly of a multigene plasmid will remove sfGFP from the vector plasmid and replace it with the desired cassette sequences. Thus, *E. coli* that take up the modified plasmid with all cassette inserts will contain the resistance gene (allowing them to grow on, e.g., LB + carbenicillin plates), but will not contain sfGFP (meaning they won't fluoresce green). GFP-positive colonies can be readily identified by placing the plate on a UV transilluminator (see Subheading 3.3.1, **Step 4**), allowing for selection of GFP-negative colonies for culturing.
5. GFP-negative colonies are cultured and plasmids are purified using a plasmid miniprep kit according to manufacturer instructions. Construct assembly efficiency is high for most sequences, so preparing two different colonies will typically ensure a correct plasmid is obtained.
6. Concentrations of purified plasmids are measured by nanodrop and the sequence of the insert and flanking restriction sites are verified by sequencing. Sanger sequencing (e.g., Functional Biosciences, GENEWIZ) from just outside the insert or whole plasmid sequencing (e.g., Plasmidsaurus) can be performed.

3.3.4 Yeast Transformation

1. *NotI digestion.* Multigene plasmids or cassettes designed for integration (*see* **Note 4**) can be NotI-digested and transformed into an appropriate yeast strain by mixing 500 ng plasmid with 0.5 μL NotI-HF, 1 μL 10x CutSmart buffer, adjusting the final volume to 10 μL, and incubating at 37 °C for at least 15 min. Alternatively, plasmids with a CEN6/ARS4 or 2micron origin can be directly transformed into an appropriate yeast strain.
2. Yeasts are transformed according to published methods [25].
   - (a) Grow up a colony of the strain to be transformed in 5 mL YPD on a roller drum at 30 °C overnight.
   - (b) Add 500 μL of the overnight culture (or 2.5×10^8 cells) to a 250 mL flask with 50 mL YPD. Grow in a shaking incubator for 3–5 h (or until the cell titer is at least 2×10^7 cells/mL).
   - (c) Centrifuge cells at 3130 *g* for 5 min and decant supernatant.

(d) Resuspend cells in 25 mL sterile water. Centrifuge again, decant supernatant, and resuspend cells in 1 mL 100 mM LiAc.

(e) Transfer cell suspension to a 1.5 mL tube, centrifuge at 3000 *g*, and discard the supernatant by pipetting.

(f) Add 400 μL 100 mM LiAc and resuspend cells by pipetting. Aliquot 50 mL cell mixture into 1.5 mL tubes (one for each transformation), centrifuge at 3000 *g* for 2 min, and remove supernatant by aspiration.

(g) Add 300 μL transformation mix (240 μL PEG, 35 μL 1.0 M LiAc, 25 μL 2 mg/mL ssDNA) and 70 μL sterile water with 0.1–10 μg (*see* **Note 6**) plasmid DNA (NotI-digested if constructs are to be genomically integrated) to each cell pellet.

(h) Resuspend cells by vortex mixing. Incubate for 30 min at 30 °C (without shaking).

(i) Perform a heat shock by placing the tubes in a water bath at 42 °C for 20–25 (up to 40) minutes.

(j) Centrifuge tubes at 3000 *g* for 15 s and discard supernatant by pipetting.

(k) If cells are being transformed with a drug selection marker (e.g., G418, clonNat, hygromycin, or 5-FOA), perform a recovery step prior to plating: resuspend the cell pellet in 1 mL YPD and incubate at 30 °C for 2–18 h. Centrifuge tubes at 3000 *g* for 15 s and discard supernatant by pipetting. Disregard this recovery step if plating onto an amino acid deficiency selective plate.

(l) Resuspend cells in 200 μL sterile water. To ensure single colonies, plate 150 μL sterile water and 20 μL cell suspension in one selection plate and the remaining 180 μL cell suspension in a second selection plate.

(m) Spread the culture on the plates with sterile glass beads and incubate at 30 °C for 2–4 days.

3. *gDNA preparation.* Construct integration can be verified by PCR (for optogenetic constructs that generate a readily detectable reporter, such as a fluorescent protein, this step can be omitted and transformants can be directly screened for activity as described in Subheading 3.3.1; *see* **Note 7**). gDNA is prepared from colonies according to an established LiAc SDS protocol [26] (colony PCR can be directly performed from transformant colonies, but is generally less reliable).

   (a) Add 50 μL LiAc SDS buffer to a 1.5 mL tube and pick a colony by lightly touching it with a pipette tip and swirling it in the buffer.

   (b) Heat the tube to 70 °C for 5 min and add 150 μL EtOH.

(c) Centrifuge tube at 15,000 *g* for 3 min and aspirate the supernatant.

(d) Resuspend the pellet in 100 μL 70% EtOH. Pellet and aspirate as before.

(e) Resuspend pellet in 20 μL sterile deionized water. Centrifuge at 15,000 *g* for 30 s to pellet cell debris.

4. *PCR Validation of Construct Integration.* Use 1 μL of the gDNA supernatant (from Subheading 3.3.4, **Step 3.e**) for PCR. Integration of each construct should be validated using two pairs of primers (one for 5′ integration and one for 3′ integration). For each primer pair, one primer is designed to bind the homologous region, and one primer is designed to bind in the insert, generating, for example, a 500 bp PCR product that can be verified by gel electrophoresis. Primer sequences for PCR verification of URA3, LEU2, and HO integrations can be found in Supplementary Table S4 of Lee et al. 2015 [19].

3.3.5 Screening and Validation of Optogenetic Tools

Once optogenetic systems have been transformed into yeast and integration checked by colony PCR, their functionality (i.e., response to light) should be measured (Fig. 2). In this protocol, we will focus on using a commonly used open-source illumination solution called the optoPlate [27]. However, many additional illumination solutions are available (*see* **Note 8**).

3.3.6 Illumination with optoPlate

1. Colonies are picked from YPD plates to test tubes with 3 mL SC media and grown overnight at 30 °C on a roller drum.
2. Taking care to prevent light exposure of optogenetic strains (*see* **Note 9**), dilute overnight cultures to desired density (e.g., OD600 = 0.01–0.5), pipette 100–200 μL into each well of a 96-well plate (black-walled, glass bottom), and incubate for 3–5 h at 30 °C, shaking. Blank media and a nonfluorescent strain should be included as controls if a plate reader will be used for measurements. Each condition is typically performed in triplicate to ensure reliability of results (*see* Fig. 2a; **Note 7**).
3. Light induction is performed for the length of time desired (typically 1–16 h; *see* **Note 10**). Shaking is important for continued growth of cells.

3.3.7 Screening via Plate Reader

1. The plate reader should first be calibrated to determine the optimal measurement parameters for the desired reporter protein and well plate. This is most easily accomplished using two control strains: a strain with the desired reporter under constitutive expression and a nonfluorescent strain. The Z value (distance between the plate reader and the plate), where applicable, for each well plate is determined by measuring both

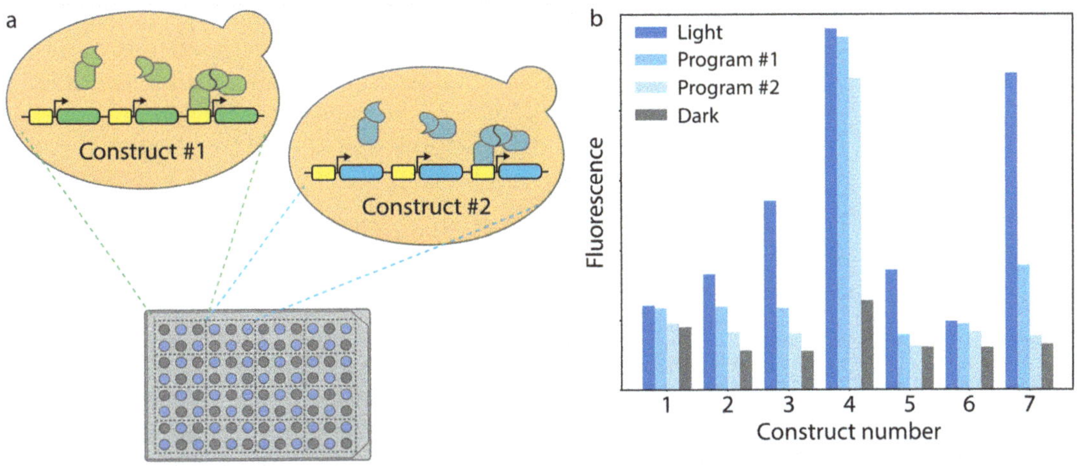

Fig. 2 (**a**) Strains with different constructs are cultured, diluted back, and added to different sections of a 96-well plate with checkerboarded light conditions (to reduce edge effects) performed in triplicate. (**b**) Sample fluorescence means for each construct and light condition

strains at a range of Z values and determining which distance has the largest difference between the reporter and nonfluorescent controls.

2. For strains that produce a readily measurable output, such as a fluorescent or bioluminescent reporter, screening of activity can be readily performed in a plate reader. Cultures should be resuspended by shaking prior to measurement (e.g., 1 min shaking at 1000 rpm with 2 mm orbital), to ensure accuracy of measurement. Inclusion of triplicate (or higher) sample numbers allows for determination of statistical significance between different illumination conditions (such as varying light intensity or duty cycle).

3.3.8 Scanning Light Intensity and Dynamics

1. Level of gene expression can be readily controlled using light intensity and light-duty cycle modulation [28]. The first experiment with a new optogenetic system should use constant illumination to determine whether the system is functioning as desired (*see* **Note 10**). Once functionality of the optogenetic system has been established, light intensity, period, and duty cycle can be optimized to achieve desired characteristics (Fig. 2b).

4 Notes

1. For higher efficiency Golden Gate assembly reactions (especially with multiple inserts), it is recommended to use the longer incubation time protocol: (37 °C or 42 °C, 5 min → 16 °C, 5 min) × 30 → 60 °C, 5 min. Note that the 60 °C denaturation

step will denature the ligase, but not most Type IIS restriction enzymes (consult enzyme denaturation temperatures from the manufacturer). This is important for cloning plasmids with internal restriction sites (such as when making a GFP-dropout vector), and the 60 °C denaturation step should be skipped for such assemblies.

2. DH5α is used here as an example. Other cloning-competent *E. coli* strains, such as TOP10 (ThermoFisher Scientific) may also be used, but manufacturer-recommended transformation protocols should be consulted.

3. GFP-positive colonies can be readily identified by briefly placing the plate on a UV transilluminator and noting which colonies fluoresce brightly (using proper safety equipment to avoid any UV exposure). Alternatively, the difference between GFP-positive and negative colonies will become more visible after 8–36 additional hours (though plates should be placed at either 4 °C or 21 °C after 24 h post-plating to prevent colony overgrowth). This allows selection of GFP-negative colonies for culturing.

4. If cassettes are intended to be transformed individually, individual toolkit parts 6–8 can be selected (pYTK074-pYTK094). If multiple cassettes are to be assembled into a multigene plasmid, assembly vector pYTK095 can be used and the resulting transformant colonies selected on carbenicillin media and screened for GFP dropout. Cassettes to be assembled into a multigene plasmid should be designed with appropriate connectors in mind (e.g., the cassette in the first position of a multigene plasmid should have pYTK002 ConLS and pYTK067 ConR1 in positions 1 and 5). Constructs on plasmids will exhibit greater variability in copy number but can be maintained at higher copy numbers. Clonal populations of cells with integrated constructs will exhibit more consistent activity and don't need to be grown in selective media to prevent curing of the construct. Strains using fluorescent reporters can be used to rapidly screen the efficacy of different promoter combinations.

5. Expression levels for two-component split transcription factors are critical for determining level of activity [1, 29]. More optimal activity will typically be obtained if the AD component is expressed at an equal or higher level than the DBD component, as binding sites in the genome are limited and only localized AD will have the desired effect.

6. The amount of DNA may need to be optimized for the specific strain and construct, but 500 ng is usually sufficient to yield tens of colonies for standard strains and constructs.

7. For an initial screening of transformants of optogenetic construct that generate a fluorescent or bioluminescent output, transformants can be picked to wells of a 96-well plate and grown overnight at 30 °C, shaking. Each culture can be diluted by the same factor. For example, 25 μL cell culture can be transferred to 175 μL SC media, mixing, then transferring 50 μL of that dilution to 150 μL SC media (once each for a light and dark control) in a 96-well plate. This plate can then be incubated in the dark at 30 °C, shaking for 5 h before beginning light induction. It is important to screen multiple transformants from each integration transformation as integration copy number variation can occur and can have a significant effect on the activity of the optogenetic system [29].
8. Various illumination platforms are available for light induction in microwell plates. The LPA [30] and optoWELL-24 [31] (commercially available) are suitable for illumination in 24-well plates. The optoWELL-96 [31], LITOS [32], Lumiplate and optoPlate [27] (commercially available) are suitable for induction in higher-capacity plates, such as 96-well plates. There are also more advanced options that allow for control of illumination in chemostat or turbidostat conditions [33–35] and could potentially allow for continuous measurement. However, these systems are not preferred for screening due to low throughput.
9. It is ideal to culture and prepare strains for optogenetic experiments exploring light induction response in a dark environment. For blue light-sensitive optogenetic strains, red lights can be used for visibility without activating the optogenetic system. Blackout curtains can be used to protect specific areas or pieces of equipment from light, and aluminum foil can be used to cover glass windows in incubators, etc. Some optogenetic proteins, such as CRY2PHR, are highly sensitive to light [29].
10. Most strains and optogenetic systems will show the most measurable change in gene expression while cultures are in an exponential phase of growth. For the experimental conditions described here, this will typically occur around 5–8 h after the start of induction (for strains lacking a significant growth defect). Determining the ideal time to induce and measure is simpler with an automated device that controls illumination and takes regular measurements, such as the optoPlateReader [36], Lustro [29, 37], or other devices [38–40]. When such a device is not available, 12 h of constant light induction at 100 $\mu W/cm^2$ should allow for measurement of a fluorescent reporter in order to determine whether the optogenetic system is functioning as desired. Excess light is mildly phototoxic to yeast [41] and should be avoided for more sensitive experiments.

References

1. An-adirekkun J (My), Stewart CJ, Geller SH, Patel MT, Melendez J, Oakes BL, Noyes MB, MN MC (2020) A yeast optogenetic toolkit (yOTK) for gene expression control in *Saccharomyces cerevisiae*. Biotechnol Bioeng 117: 886–893. https://doi.org/10.1002/bit.27234
2. Geller SH, Antwi EB, Di Ventura B, McClean MN (2019) Optogenetic repressors of gene expression in yeasts using light-controlled nuclear localization. Cell Mol Bioeng 12:511–528. https://doi.org/10.1007/s12195-019-00598-9
3. Yang X, Jost AP-T, Weiner OD, Tang C (2013) A light-inducible organelle-targeting system for dynamically activating and inactivating signaling in budding yeast. Mol Biol Cell 24: 2419–2430. https://doi.org/10.1091/mbc.e13-03-0126
4. Niopek D, Wehler P, Roensch J, Eils R, Di Ventura B (2016) Optogenetic control of nuclear protein export. Nat Commun 7: 10624. https://doi.org/10.1038/ncomms10624
5. Niopek D, Benzinger D, Roensch J, Draebing T, Wehler P, Eils R, Di Ventura B (2014) Engineering light-inducible nuclear localization signals for precise spatiotemporal control of protein dynamics in living cells. Nat Commun 5:4404. https://doi.org/10.1038/ncomms5404
6. Lu H, Mazumder M, Jaikaran ASI, Kumar A, Leis EK, Xu X, Altmann M, Cochrane A, Woolley GA (2019) A yeast system for discovering optogenetic inhibitors of eukaryotic translation initiation. ACS Synth Biol 8:744–757. https://doi.org/10.1021/acssynbio.8b00386
7. Rojas V, Larrondo LF (2023) Coupling cell communication and optogenetics: implementation of a light-inducible intercellular system in yeast. ACS Synth Biol 12:71–82. https://doi.org/10.1021/acssynbio.2c00338
8. Stewart-Ornstein J, Chen S, Bhatnagar R, Weissman JS, El-Samad H (2017) Model-guided optogenetic study of PKA signaling in budding yeast. Mol Biol Cell 28:221–227. https://doi.org/10.1091/mbc.e16-06-0354
9. Moreno Morales N, Patel MT, Stewart CJ, Sweeney K, McClean MN (2021) Optogenetic tools for control of public goods in *Saccharomyces cerevisiae*. mSphere 6:e00581-21. https://doi.org/10.1128/mSphere.00581-21
10. Zhao EM, Suek N, Wilson MZ, Dine E, Pannucci NL, Gitai Z, Avalos JL, Toettcher JE (2019) Light-based control of metabolic flux through assembly of synthetic organelles. Nat Chem Biol 15:589–597. https://doi.org/10.1038/s41589-019-0284-8
11. Pouzet S, Cruz-Ramón J, Le Bec M, Cordier C, Banderas A, Barral S, Castaño-Cerezo S, Lautier T, Truan G, Hersen P (2023) Optogenetic control of beta-carotene bioproduction in yeast across multiple lab-scales. Front Bioeng Biotechnol 11: 1085268. https://doi.org/10.3389/fbioe.2023.1085268
12. Sweeney K, McClean MN (2022) Transcription factor localization dynamics and DNA binding drive distinct promoter interpretations. 2022.08.30.505887
13. Chen SY, Osimiri LC, Chevalier M, Bugaj LJ, Nguyen TH, Greenstein RA, Ng AH, Stewart-Ornstein J, Neves LT, El-Samad H (2020) Optogenetic control reveals differential promoter interpretation of transcription factor nuclear translocation dynamics. Cell Syst 11: 336–353.e24. https://doi.org/10.1016/j.cels.2020.08.009
14. Zhao EM, Zhang Y, Mehl J, Park H, Lalwani MA, Toettcher JE, Avalos JL (2018) Optogenetic regulation of engineered cellular metabolism for microbial chemical production. Nature 555:683–687. https://doi.org/10.1038/nature26141
15. Aditya C, Bertaux F, Batt G, Ruess J (2021) A light tunable differentiation system for the creation and control of consortia in yeast. Nat Commun 12:5829. https://doi.org/10.1038/s41467-021-26129-7
16. Andreou AI, Nakayama N (2018) Mobius assembly: a versatile golden-gate framework towards universal DNA assembly. PLoS One 13:e0189892. https://doi.org/10.1371/journal.pone.0189892
17. Larroude M, Park Y-K, Soudier P, Kubiak M, Nicaud J-M, Rossignol T (2019) A modular Golden Gate toolkit for *Yarrowia lipolytica* synthetic biology. Microb Biotechnol 12: 1249–1259. https://doi.org/10.1111/1751-7915.13427
18. Prielhofer R, Barrero JJ, Steuer S, Gassler T, Zahrl R, Baumann K, Sauer M, Mattanovich D, Gasser B, Marx H (2017) GoldenPiCS: a Golden Gate-derived modular cloning system for applied synthetic biology in the yeast *Pichia*

pastoris. BMC Syst Biol 11:123. https://doi.org/10.1186/s12918-017-0492-3

19. Lee ME, DeLoache WC, Cervantes B, Dueber JE (2015) A highly characterized yeast toolkit for modular, multipart assembly. ACS Synth Biol 4:975–986. https://doi.org/10.1021/sb500366v
20. Weber E, Engler C, Gruetzner R, Werner S, Marillonnet S (2011) A modular cloning system for standardized assembly of multigene constructs. PLoS One 6:e16765. https://doi.org/10.1371/journal.pone.0016765
21. Moore SJ, Lai H-E, Kelwick RJR, Chee SM, Bell DJ, Polizzi KM, Freemont PS (2016) EcoFlex: a multifunctional MoClo kit for *E. coli* synthetic biology. ACS Synth Biol 5:1059–1069. https://doi.org/10.1021/acssynbio.6b00031
22. Fonseca JP, Bonny AR, Kumar GR, Ng AH, Town J, Wu QC, Aslankoohi E, Chen SY, Dods G, Harrigan P, Osimiri LC, Kistler AL, El-Samad H (2019) A toolkit for rapid modular construction of biological circuits in mammalian cells. ACS Synth Biol 8:2593–2606. https://doi.org/10.1021/acssynbio.9b00322
23. Kolar K, Knobloch C, Stork H, Žnidarič M, Weber W (2018) OptoBase: a web platform for molecular optogenetics. ACS Synth Biol 7:1825–1828. https://doi.org/10.1021/acssynbio.8b00120
24. Figueroa D, Rojas V, Romero A, Larrondo LF, Salinas F (2021) The rise and shine of yeast optogenetics. Yeast Chichester Engl 38:131–146. https://doi.org/10.1002/yea.3529
25. Gietz RD, Schiestl RH (2007) High-efficiency yeast transformation using the LiAc/SS carrier DNA/PEG method. Nat Protoc 2:31–34. https://doi.org/10.1038/nprot.2007.13
26. Lõoke M, Kristjuhan K, Kristjuhan A (2011) Extraction of genomic DNA from yeasts for PCR-based applications. BioTechniques 50:325–328. https://doi.org/10.2144/000113672
27. Bugaj LJ, Lim WA (2019) High-throughput multicolor optogenetics in microwell plates. Nat Protoc 14:2205–2228. https://doi.org/10.1038/s41596-019-0178-y
28. Benzinger D, Khammash M (2018) Pulsatile inputs achieve tunable attenuation of gene expression variability and graded multi-gene regulation. Nat Commun 9:3521. https://doi.org/10.1038/s41467-018-05882-2
29. Harmer ZP, McClean MN (2023) Lustro: high-throughput optogenetic experiments enabled by automation and a yeast optogenetic toolkit. ACS Synth Biol. https://doi.org/10.1021/acssynbio.3c00215
30. Gerhardt KP, Olson EJ, Castillo-Hair SM, Hartsough LA, Landry BP, Ekness F, Yokoo R, Gomez EJ, Ramakrishnan P, Suh J, Savage DF, Tabor JJ (2016) An open-hardware platform for optogenetics and photobiology. Sci Rep 6:35363. https://doi.org/10.1038/srep35363
31. optoWELL › opto biolabs. https://optobiolabs.com/products/optowell/. Accessed 30 Mar 2023
32. Höhener TC, Landolt AE, Dessauges C, Hinderling L, Gagliardi PA, Pertz O (2022) LITOS: a versatile LED illumination tool for optogenetic stimulation. Sci Rep 12:13139. https://doi.org/10.1038/s41598-022-17312-x
33. Steel H, Habgood R, Kelly CL, Papachristodoulou A (2020) In situ characterisation and manipulation of biological systems with Chi.Bio PLOS Biol 18:e3000794. https://doi.org/10.1371/journal.pbio.3000794
34. Wong BG, Mancuso CP, Kiriakov S, Bashor CJ, Khalil AS (2018) Precise, automated control of conditions for high-throughput growth of yeast and bacteria with eVOLVER. Nat Biotechnol 36:614–623. https://doi.org/10.1038/nbt.4151
35. Stewart CJ, McClean MN (2017) Design and implementation of an automated illuminating, culturing, and sampling system for microbial optogenetic applications. JoVE J Vis Exp: e54894. https://doi.org/10.3791/54894
36. Datta S, Benman W, Gonzalez-Martinez D, Lee G, Hooper J, Qian G, Leavitt G, Salloum L, Ho G, Mhatre S, Magaraci MS, Patterson M, Mannickarottu SG, Malani S, Avalos JL, Chow BY, Bugaj LJ (2022) High-throughput feedback-enabled optogenetic stimulation and spectroscopy in microwell plates. 2022.07.13.499906
37. Harmer ZP, McClean MN (2023) High-throughput optogenetics experiments in yeast using the automated platform Lustro. JoVE J Vis Exp:e65686. https://doi.org/10.3791/65686
38. Melendez J, Patel M, Oakes BL, Xu P, Morton P, McClean MN (2014) Real-time optogenetic control of intracellular protein concentration in microbial cell cultures. Integr Biol 6:366–372. https://doi.org/10.1039/c3ib40102b
39. Bertaux F, Sosa-Carrillo S, Gross V, Fraisse A, Aditya C, Furstenheim M, Batt G (2022) Enhancing bioreactor arrays for automated

measurements and reactive control with ReacSight. Nat Commun 13:3363. https://doi.org/10.1038/s41467-022-31033-9

40. Benisch M, Benzinger D, Kumar S, Hu H, Khammash M (2023) Optogenetic closed-loop feedback control of the unfolded protein response optimizes protein production. Metab Eng. https://doi.org/10.1016/j.ymben.2023.03.001

41. Robertson JB, Davis CR, Johnson CH (2013) Visible light alters yeast metabolic rhythms by inhibiting respiration. Proc Natl Acad Sci USA 110:21130–21135. https://doi.org/10.1073/pnas.1313369110

Chapter 3

Optoribogenetic Modulation of Transcription

Christian Renzl and Günter Mayer

Abstract

Light can be used as a precise and reversible trigger for the activation of optogenetic tools with subcellular resolution. The interaction of the photoreceptor PAL and aptamer 53 was integrated into a CRISPR/dCas9 system, which can be applied for light-controlled activation of gene expression. Here, we describe a protocol for in vitro application of light-dependent overexpression using eBFP as a proof of concept. The experiment can be done in 3 days, which is split into cell seeding, transfection, and evaluation by flow cytometry. The method is broadly applicable including the upregulation of endogenous genes.

Key words Gene regulation, CRISPRa, Optoribogenetics, Photoreceptor protein, Aptamers

1 Introduction

Clustered regularly interspaced short palindromic repeats (CRISPR) loci and its CRISPR-associated (cas) proteins are an integral part of a prokaryotic RNA-guided adaptive immune system [1]. Cas9, an RNA programmable nuclease, specifically binds and cleaves DNA sequences and has become an important tool in molecular biological research [2–4]. A mutated variant of Cas9, which is named dCas9, omits the cleavage reaction and therefore can be used as a biological navigation system, when a single guide RNA (sgRNA) is co-expressed. The CRISPR/dCas9 complex can be extended by generating effector-dCas9 fusion proteins. CRISPR-activation (CRISPRa) of transcription can be obtained by generating fusion proteins of transactivation domains, for example, P65 from the NF-kB pathway or Heat Shock Factor 1 (HSF1) to dCas9 [5–9]. When photodimerizable components like CIB1 and CRY2 are fused to the transactivation domains, activation of transcription can be rendered light dependent [10–12]. The generation of on-switches based on the direct and light-dependent interaction of the photoreceptor PAL and aptamer 53 [13] enables the overexpression of user-defined genes under blue light, whereas in

Armin Baumschlager (ed.), *Optogenetics: Methods and Protocols*, Methods in Molecular Biology, vol. 2840,
https://doi.org/10.1007/978-1-0716-4047-0_3,

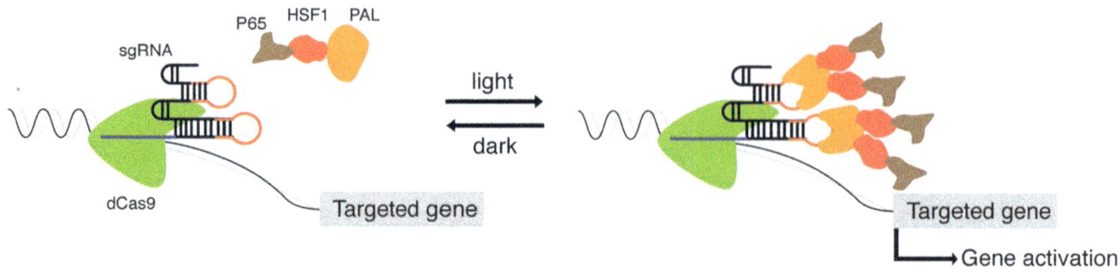

Fig. 1 *Light-dependent activation of transcription.* The CRISPR/dCas9 complex is programmed to bind upstream of the promoter of a targeted gene. Upon irradiation with blue light, the P65-HSF1-PAL fusion protein is recruited to the complex, and the transcription rate of the gene is increased. In darkness, the fusion protein detaches from the complex, and the transcription rate is reduced to a basal level

darkness, transcription levels are reverted to the basal level. We demonstrate a light-dependent RNA–protein interface that reversibly activates transcription of a targeted reporter gene by generating a fusion-variant of the activation domains p65 and HSF1 with PAL (P65-HSF1-PAL, dubbed PHP) (*see* Fig. 1). Upon irradiation with blue light, PHP is recruited to the sgRNA/dCas9 complex and modulates the rate of gene expression.

2 Materials

2.1 Reagents for Flow Cytometry

1. FACS Clean Solution.
2. FACS Flow Solution.
3. FACS Shutdown Solution.
4. Illumination device: blue light (465 nm, 106 μW/cm^2, pulsing with 30 s on-time, 60 s off-time).
5. 0.05% Trypsin-EDTA (1x). Before usage, the reagent can be stored at 4 °C.
6. DMEM (1x) + GlutaMAX-I. Store at 4 °C.
7. DPBS (1x). Store at 4 °C.
8. Fetal Calf Serum (FCS). Store at—20 °C.
9. Lipofectamine 2000. Store at 4 °C.
10. MEM NEAA (100x). Store at 4 °C.
11. Opti-MEM I (1x). Store at 4 °C.
12. Penicillin/Streptomycin (PenStrep) (100x). Store aliquots of 6 mL at −20 °C.
13. 100 mM Sodium Pyruvate (100x). Store at 4 °C.
14. D10 medium: DMEM (1x) + GlutaMAX-I supplied with 10% FCS, 1x NEAA, 1x Sodium Pyruvate and 1x PenStrep. Store at 4 °C.
15. HeLa cells (*see* **Note 1**).

2.2 Plasmids

The plasmids can be obtained from the Mayer Lab (*see* **Note 2**).

1. For the expression of dCas9, the plasmid pdCas9_eGFP is used. The plasmid encodes for an NLS-tagged dCas9 and co-expression of eGFP as transfection control using a CMV promoter.
2. For the expression of P65-HSF1-PAL, the plasmid pNLS-P65-HSF1-PAL-T2A-mCherry is used. NLS-tagged P65-HSF1-PAL is expressed using a CMV promoter. mCherry is co-expressed as transfection control.
3. For the expression of MS2-P65-HSF1, the plasmid pMS2-P65-HSF1-T2A-eGFP is used. NLS-tagged MS2-P65-HSF1 is expressed using a CMV promoter. eGFP is co-expressed as a transfection control.
3. For the expression of the eBFP reporter, the plasmid pGL3-Basic-8xsgRNA-eBFP is used. The plasmid contains a minimal CMV promoter, with a low basal expression of eBFP. An 8x sgRNA binding site is located upstream of the promoter, which can be targeted by the CRISPR/dCas9 complex.
4. The following plasmids express the guide RNAs:
   (a) Control sgRNA: pENTR.hU6.sgRNA.CMV.2.0. The plasmid expresses the sgRNA 2.0, which contains two MS2 aptamers.
   (b) sgRNA SG9: pENTR.hU6.sgRNA.CMV.M2S9. The plasmid expresses the sgRNA SG9, which contains two copies of aptamer 53.
   (c) sgRNA SG9M21: pENTR.hU6.sgRNA.CMV.M2S9M14. The plasmid expresses the sgRNA SG9M21, which contains two copies of the mutated aptamer 53.

2.3 Software

1. FlowJo Version 9.6.3.
2. GraphPad Prism Version 8.0.2.

3 Methods

Unless otherwise specified, carry out all procedures at room temperature.

3.1 Cultivation and Seeding of HeLa Cells

1. HeLa cells (*see* **Note 3**) are cultivated at 37 °C and 5% CO_2 using the cell culture medium D10 medium, which needs to be prepared in advance. The cell culture medium can be stored at 4 °C (*see* **Note 4**). The cells are grown until 80–90% confluency in, for example, T75 flasks. HeLa cells need to be passaged 2–3 times per week using 0.05% Trypsin-EDTA. For passaging, the cells are first rinsed with 5 mL prewarmed DPBS (prewarm to

37 °C). Next, 2 mL of 0.05% Trypsin-EDTA is added to the cells, and the cells are incubated for 5 min in the cell culture incubator at 37 °C. The reaction is stopped by adding 8 mL D10. Centrifuge the sample at 200× *g* for 5 min at 20 °C. The supernatant is removed, and the cells are gently resuspended in 5 mL D10. The cells can be either used for passaging or for seeding onto 24-well plates. For passaging, 2 mL cell suspension is diluted with 10 mL D10 and transferred into a new T75 cell culture flask.

2. For seeding, a 1:10 dilution of the cell suspension is prepared using D10.
3. The cells are counted in a Neubauer counting chamber (*see* **Note 5**).
4. Prepare 15 mL HeLa cell suspension with 140.000 N/mL.
5. Seed the cells onto the 24-well plates using the prepared cell suspension with 140.000 N/mL, and use 500 μL per well. After seeding, incubate the cells for 5 min in the cell culture incubator, take them out again, and move the plate gently to avoid the cells settling in the middle of the well (*see* **Note 6**).
6. Incubate HeLa cells in the cell culture incubator overnight.

3.2 Transfection Preparation

1. Thaw plasmids on ice and prewarm D10 medium, DPBS, and OptiMEM at 37 °C.
2. Calculate the amount of plasmid needed for each transfection. For every sample (e.g., MS2, SG9, or SG9M21), four plasmids are needed. The sgRNAs SG9 and SG9M21 are used together with the effector P65-HSF1-PAL and the sgRNA 2.0 is used with the effector MS2-P65-HSF1. The total amount of plasmid is 200 fmol and the molar ratio between dCas9, effector, gRNA, and reporter is 2:1:1:1. Pipette 80 fmol of dCas9 plasmid and add 40 fmol of effector, sgRNA and reporter plasmid, which in summary will yield the 200 fmol. Per sample 200 fmol plasmids is diluted in OptiMEM in a final volume of 25 μL.
3. Prepare one stock of lipofection reagent mix for all samples. Per sample, 1.5 μL of Lipofectamine 2000 reagent is used in 24 μL OptiMEM. Pipette 329 μL OptiMEM into the lipofection reagent mix tube and add 21 μL Lipofectamine 2000 reagent. Vortex sample for 5 s and spin down tube. The prepared mix is sufficient for 12 samples (*see* **Note 7**).
4. Plasmid mixes are prepared per sample set by first adding 50 μL of OptiMEM to each reaction tube and then adding the appropriate amount of plasmid (*see* **Note 8**). Finally, the volume is filled up to 110 μL using OptiMEM. Vortex for 5 s and spin down tubes.

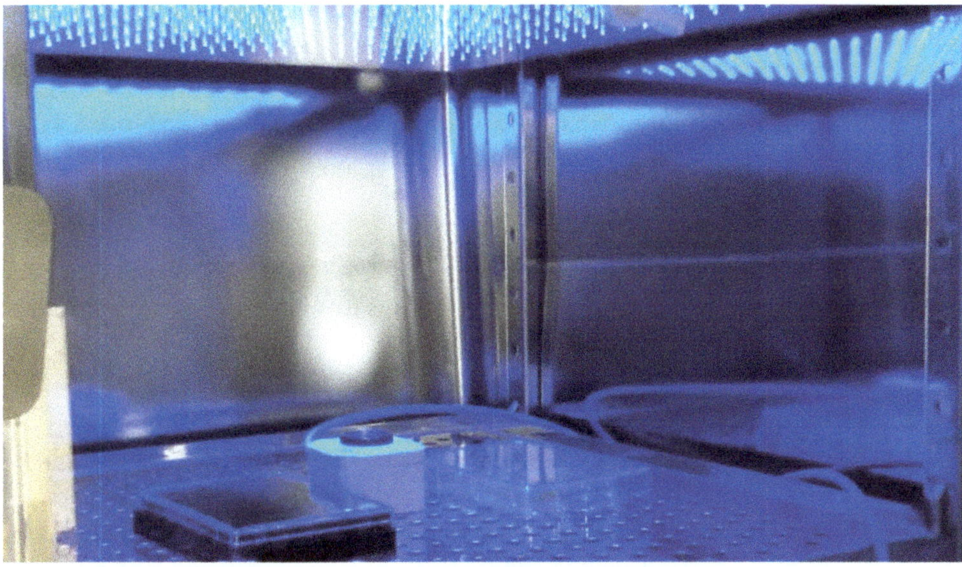

Fig. 2 *A custom-made LED panel can be installed in the cell culture incubator.* A light sensor below the panel constantly measures the light intensity and feeds the data into an external controller unit

5. The plasmid mix and the lipofection reagent mix are combined in a 1:1 ratio. Per well, 25 μL of plasmid mix and 25 μL of lipofection reagent mix are prepared. To each 110 μL of plasmid mix, add 110 μL of lipofection mix (*see* **Note 9**).
6. Vortex all samples for 5 s and spin them down. Incubate the combined transfection mixes for 30 min at 20 °C.
7. Remove the supernatant from the cells and carefully add 500 μL OptiMEM to the cells (*see* **Note 10**). The temperature of the OptiMEM should be at 37 °C.
8. Switch on the illumination device in the cell culture incubator and program the following settings: Light Intensity = 100 μW/cm^2, Mode = Pulse with 60 s off-time and 30 s on-time. Start the program and check if the illumination device is working as expected. An example of an illumination setup is depicted in Fig. 2.
9. Add 50 μL transfection mix into each well by holding the pipette tip close over the liquid surface of the well. Carefully transfer the transfection mix into the wells (*see* **Note 11**).
10. After the addition of transfection mix, move the plate 4–5 times to gently distribute the transfection reagent within the wells.
11. Incubate one of the plates in an incubator with and the other plate in an incubator without illumination. Incubation time should be for 4 h.

12. Carefully remove and discard the supernatant and replace it with D10 medium (*see* **Note 10**).
13. Place the cell culture plates back into either the dark or the illuminated incubator and wait for 20 h (*see* **Note 12**).

3.3 Flow Cytometer Measurement

1. Switch on the flow cytometer and perform all startup procedures.
2. Label FACS tubes according to the samples.
3. Take out the 24-well plates from the incubator and remove the supernatant (*see* **Notes 10**).
4. Wash cells once gently with 500 μL DPBS (at 37 °C) (*see* **Note 10**).
5. Add 200 μL 0.05% Trypsin-EDTA to the cells and place them for 10 min in the dark cell culture incubator.
6. Add 300 μL D10 medium to the cells to stop the reaction and transfer the cell suspension into the labeled FACS tubes.
7. Wash the cells using 1 mL DPBS and centrifuge the tubes for 5 min at 200× *g*. Remove the supernatant but leave ~100 μL of liquid in the tubes for the flow cytometer measurement.
8. Measure 30.000 cells per sample. For optimal visualization, plot the reporter channel cytogram (e.g., PacificBlue-A) against the SSC-A and display 1000 events per second.

3.4 Flow Cytometer Data Analysis

1. Open the data files with the FlowJo software. Load the cytogram, which displays the untransfected cells, and set the SSC-H vs. SSC-A plot to isolate single cells by excluding cell debris at the same time. Next, apply a quadrant gate to the cytogram of the untransfected cells and set the PacificBlue-A autofluorescence to 1%. Apply these settings to all samples and export a table with the quadrant 2 shift data (*see* Fig. 3a).
2. Normalize the data using sgRNA 2.0 (dark) as positive control (maximum value) and untransfected cells (dark) as negative control (minimum value). The normalization is done by applying the feature scaling method on each sample.
3. Plot the normalized data in the Prism software (*see* Fig. 3b).

4 Notes

1. The passage number should not be higher than 30, after the first inoculation.
2. The plasmids should be freshly prepared from a plasmid extraction of an overnight liquid culture. The purity of the plasmids should be verified at least by photospectroscopy and by visualization on an agarose gel. The yields for the pENTER constructs usually are lower compared to the other constructs.

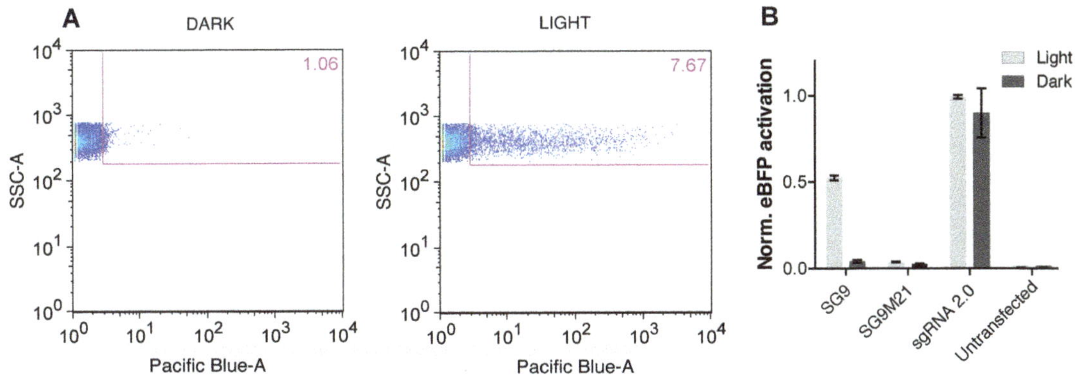

Fig. 3 *Light-dependent activation of eBFP expression using flow cytometry.* (**a**) Scatter plots for SG9 in the dark and the light. Amount of shift for quadrant 2 is depicted in the top right corner. (**b**) Normalized eBFP activation as measured by flow cytometry. SG9 shows eBFP-activated cells in light but not in darkness. The aptamer mutant SG9M21 induces eBFP expression neither in light nor in the darkness. The sgRNA 2.0 combined with MPH leads to a strong induction of eBFP activation, which is light independent. Control sgRNA without any aptamers remains at background level comparable to the SG9M21 mutant. Untransfected cells do not exhibit eBFP fluorescence. The data was normalized to sgRNA 2.0 using Min-Max scaling

3. In our lab, we used HeLa cells but we expect the protocol also to work with different easy-to-transfect cell lines such as HEK293 cells.
4. All additional reagents are directly added to the bottle. The medium can be prewarmed in a water bath. Avoid longer prewarming times than 20 min at 37 °C.
5. It is also useful, to stain the cells, for example, with trypan blue, to indicate the ratio between dead and alive cells.
6. Sometimes, the cells tend to settle in the middle of the well. Moving the plate as described can help to distribute the cells in the well more evenly.
7. Always calculate for more samples, to exclude pipette errors.
8. The plasmids should be in a volume range between 1 and 10 μL.
9. Pipette the lipofection reagent mix below the liquid surface of the plasmid mix and quickly pipette up and down three times.
10. Make sure that the cells are not detaching from the well surface.
11. Pipette in a way, that drops are formed at the tip and merge with the liquid surface. There should be approximately three drops per transfection.
12. If the dark incubator is used by colleagues in parallel, it is sufficient to cover the cells with a sterile paper box, to shield light from the plate.

Acknowledgments

This work was supported by funds from the European Union ERC ("OptoRibo", 615381) to G.M. and the German Research Council (grants MA3442/5-1 and 5-2 and SFB1089-P1 to G.M.).

References

1. Barrangou R, Fremaux C, Deveau H et al (2007) CRISPR provides acquired resistance against viruses in prokaryotes. Science 315(5819):1709–1712
2. Leenay RT, Beisel CL (2017) Deciphering, communicating, and engineering the CRISPR PAM. J Mol Biol 429(2):177–191
3. Hsu PD, Lander ES, Zhang F (2014) Development and applications of CRISPR-Cas9 for genome engineering. Cell 157(6):1262–1278
4. Doudna JA, Charpentier E (2014) Genome editing. The new frontier of genome engineering with CRISPR-Cas9. Science 346(6213): 1258096
5. Gilbert LA, Larson MH, Morsut L et al (2013) CRISPR-mediated modular RNA-guided regulation of transcription in eukaryotes. Cell 154(2):442–451
6. Bassett AR, Liu JL (2014) CRISPR/Cas9 and genome editing in Drosophila. J Genet Genomics 41(1):7–19
7. Chavez A, Scheiman J, Vora S et al (2015) Highly efficient Cas9-mediated transcriptional programming. Nat Methods 12(4):326–328
8. Konermann S, Brigham MD, Trevino AE et al (2015) Genome-scale transcriptional activation by an engineered CRISPR-Cas9 complex. Nature 517(7536):583–588
9. Jia Y, Xu RG, Ren X et al (2018) Next-generation CRISPR/Cas9 transcriptional activation in Drosophila using flySAM. Proc Natl Acad Sci USA 115(18):4719–4724
10. Nihongaki Y, Yamamoto S, Kawano F et al (2015) CRISPR-Cas9-based photoactivatable transcription system. Chem Biol 22(2): 169–174
11. Polstein LR, Gersbach CA (2015) A light-inducible CRISPR-Cas9 system for control of endogenous gene activation. Nat Chem Biol 11(3):198–200
12. Nihongaki Y, Furuhata Y, Otabe T et al (2017) CRISPR-Cas9-based photoactivatable transcription systems to induce neuronal differentiation. Nat Methods 14(10):963–966
13. Weber AM, Kaiser J, Ziegler T et al (2019) A blue light receptor that mediates RNA binding and translational regulation. Nat Chem Biol 15(11):1085–1092

Chapter 4

A Red Light–Activatable Endogenous Gene Transcription System with Red-CPTS

Takahiro Nakajima, Yuto Kuwasaki, Shota Yamamoto, Takahiro Otabe, and Moritoshi Sato

Abstract

Red light penetrates deep into mammalian tissues and has low phototoxicity. We developed a red light–activatable photoswitch (MagRed) for deep tissue optogenetics. Using MagRed, we developed a red light–activatable endogenous gene transcription system (Red-CPTS) based on CRISPR–Cas9. Here we provide a detailed protocol for endogenous gene activation using Red-CPTS in cultured mammalian cells and living mice in vivo.

Key words Photoswitch, Optogenetics, Red light, Endogenous gene activation, CRISPR–Cas9

1 Introduction

After the demonstration of optogenetic control of neurons using microbial opsins [1], blue light–activatable photoswitches emerged as a powerful core technology for direct manipulation of protein activity with high spatiotemporal resolution using blue light [2–6]. The versatility and regulatability of blue light–activatable photoswitches have facilitated biological applications in a variety of research fields beyond neuroscience [7–18]. However, blue light is easily scattered and absorbed in mammalian tissues, which greatly hinders the application of blue light–activatable photoswitches to deep tissues [19].

Compared to blue light, red light is advantageous for optogenetic use in mammalian deep tissues due to its high tissue transparency, low invasiveness, and low light scattering [20]. Recently, we developed a red light–activatable photoswitch that has high versatility and regulatability for enabling optogenetic manipulation using red light, named MagRed [21]. MagRed consists of a red light–absorbing bacterial phytochrome incorporating a mammalian

Armin Baumschlager (ed.), *Optogenetics: Methods and Protocols*, Methods in Molecular Biology, vol. 2840,
https://doi.org/10.1007/978-1-0716-4047-0_4,

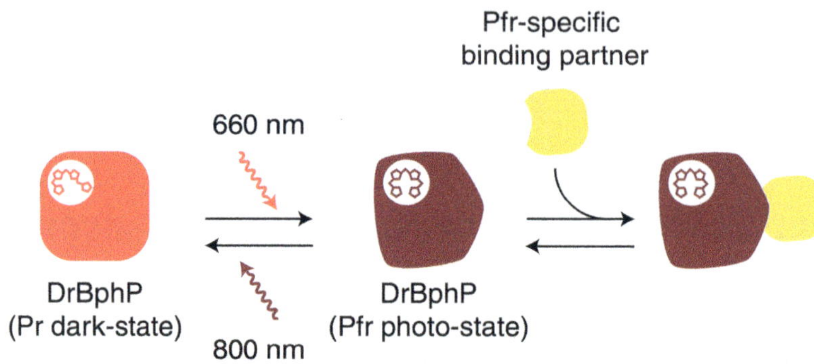

Fig. 1 Schematic representation of MagRed. DrBphP is a red light-absorbing bacterial phytochrome incorporating a mammalian endogenous chromophore, biliverdin, and reversibly photoconverts between a Pr dark-state and a Pfr photo-state with two-wavelength light illuminations. Pfr-specific binding partner is a photo-state-specific binder that we developed using Affibody library selection. (Reproduced from Ref. [21] with permission from Springer Nature)

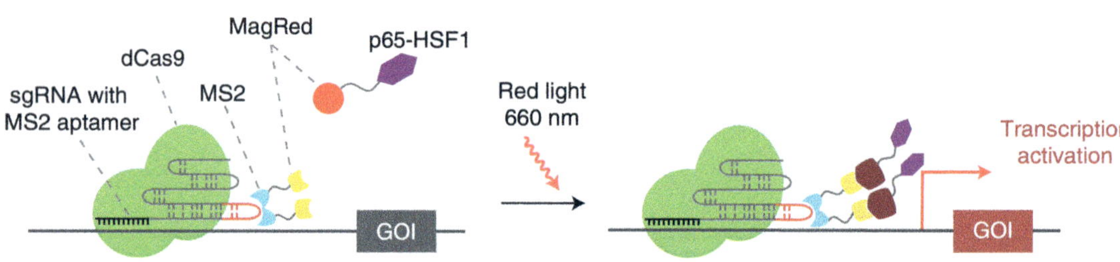

Fig. 2 Schematic representation of Red-CPTS system. Catalytically inactive Cas9 (dCas9) and sgRNA-bearing MS2 RNA aptamers work as a genomic anchor. MagRed is fused to the transcription activation domain p65-HSF1 and MS2 coat protein. Upon 660-nm light illumination, p65-HSF1 is recruited to target loci by MagRed interaction, activating transcription of targeted genes. (Reproduced from Ref. [21] with permission from Springer Nature)

endogenous chromophore, biliverdin, and a photo-state-specific binder that we developed using Affibody library selection (Fig. 1).

Using MagRed, we developed a red light–activatable endogenous gene transcription system, named Red-CPTS [21] (Fig. 2). Red-CPTS is based on CRISPR–Cas9. In this system, MagRed plays a role in the red light–dependent recruitment of transcription activation domains, p65 and HSF1, to target loci, harboring inactive Cas9 (dCas9), single-guide RNA (sgRNA)-bearing MS2 RNA aptamer and MS2 coat protein (Fig. 2). Red-CPTS enables multiplexed activation of user-defined endogenous genes using red light illumination.

In this chapter, we describe endogenous gene activation using Red-CPTS in cultured mammalian cells and living mice in vivo. First, we introduce a protocol for multiple endogenous gene activation by red light in cultured mammalian cells. In this experiment, we simultaneously activate four human genes, such as *ASCL1*,

HBG1, *IL1R2*, and *MYOD1*. Next, we provide a protocol for reporter gene activation by red light in mouse liver in vivo. Bioluminescence imaging using a luciferase reporter allows us to observe gene activation by Red-CPTS in living mice. Finally, we describe a protocol for endogenous gene activation such as mouse *ASCL1* gene by red light in mouse liver in vivo.

2 Materials

2.1 Cell Culture and Transfection

1. HEK293T.
2. Culture medium: Dulbecco's Modified Eagle Medium (DMEM) with 10% Fetal Bovine Serum (FBS), 100 unit/mL penicillin, 100 μg/mL streptomycin.
3. 96-well cell culture plate.
4. Lipofection reagent for mammalian cells.

2.2 Mice and Hydrodynamic Tail Vein Injection

1. Four-week-old female ICR mice.
2. Six-week-old female BALB/c mice.
3. TransIT-EE Hydrodynamic Delivery Solution (Mirus).
4. Animal clippers.
5. Depilatory cream.
6. Transparent acryl box (W 140 mm × D 140 mm × H 180 mm).
7. Wire mesh laid (W 130 mm × D 130 mm).
8. Water gel for animal hydration.

2.3 Plasmids of Red-CPTS (See Note 1)

1. NLS-dCas9-NLS (used in Subheadings 3.1, 3.2, and 3.3).
2. 3xNLS-DrBphP-p65-HSF1 (used in Subheadings 3.1, 3.2, and 3.3).
3. 2xAff6_V18FΔN-MS2-3xNLS (used in Subheadings 3.1, 3.2, and 3.3).
4. sgRNA_ASCL1 (used in Subheading 3.1 and *see* **Notes 2–4**).
5. sgRNA_HBG1 (used in Subheading 3.1 and *see* **Notes 2–4**).
6. sgRNA_IL1R2 (used in Subheading 3.1 and *see* **Notes 2–4**).
7. sgRNA_MYOD1 (used in Subheading 3.1 and *see* **Notes 2–4**).
8. sgRNA_GAL4UAS (used in Subheading 3.2 and *see* **Notes 2–4**).
9. GAL4UAS-luc2 (used in Subheading 3.2).
10. sgRNA_mASCL1 (used in Subheading 3.3 and *see* **Notes 2–4**).
11. sgRNA_empty (used in Subheadings 3.1, 3.2, and 3.3).

2.4 RNA Extraction Kit, Reverse Transcription Kit, and Bioluminescence Imaging Reagent

1. CellAmp Direct RNA prep Kit (TaKaRa).
2. SuperScript IV VILO Master Mix (Invitrogen).
3. RNAlater stabilization solution (Invitrogen).
4. Precellys Lysing Kit CK28 (Bertin Instruments).
5. NucleoSpin RNA kit (Macherey-nagel).
6. D-luciferin: 100 mM solution in saline.

2.5 Quantitative PCR Kit

1. TaqMan Gene Expression Master Mix (Applied Biosystems).
2. Luna Universal Probe qPCR Master Mix (New England Biolabs).
3. TaqMan Gene Expression Assays for *ASCL1*: Hs04187546_g1 (Applied Biosystems).
4. TaqMan Gene Expression Assays for *HBG1*: Hs00361131_g1 (Applied Biosystems).
5. TaqMan Gene Expression Assays for *IL1R2*: Hs01030384_m1 (Applied Biosystems).
6. TaqMan Gene Expression Assays for *MYOD1*: Hs02330075_g1 (Applied Biosystems).
7. TaqMan Gene Expression Assays for *GAPDH*: Hs99999905_m1 (Applied Biosystems).
8. TaqMan Gene Expression Assays for mouse *ASCL1*: Mm03058063_m1 (Applied Biosystems).
9. TaqMan Gene Expression Assays for mouse *GAPDH*: Mm99999915_g1 (Applied Biosystems).

2.6 Instruments

1. 660 nm LED light source (CCS).
2. Regulated DC power supply (Kikusui Electronics).
3. Arduino mega microcontroller board (Arduino).
4. StepOnePlus Real-Time PCR system (Applied Biosystems).
5. Lumazone bioluminescence imager (Shoshin EM).
6. Evolve 512 EMCCD camera (Photometrics).
7. SlideBook software (Intelligent Imaging Innovations).
8. Precellys Evolution tissue homogenizer with cooling system (Bertin Instruments).

3 Methods

3.1 Multiple Endogenous Gene Activation by Red Light in Cultured Mammalian Cells (See Fig. 3)

1. Plate HEK293T cells at 2.0×10^4 cells per well into a 96-well cell culture plate. Culture the cells for 24 h at 37 °C in 5% CO_2.
2. Transfect the cells by lipofection with plasmids encoding NLS-dCas9-NLS, 3xNLS-DrBphP-p65-HSF1, 2xAff6_V18F ΔN-MS2-3xNLS, sgRNA_ASCL1, sgRNA_HBG1, sgRNA_

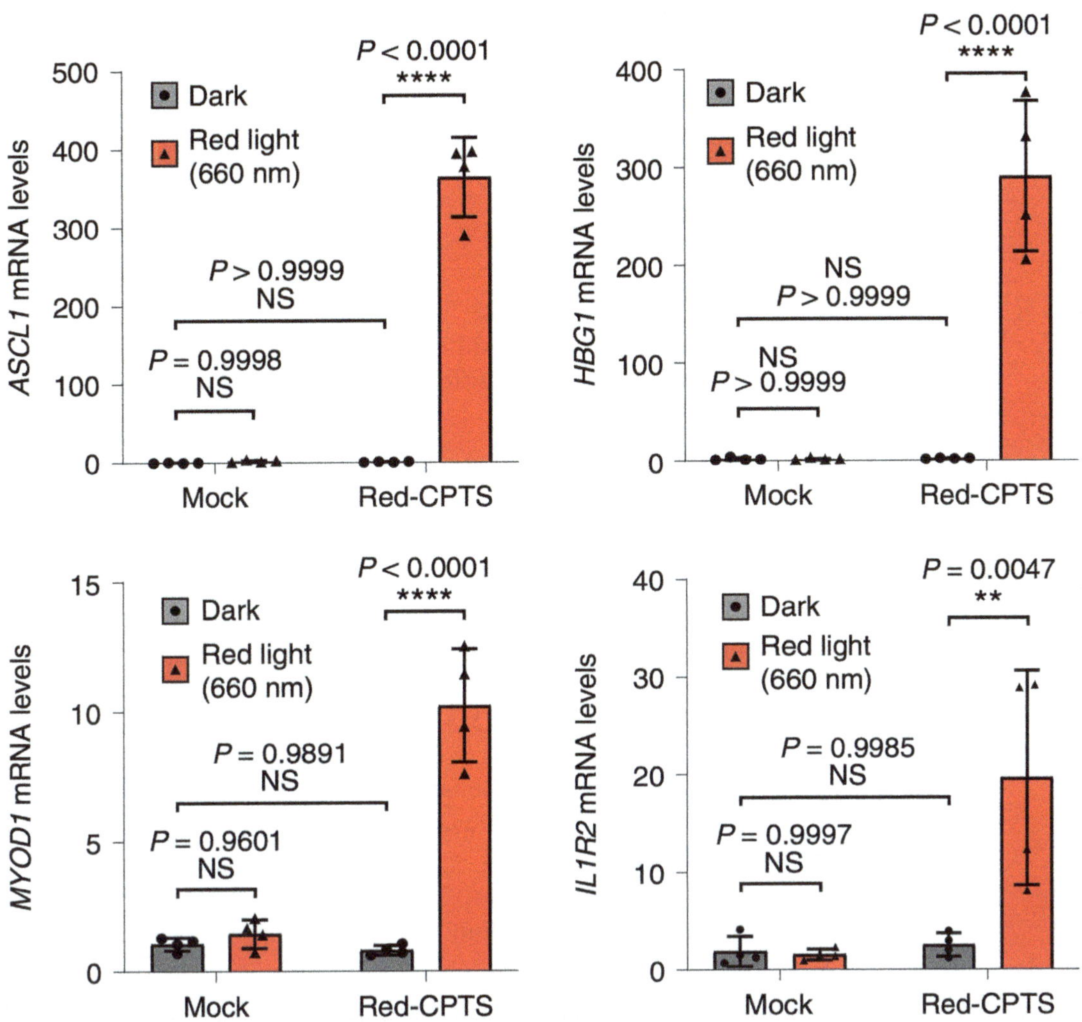

Fig. 3 Multiple endogenous gene activation by red light in cultured mammalian cells. Simultaneous activation of multiple endogenous gene targets is achieved using a mixture of sgRNAs targeted to *ASCL1*, *HBG1*, *IL1R2*, and *MYOD1* in HEK293T cells. *P* values are indicated above the bars. (*NS* not significant $P > 0.05$; $^*P < 0.05$; $^{**}P < 0.01$; $^{***}P < 0.001$; $^{****}P < 0.0001$; two-way ANOVA with multiple comparisons, $n = 4$ biologically independent samples, mean ± s.d.). a.u., arbitrary units. (Reproduced from Ref. [21] with permission from Springer Nature)

IL1R2, and sgRNA_MYOD1 at a 4:4:4:1:1:1:1 ratio. The total amount of DNA is 0.1 μg per well.

3. Twenty-four hours after the transfection, add 100 μL of fresh media to the samples and incubate them at 37 °C in 5% CO_2 under 660-nm pulsed light illumination (1 W m^{-2}) that is repeatedly switched on for 1 min and then turned off for 4 min (*see* **Note 5**). Generate the pulsed illumination pattern using an Arduino mega microcontroller board. For dark condition, wrap the sample in aluminum foil during incubation.

4. After the incubation for 24 h, extract total RNA using CellAmp Direct RNA prep Kit according to the manufacturer's protocol.
5. Reversely transcribe the total RNA into cDNA using SuperScript IV VILO Master Mix according to the manufacturer's protocol.
6. Perform quantitative PCR with StepOnePlus Real-Time PCR system using TaqMan Gene Expression Master Mix. Use TaqMan primers and probes to quantify the expression level of the target genes. As endogenous control, measure *GAPDH* gene.
7. Use the ΔΔCt method [22] to show each relative mRNA level to that from nontransfected HEK293T cells as a negative control.

3.2 Luciferase Reporter Gene Activation by Red Light in Mouse Liver (See Figs. 4, 5a, b, c and Note 6)

1. Keep 4-week-old female ICR mice under conditions with 20–26 °C, 40–60% humidity, a 14-h/10-h light/dark cycle, and free access to food and water.
2. Hydrodynamically transfect the livers of the mice by injecting plasmids via the tail vein with TransIT-EE Hydrodynamic Delivery Solution. Transfect the plasmids encoding NLS-dCas9-NLS, 3xNLS-DrBphP-p65-HSF1, 2xAff6_V18FΔN-MS2-3xNLS, sgRNA_GAL4UAS, and luciferase reporter gene at a 1:1:1:1:1 ratio. The total amount of DNA is 150 μg per mouse (*see* **Note 7**).
3. Eight hours after the injection, remove the abdominal surface fur of the mice using clippers and a depilatory cream. Randomly assign the mice to dark and red light groups (Fig. 4a). As to the assignment, blind the person performing the hydrodynamic injections.
4. Illuminate the mice of the red light group using an LED light source (660 nm, 100 W m^{-2}, continuous) in a transparent acrylic box with a wire mesh laid at the bottom and mouse diet pellets and water gel (Fig. 4b, *see* **Note 8**). Keep the mice of the dark group in the home cage that receives no light (Fig. 4a).
5. Twenty-four hours after the illumination, perform bioluminescence imaging of the mice. To perform bioluminescence imaging, inject 200 μL of 100 mM D-luciferin intraperitoneally into the mice. Then anesthetize the mice with isoflurane. Five minutes after D-luciferin injection, obtain bioluminescence images of the mice using Lumazone bioluminescence imager equipped with Evolve 512 EMCCD camera controlled by SlideBook software.

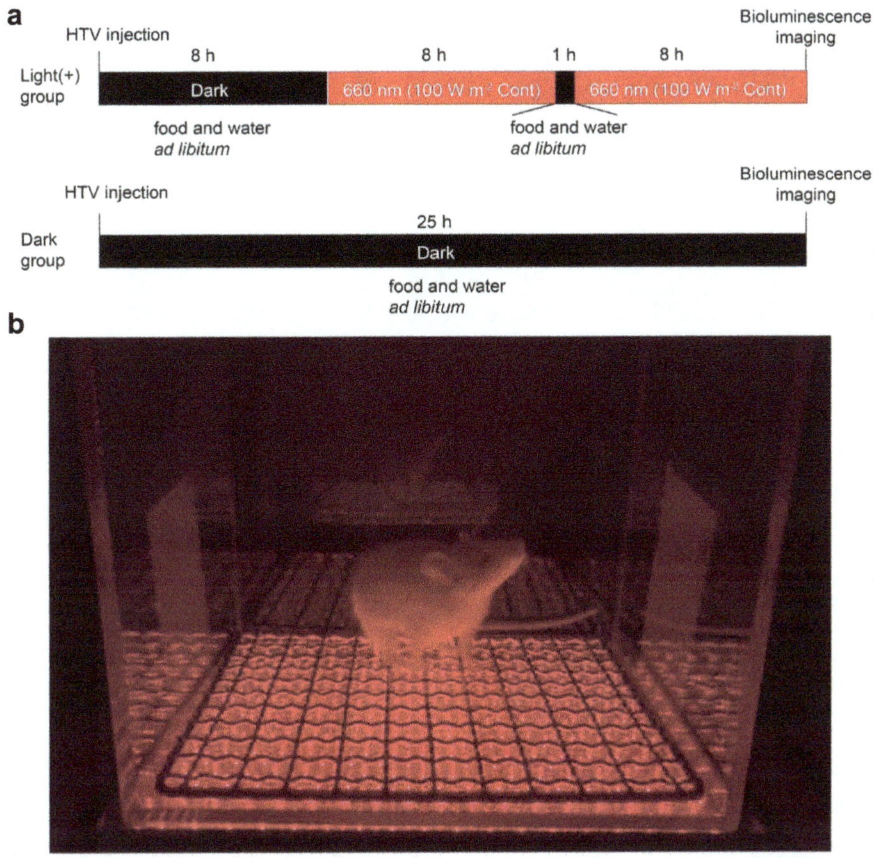

Fig. 4 Noninvasive illumination used for optogenetic gene activation by Red-CPTS in living mice. (**a**) Experimental time course for the activation of Red-CPTS in the mouse liver. ICR mice and BALB/c mice are transfected with plasmids by the hydrodynamic tail vein (HTV) injection method. After HTV injection, the mice are illuminated with red light (660 nm, 100 W m^{-2}, continuous) for 16 h or kept in the dark. (**b**) A photograph to illustrate external and noninvasive illumination of the mice with an LED light source. The transfected mice are kept in a transparent acrylic box with a wire mesh laid at the bottom. (Reproduced from Ref. [21] with permission from Springer Nature)

3.3 Endogenous Gene Activation by Red Light in Mouse Liver (See Figs. 4, 5d, and Note 6)

1. Keep 6-week-old female BALB/c mice under conditions with 20–26 °C, 40–60% humidity, a 14-h/10-h light/dark cycle, and free access to food and water.
2. Hydrodynamically transfect the livers of the mice by injecting plasmids via the tail vein with TransIT-EE Hydrodynamic Delivery Solution. Transfect the plasmids encoding NLS-dCas9-NLS, 3xNLS-DrBphP-p65-HSF1, 2xAff6_V18FΔN-MS2-3xNLS, and sgRNA_mASCL1 at a 1:1:1:1 ratio. The total amount of DNA is 120 μg per mouse.
3. Eight hours after the injection, remove the abdominal surface fur of the mice using clippers and a depilatory cream. Randomly assign the mice to dark and red light groups (Fig. 4a). As to the assignment, blind the person performing the hydrodynamic injections.

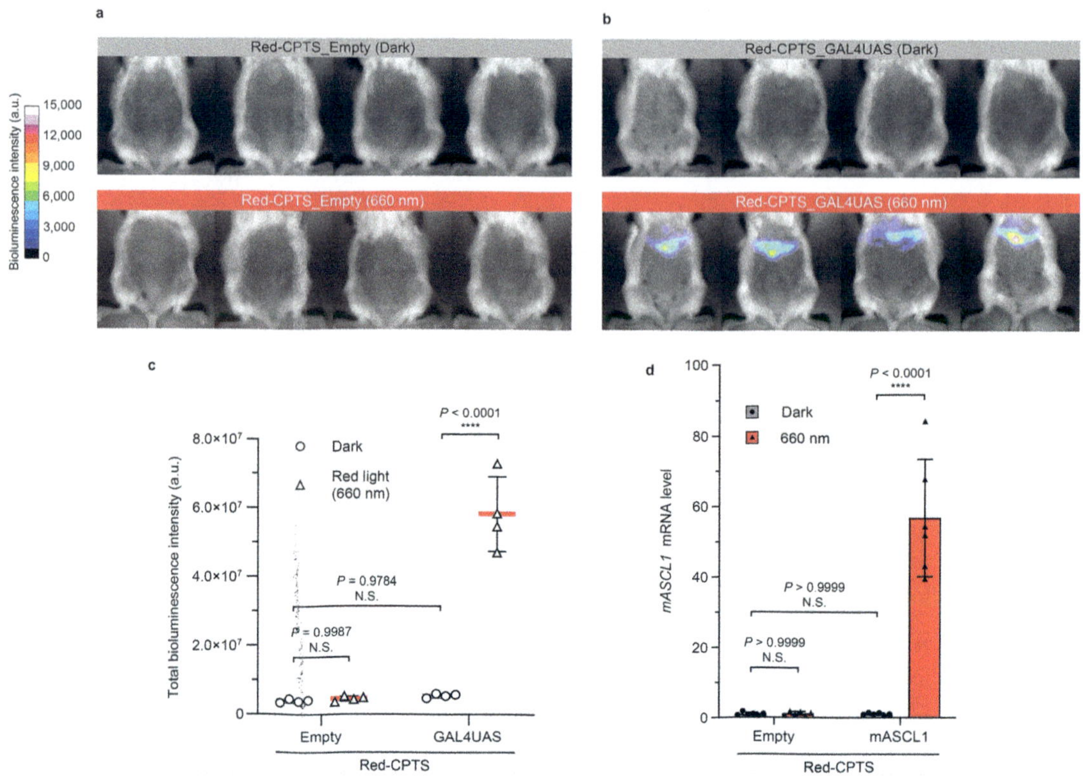

Fig. 5 In vivo gene activation by Red-CPTS upon noninvasive red light illumination. (**a**, **b**) ICR mice are transfected with plasmids encoding Red-CPTS and luciferase reporter together with a plasmid encoding unrelated sgRNA as a negative control (**a**, Empty) or sgRNA targeting GAL4UAS (**b**, GAL4UAS). After the transfection, the mice are noninvasively illuminated at 660 nm or kept in the dark, as shown in Fig. 4a, and then bioluminescence imaging of the mice is performed. (**c**) Total bioluminescence intensities of the mice shown in (**a**) and (**b**). Gray and red bars represent the mean ± s.d., and dots represent the total bioluminescence intensity of each mouse ($n = 4$ mice per group). (*N.S.* not significant $P > 0.05$; ****$P < 0.0001$; using two-way ANOVA with multiple comparisons). (**d**) Red light-dependent endogenous gene activation by Red-CPTS with unrelated sgRNA as a negative control (Empty) or sgRNA targeting mouse *ASCL1* (m*ASCL1*) in vivo in living BALB/c mice. Data are represented as the relative mRNA level to the nontransfected negative control ($n = 6$ mice per group). Gray and red bars represent the mean ± s.d., and dots represent individual data points. No difference can be observed in the appearance between the mice maintained in the dark and the ones illuminated with red light at 660 nm for 16 h. *P* values are indicated above the bars. (*N.S.* not significant $P > 0.05$; ****$P < 0.0001$; using two-way ANOVA with multiple comparisons)

4. Illuminate the mice of the red light group with an LED light source (660 nm, 100 W m^{-2}, continuous) in a transparent acrylic box with a wire mesh laid at the bottom and mouse diet pellets and water gel (Fig. 4b, *see* **Note 8**). Keep the mice of the dark group in the home cage that receives no light (Fig. 4a).
5. Twenty-four hours after the illumination, euthanize the mice. Then dissect the mice and remove the livers. Cut the livers into pieces and collect them in RNAlater stabilization solution.

6. Homogenize the pieces of the livers using Precellys Lysing Kit CK28 and Precellys Evolution tissue homogenizer with cooling system. Extract and purify total RNA from the homogenates of the livers using NucleoSpin RNA kit.
7. Reversely transcribe the total RNA into cDNA using SuperScript IV VILO Master Mix according to the manufacturer's protocol.
8. Perform quantitative PCR with StepOnePlus Real-Time PCR system using Luna Universal Probe qPCR Master Mix. Use TaqMan primers and probes to detect mouse *ASCL1* and mouse *GAPDH* genes. As endogenous control, measure *GAPDH* gene.
9. Use the ΔΔCt method [22] to show each relative mRNA level to that from the liver of nontransfected mice as a negative control.

4 Notes

1. Contact Moritoshi Sato (moritoshisato@g.ecc.u-tokyo.ac.jp) to request plasmids related to Red-CPTS.
2. When designing sgRNAs, the following three points should be considered. The first point is the genomic region to which the sgRNA is bound. Although it varies depending on the target gene, in our experience, gene expression is highly efficient if the sgRNA is bound to the upstream region up to about 200 bases from the transcription start point. Second, the target sequence of the genome to which the sgRNA binds must be in close proximity to a nucleotide sequence called PAM. Because Red-CPTS uses Cas9 from *Streptococcus pyogenes*, the PAM sequence is NGG. The third point is the binding selectivity of sgRNA to the genome. If the selectivity is low, off-targeting may be a problem. The 5′ end sequence of sgRNA (about 20 bases) is designed in consideration of the above three points.
3. To construct the sgRNA vector, we introduced the MS2 RNA aptamer into the pSPgRNA vector (plasmid #47108) available from Addgene and used *Bbs*I restriction enzyme site to introduce an arbitrary sequence at the 5′ end of the sgRNA.
4. The expression of a target gene can be induced more strongly by simultaneously binding several different sgRNAs to the upstream regions of the gene.
5. The pulsed illumination pattern of 1-min light-on and 4-min light-off is used to avoid temperature increase in the incubator. If the reader changes the illumination conditions, the temperature in the incubator should be measured.

6. All procedures involving animals should be approved by your Institutional Animal Care and Use Committee (IACUC).
7. The manufacturer's protocol for TransIT-EE Hydrodynamic Delivery Solution recommends a DNA amount of 1–50 μg. We found that the total DNA amount of 150 μg gave better results than the total DNA amount of 50 μg in the mice experiments using Red-CPTS.
8. To illuminate the mice using an LED light source, use a fan or an air circulator to avoid temperature increase.

Acknowledgments

This work was supported by a CREST grant (JPMJCR1653) from the Japan Science and Technology Agency; a project grant from the Kanagawa Institute of Industrial Science and Technology; and the UTEC-UTokyo FSI Research Grant Program.

References

1. Boyden ES, Zhang F, Bamberg E, Nagel G, Deisseroth K (2005) Millisecond-timescale, genetically targeted optical control of neural activity. Nat Neurosci 8:1263–1268
2. Kennedy MJ, Hughes RM, Peteya LA, Schwartz JW, Ehlers MD, Tucker CL (2010) Rapid blue-light-mediated induction of protein interactions in living cells. Nat Methods 7:973–975
3. Guntas G, Hallett RA, Zimmerman SP, Williams T, Yumerefendi H, Bear JE, Kuhlman B (2015) Engineering an improved light-induced dimer (iLID) for controlling the localization and activity of signaling proteins. Proc Natl Acad Sci USA 112:112–117
4. Kawano F, Suzuki H, Furuya A, Sato M (2015) Engineered pairs of distinct photoswitches for optogenetic control of cellular proteins. Nat Commun 6:6256
5. Zhang K, Cui B (2015) Optogenetic control of intracellular signaling pathways. Trends Biotechnol 33:92–100
6. Losi A, Gardner KH, Möglich A (2018) Blue-light receptors for optogenetics. Chem Rev 118:10659–10709
7. Nihongaki Y, Yamamoto S, Kawano F, Suzuki H, Sato M (2015) CRISPR–Cas9-based photoactivatable transcription system. Chem Biol 22:169–174
8. Nihongaki Y, Kawano F, Nakajima T, Sato M (2015) Photoactivatable CRISPR–Cas9 for optogenetic genome editing. Nat Biotechnol 33:755–760
9. Kawano F, Okazaki R, Yazawa M, Sato M (2016) A photoactivatable Cre-loxP recombination system for optogenetic genome engineering. Nat Chem Biol 12:1059–1064
10. Nihongaki Y, Furuhata Y, Otabe T, Hasegawa S, Yoshimoto K, Sato M (2017) CRISPR–Cas9-based photoactivatable transcription systems to induce neuronal differentiation. Nat Methods 14:963–966
11. Furuya A, Kawano F, Nakajima T, Ueda Y, Sato M (2017) Assembly domain-based optogenetic system for the efficient control of cellular signaling. ACS Synth Biol 16:1086–1095
12. Nihongaki Y, Otabe T, Ueda Y, Sato M (2019) A split CRISPR–Cpf1 platform for inducible genome editing and gene activation. Nat Chem Biol 15:882–888
13. Aoki K, Kumagai Y, Sakurai A, Komatsu N, Fujita Y, Shionyu C, Matsuda M (2013) Stochastic ERK activation induced by noise and cell-to-cell propagation regulates cell density-dependent proliferation. Mol Cell 52:529–540
14. Lee D, Creed M, Jung K, Stefanelli T, Wendler DJ, Oh WC, Mignocchi NL, Lüscher C, Kwon HB (2017) Temporally precise labeling and control of neuromodulatory circuits in the mammalian brain. Nat Methods 14:495–503
15. Pu J, Zinkus-Boltz J, Dickinson BC (2017) Evolution of a split RNA polymerase as a

versatile biosensor platform. Nat Chem Biol 13:432–438

16. Jung H, Kim SW, Kim M, Hong J, Yu D, Kim JH, Lee Y, Kim S, Woo D, Shin HS, Park BO, Heo WD (2019) Noninvasive optical activation of Flp recombinase for genetic manipulation in deep mouse brain regions. Nat Commun 10: 314
17. Yu D, Lee H, Hong J, Jung H, Jo YJ, Oh BH, Park BO, Heo WD (2019) Optogenetic activation of intracellular antibodies for direct modulation of endogenous proteins. Nat Methods 16:1095–1100
18. Liu Q, Sinnen BL, Boxer EE, Schneider MW, Grybko MJ, Buchta WC, Gibson ES, Wysoczynski CL, Ford CP, Gottschalk A, Aoto J, Tucker CL, Kennedy MJ (2019) A photoactivatable botulinum neurotoxin for inducible control of neurotransmission. Neuron 101: 863–875
19. Ziegler T, Möglich A (2015) Photoreceptor engineering. Front Mol Biosci 2:30
20. Weissleder R (2001) A clearer vision for in vivo imaging biological imaging. Nat Biotechnol 19:316–317
21. Kuwasaki Y, Suzuki K, Yu G, Yamamoto S, Otabe T, Kakihara Y, Nishiwaki M, Miyake K, Fushimi K, Bekdash R, Shimizu Y, Narikawa R, Nakajima T, Yazawa M, Sato M (2022) A red light–responsive photoswitch for deep tissue optogenetics. Nat Biotechnol 40:1672–1679
22. Livak KJ, Schmittgen TD (2001) Analysis of relative gene expression data using real-time quantitative PCR and the 2-ΔΔCT method. Methods 25:402–408

Chapter 5

Targeted Activation of Programmed Cell Death Pathways by Optogenetics

Kateryna Shkarina and Petr Broz

Abstract

Regulated cell death is an important biological process by which an organism removes unwanted, malignant, or infected cells. Although it has become clear that different forms of regulated cell death exist, it remains difficult to compare their consequences at the cellular and tissue level as they are induced by different stimuli and proceed with different kinetics. Moreover, it was so far difficult to target and induce cell death in selected cells within cell populations or complex tissues without affecting its neighbors. To overcome these obstacles researchers developed novel synthetic biology tools that use chemical or optogenetic approaches to specifically induce PCD in selected cells. Here, we describe a protocol for optogenetic activation of three major types of regulated cell death: apoptosis, necroptosis, and pyroptosis, using light-induced forced oligomerization of their major effector proteins (caspases or kinases).

Key words Optogenetics, Regulated cell death, Pyroptosis, Necroptosis, Apoptosis

1 Introduction

Regulated or programmed cell death (RCD/PCD) is an essential biological process that allows the removal of superfluous, irreversibly damaged, or potentially harmful cells. While basic forms of RCD can be observed in bacteria and unicellular eukaryotes [1, 2], it has been best studied in mammals and shown to play key roles in immunity, development, and homeostasis. In contrast to accidental necrosis that occurs upon exposure of cells to physical, chemical, or mechanical insults, RCD relies on dedicated signaling pathways that can be engaged by external or internal signals. Research over the last 20 years has identified over a dozen distinct forms of RCD and characterized the respective signaling pathways [3]. The most well-studied form of RCD, apoptosis, is induced by different multiprotein signaling complexes that integrate external or internal signals to control the activation of apoptotic caspases, a family of

Armin Baumschlager (ed.), *Optogenetics: Methods and Protocols*, Methods in Molecular Biology, vol. 2840,
https://doi.org/10.1007/978-1-0716-4047-0_5,

cysteine proteases. Examples of these complexes are the death-inducing signaling complex (DISC) or the TNF complex II that controls activation of caspase-8 (extrinsic pathway of apoptosis) [4], or the apoptosome that controls activation of caspase-9 (intrinsic pathway) [5]. These caspases are expressed as inactive precursors (zymogens) that require recruitment into the complex via protein–protein interaction domains, and oligomerization-dependent autoactivation, which results in the formation of the active caspase hetero-tetramer composed of the catalytic p20 and p10 subunits [6]. Active caspase-8 and -9 process and activate executioner caspases-3 and-7 to cleave downstream substrates and induce cell death [7]. Pyroptosis, another major form of RCD, is induced by activation of inflammatory caspases-1 or -11 in mice or -1, -4, or -5 in humans. These caspases are activated within macromolecular inflammasome complexes that respond to host or pathogen-derived danger signals as part of the innate immune response to infection [8]. To induce cell death these caspases cleave and activate gasdermin-D (GSDMD), a pore-forming protein that permeabilizes the plasma membrane, resulting in necrotic cell death [9]. Cell death can also involve the activation of kinases, as shown for necroptosis. Under conditions of compromised caspase-8 activity, activation of the kinase RIPK3 (RIPK3 receptor-interacting serine/threonine kinase 3) within the necrosome complex results in the phosphorylation of MLKL (mixed lineage kinase like), which targets the plasma membrane and induces necroptosis [10].

While the signaling pathways underlying the induction of different cell death pathways have been well characterized, less is known about how different forms of cell death impact the surrounding tissue and the organism as a whole. It is becoming increasingly clear that different forms of RCD produce diverse and distinct sets of signals and immunological outcomes, ranging from the immunological silent apoptosis to highly proinflammatory forms of cell death like pyroptosis and necroptosis [11], but the exact nature of these signals, and their sensing and processing by surrounding tissues remain to be further elucidated. Studies of such responses typically require selective induction of these forms of RCD combined with the simultaneous monitoring of the behavior of their neighbors. However, they are often limited by varying cell death kinetics, the inability to selectively target single cells in multicellular populations, as well as to the pleiotropic effects of commonly used "natural" cell death triggers both on the dying cells and their neighbors.

Over the past years, multiple strategies have been developed to specifically ablate cells both in vitro and in living animals. Nowadays, the inducible cell death toolbox includes immunological, genetics, optical, and chemo/optogenetic methods which enable selective depletion of either specific cells or cell populations in vitro and in live animals. The examples range from antibody-mediated depletion of specific cell types and inducible expression of the toxins to the laser ablation or chemo−/optogenetic engagement of specific RCD [12]. Optogenetic approaches, in particular, provide several major advantages compared to the other techniques, including superior spatiotemporal control and pathway selectivity, combined with the easy and relatively noninvasive trigger delivery and ability to precisely dose its intensity [13], allowing to shift between lethal and sub-lethal RCD modalities. The current optogenetic RCD toolbox includes tools for induction of cell death signaling via receptor or adaptor protein activation, or direct activation of various cell death effectors [12]. Many of them take advantage of the central role that di- or oligomerization plays in cell death signaling by artificially driving such multimerization using light-activated protein domains (photoactuators), although several employ other types of protein manipulation, such as light-dependent protein recruitment, proteolysis, or degradation. Recently, we further expanded it by developing a set of optogenetically activated cell death effectors (optoCDEs) [14] that enable blue light-induced activation of the major effectors of the three most studied cell death pathways: apoptosis, necroptosis, and pyroptosis. Blue light illumination of the cells expressing these constructs results in activation of these artificial cell death switches, which is in most cell types sufficient to trigger the first signs of cell death within minutes post illumination.

The diverse applications of optoCDEs and other optogenetic cell death systems include selective and highly specific elimination of chosen cells or cell populations both in cell culture settings and in live animals, as well as longitudinal tracking of the neighboring cell responses to these events [12]. They can be used to further decipher the fates of the different types of dying cells in a complex multicellular environment, as well as provide valuable insights into the following niche remodeling and long-term consequences [14, 15]. Besides, the precise control over the illumination parameters, such as light intensity and duration, provides new means for cellular and mechanistic studies of these forms of cell death, as well as probing cell survival mechanisms that limit cellular damage downstream of these effectors. Here we discuss considerations for the design or optoCDEs and describe a protocol for the generation of transgenic cells expressing inducible optoCDEs, the activation of optoCDEs by blue-light illumination, and how to analyze the induction apoptosis, necroptosis, and pyroptosis in cultured cells and how to set up experiment aimed at analyzing the response of live bystander cells.

2 Materials

2.1 Common Materials and Equipment

1. General tissue culture equipment (cell culture incubator with CO_2 supply, sterile laminar flow cabinet).
2. Centrifuge, suitable for tubes (up to 50 mL) and plates.
3. Microcentrifuge.
4. Transmitted light microscope.
5. Confocal or epifluorescent microscope equipped with a suitable set of lasers or filters, humidified incubation chamber, and CO_2 supply.
6. Light Plate Apparatus for 24-well plates, equipped with blue light (450 nm) LEDs.
7. Common cell culture consumables (serological pipettes, pipette tips, petri dishes or flasks, cell scrapers, sterilized microcentrifuge tubes, laboratory pens, etc.)
8. Water bath or heating block suitable for warming up the medium.
9. Cell counting equipment.
10. Aluminum foil.
11. Sterile PBS, cell culture grade.
12. Cell culture medium (depending on the cell line of choice).
13. Trypsin or accutase.

2.2 Reagents and Equipment for Cell Line Generation

1. OptoCDE-expressing lentiviral or retroviral vectors.
2. Packaging vectors.
3. Designated biosafety level 2 area and equipment (including laminar flow cabinet and tissue culture incubator and plate centrifuge).
4. Appropriate safety equipment.
5. Disinfection reagents (such as DESCOGEN®-I for surface disinfection and ULTRASOL® active for spill neutralization).
6. HEK cell medium: DMEM supplemented with GlutaMAX, 5% fetal calf serum (FCS) and 1% Penicillin-Streptomycin mix.
7. Opti-MEM (Gibco) or equivalent serum-free medium suitable for transfection mix assembly.
8. Transfection reagent (such as Lipofectamine 2000, JetPRIME, or PEI).
9. Endotoxin-free plasmid DNA preps (midi or maxi prep format) of optoCDEs and packaging vectors.
10. Polybrene.

11. Tissue culture treated 6-well plates.
12. 0.45 μm filters.
13. Syringes.
14. Blunt 18G syringe needles.
15. Laboratory tape.
16. Antibiotic suitable for transduced cell selection.

2.3 Reagents and Equipment for Microscopy-Based Experiments

1. Cell lines expressing optoCDEs and, for coculture experiments, reporters of choice.
2. Microscopy-suitable cell culture vessels (such as μ-bottom dishes or multiwell chambers (ibidi), CellCarrier Ultra 96-well microplates (PerkinElmer) or equivalent).
3. Opti-MEM (Phenol Red-free) or other live imaging-suitable medium of choice.
4. Dyes for detecting membrane permeabilization: SYTOX™ Green, Propidium iodide, DRAQ7™, or equivalent.
5. Reagents for detecting membrane scrambling: fluorescently conjugated Annexin V, lactadherin, or others.
6. Reagents for total cell visualization and quantification: membrane-permeable DNA dyes (Hoechst 33342, DRAQ5™, or SiR-DNA) or lipophilic membrane dyes (CellMask).
7. Confocal or epifluorescent microscope equipped with a suitable set of lasers or filters, humidified incubation chamber, and CO_2 supply.
8. For single-cell targeting: confocal or two-photon microscope with ROI excitation mode (or microscope equipped with a separate digital mirror device module).
9. Imaging software.
10. Image analysis software (Fiji, CellProfiler, or equivalent).

2.4 Reagents and Equipment for Cell Stimulation Using Light Plate Apparatus

1. Light Plate Apparatus for 24-well plates, equipped with blue light (450 nm) LEDs.
2. Power source.
3. Black 24-well plates, tissue culture-treated (4titude 4TI-0262 or equivalent).
4. If required: serum-free medium for supernatant collection.
5. Other necessary reagents and equipment for the sample collection and processing (e.g., LDH assay kit, cytokine secretion assays, Western Blot equipment, and reagents).
6. Iris webtool (http://taborlab.github.io/Iris/).

3 Methods

3.1 Generation of Transgenic Cell Lines Stably Expressing Optogenetically Activated Cell Death Effectors (optoCDEs)

While transient transfection of optoCDEs can be used, we recommend generating transgenic cell lines that stably express the optoCDE constructs under an inducible promoter. This allows for the selection of a cell population that uniformly expresses the optoCDE constructs, and inducible expression reduces the risk of construct activation by ambient light conditions (*see* **Note 1**).

1. Assemble your optogenetic constructs for cell death induction using an appropriate destination vector, and standard molecular biology techniques. Consider that an effector module should ideally consist of a full-length protein or its functional domain(s) responsible for cell death execution, an *Arabidposis thaliana*-derived Cry2^{E490G} (Cry2olig) module [16] that enables light-induced oligomerization of the effector module, and, in most of the cases, a monomeric red or far-red fluorescent tag (such as mCherry, miRFP or mKate2) for monitoring protein expression and its intracellular localization and activation states, all connected via flexible or semi-flexible linkers (Fig. 1). For consideration on effector module design and delivery vectors refer to *see* **Notes 2** and **3**.
2. Thaw a vial of low-passage HEK cells and expand them to achieve desired cell numbers and viability.
3. One day before the transfection, seed HEK cells into 6-well plates at the density of 5 × 10^5 cells/well. For hard-to-transduce cell types or in the case of low transduction efficiency, viral particle production can be scaled up.

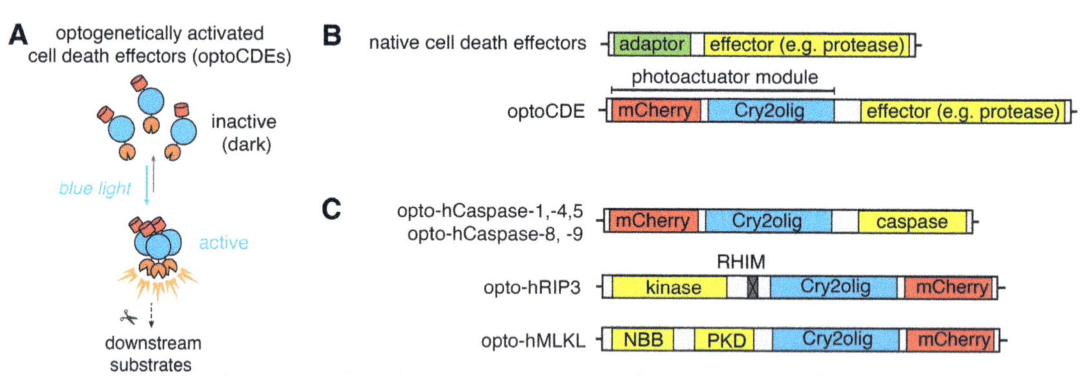

Fig. 1 *Schematic showing optogenetically activated cell death effectors (optoCDEs).* (**a**) Inactive monomeric optoCDEs are activated by illumination with blue light, which induces the clustering of the Cry2olig domain (blue) thus activating effectors domains (red) to initiate signaling and cell death. (**b**) General domain architecture of optoCDEs in comparison to native cell death effectors. OptoCDEs consist of mCherry (to visualize expression), Cry2olig (photoactuator), and an effector module. (**c**) Domain architecture of optoCDEs for pyroptosis, apoptosis, and necroptosis induction

4. Allow the cells to adhere overnight.
5. On the day of the transfection: bring the OptiMEM and transfection reagents to room temperature (RT).
6. Assemble the transfection mix: For each well of HEK cells, prepare two tubes, one containing 150 μL of Opti-MEM and 10 μL of Lipofectamine 2000, and one containing the vector DNA (1.5 μg of optoCDE inducible expression vector (such as pRetroX-TetOne-Puro), 1.5 μg of PsPax2 and 0.3 μg of VSV-g). When preparing the tubes, make sure to mix the contents well and incubate for 5 min at RT before proceeding. If using alternative transfection reagents or packaging vectors, refer to the manufacturer's manual for instructions.
7. Combine the contents of both tubes, mix, and incubate the transfection mix for 15–30 min at RT.
8. Add the transfection mix dropwise to the cells, covering the whole well surface. Mix by gently tapping the plate.
9. To facilitate the transfection, centrifuge the plate for 5 min at 500 *g* at 37 °C.
10. Place the plate back to the cell culture incubator, and incubate for 6 h at 37 °C.
11. Gently replace the transfection medium with 2 mL of collection medium (suitable for the cell line to be transfected).
12. To reduce the risk of accidentally spilling virus-containing medium, seal the plate lid with laboratory tape.
13. Incubate the cells for 24–48 h to allow for sufficient production of viral particles.
14. Under the microscope, check the morphology of the cells. Changes in the cell shape (rounding up and accidental detachment of the single cells or cell groups) indicate a successful production of viral particles.
15. Harvest virus-containing supernatant using a 10- or 20-mL luer-lock syringe equipped with a blunt 18G needle. Aspirate the small amount of air to ensure that no medium remains inside the syringe.
16. Remove the needle and tightly attach a 0.45 μm filter to the syringe using luer lock system.
17. Filter the virus-containing medium into the 15-mL falcon tube. At this step, supernatant from several wells transfected with the same construct can be combined.
18. Dispose of the syringe and the needle using the appropriate safety procedures.
19. Add 5–10 μg/mL polybrene to facilitate viral adhesion and infection.

20. For the transduction of adherent cells: add up to 8 mL of virus-containing supernatant (transduction medium) to up to 1×10^6 cells/well of target cells seeded at least 6–8 h before transduction.
21. To facilitate the infection, seal the lid and centrifuge the plate for 60 min at 1.500 *g* at 37 °C. Proceed to **Step 24**.
22. For the transduction of suspension cells: collect the desired number of cells in a 50 mL flask, pellet, and resuspend in transduction medium.
23. Centrifuge the cell suspension for 60 min at 1.500 *g* at 37 °C. Proceed to **Step 24**.
24. Using a serological pipette, resuspend the cells in the transduction medium and transfer them to 6-well plate (or other vessel depending on the initial cell numbers).
25. Incubate the cells with the transduction medium for 6 h to overnight.
26. Replace the transduction medium with the fresh medium and let the cells recover for an additional 24–48 h. If confluency is reached, passage the cells into the bigger flask.
27. For antibiotic selection: Add the appropriate amount of antibiotic to the cells. Include nontransduced cells as a viability control. We recommend performing a separate antibiotic titration on nontransduced cells beforehand to determine the optimal selection conditions.
28. Incubate for 48–72 h (or until no viable cells remain in the control well).
29. Expand and cryopreserve the surviving cells.
30. To assess the transduction efficiency, induce optoCDE expression for 24–48 h and quantify the fraction of reporter-positive cells using fluorescence microscopy or flow cytometry.

3.2 Determination of the Suitable Illumination Intensity and Phototoxicity Limits

Before performing experiments using optogenetic cell death effectors, it is important to determine optimal illumination conditions and phototoxicity limits (*see* **Note 4**). Here we describe a protocol for microscopy-based determination of these parameters.

1. Seed wild-type cells or cells expressing Cry2olig-mCherry into microscopy-suitable dishes (*see* **Note 5**). Incubate overnight (or for the period that is suitable to reach desired cell density and stable-state morphology).
2. Next day, exchange the culture medium for the prewarmed microscopy-suitable medium (OptiMEM or equivalent).
3. Preheat the microscope and set up the photostimulation and image acquisition parameters. We recommend selecting several fields of view per illumination condition, with at least 30–50

cells to be analyzed (not considering high-resolution single-cell studies, which are typically performed with lower total cell numbers).

4. Set up the photostimulation and image acquisition parameters. These should closely mimic the conditions that will be used in the real experiment in terms of light intensity, resolution, frame rate, etc. When using confocal laser point-scanning microscopes, the typical sub-cytotoxic laser intensity range typically varies between 1% and 20% of maximum laser intensity, depending on the laser model and lifetime, and laser intensity below 25 mW/cm^2 is usually noncytotoxic.
5. Perform the photostimulation and timelapse imaging. The timelapse acquisition duration should exceed the planned duration of the experiment to ensure that no signs of phototoxicity are detected throughout the whole course of the experiment.
6. Repeat until the whole desired range of illumination conditions is thoroughly tested.
7. Perform image analysis or biochemical analysis to evaluate illumination-induced changes in cell fitness and viability in your cells. Note the illumination parameters, which cause significant changes in these characteristics (i.e., phototoxicity threshold). Ensure that in further experiments, illumination parameters are always set up below this threshold to avoid nonspecific phototoxicity-induced effects.

A similar phototoxicity assessment should be performed when using Light Plate Apparatus for cell population-based assays. In this case, different light intensities can be assigned to the separate wells and tested in parallel. Phototoxicity can be assessed using the same set of characteristics.

3.3 Optogenetic Induction and Biochemical Assessment of Cell Death Using Multiwell Illumination Devices

Cell population-level experiments, such as biochemical, metabolic, or transcriptional characterization of cell death and cell survival-associated events and responses, are typically performed in multiwell-plate format, which enables simultaneous photostimulation of the multiple conditions and collecting sufficient number of samples. Below, we describe a protocol for optogenetic activation of regulated cell death in 24-well plates using a custom-built Light Plate Apparatus (LPA) [17]. However, recently many other custom [18–20] or commercial devices became available, providing possibilities to perform such experiments in other formats.

1. About 24–48 h before an experiment, seed the cells into suitable black 24-well plates to achieve 80–90% cell confluency on the day of the experiment.
2. Induce optoCDE construct expression by supplementing the medium with 1 μg/mL doxycycline.

3. On the day of the experiment: Prewarm all the necessary reagents to 37 °C (unless required otherwise).
4. Assemble the Light Plate Apparatus 24-well plate illumination device, equipped with high-intensity blue light LEDs (~450 nm) and calibrated as described previously [17, 21] and connect it to the power source.
5. For prolonged stimulations (15 min or more), securely place the device at 37 °C in a cell culture incubator. Cover the plate with a tightly fitted custom-made plastic lid or aluminum foil to prevent undesired light exposure and cross-illumination of the wells.
6. Generate a custom illumination layout using an appropriate program, such as the Iris webtool (http://taborlab.github.io/Iris/). This should typically include the illuminated and non-illuminated wells, in duplicates or triplicates, treated similarly. Additional nonilluminated wells can be included for the total cell lysis (when performing LDH assay) or for the cells treated with the conventional cell death triggers (when comparing the efficiency and dynamics of optoCDE-induced cell death to endogenous pathway engagement). The examples of such layouts are shown in Fig. 2. To minimize the position bias, the well position can be randomized.
7. Export the layout to the SD card. Insert into the device and perform a test illumination run to ensure that the device works as expected.
8. Using ambient light or red-light conditions, replace the culture medium with a suitable experimental medium. In cases where supernatants are collected and retained for further biochemical analysis, we recommend using serum-free medium, such as Opti-MEM.
9. Place the plate in LPA and cover it with the lid. Ensure that the well positions are well-aligned with the LED positions to prevent cross-illumination.
10. Turn on the illumination program. Pay attention to the green light next to SD slot, which indicates that the program is running successfully.
11. After the desired amount of time: turn off the device and gently remove the plate.
12. Perform the sample collection and processing as required for further analysis.

Optogenetic activation of cell death using microscopy has the advantage that it allows for simultaneous monitoring of morphological changes commonly associated with cell death (*see* **Note 6**).

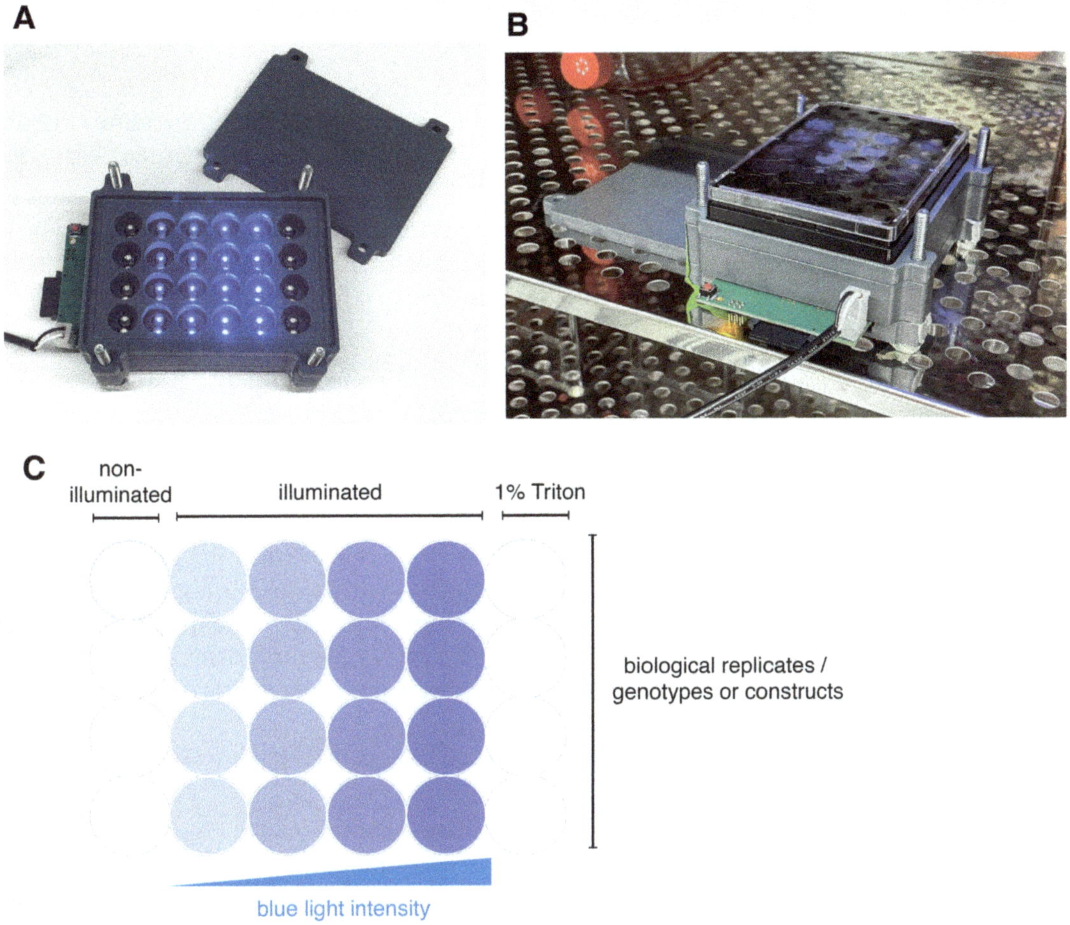

Fig. 2 *Light plate apparatus setup.* (**a**) Image of a custom-built light plate apparatus. (**b**) Setting up the device in the cell culture incubator for long-term experiments. (**c**) Example of the illumination layout used for determining the optimal illumination parameters

3.4 Optogenetic Induction of Apoptosis, Pyroptosis, and Necroptosis Using a Microscope

1. One day before the experiment, seed the cells transduced with different optogenetically activated cell death effectors into the suitable cell culture dish. The seeding density should be adjusted to achieve confluency of 50–70% on the day of the experiment.
2. To induce optoCDE expression, supplement the medium with 1–5 μg/mL doxycycline for 16–24 h. The optimal doxycycline concentration and induction duration should be adjusted based on the experimental setup and the desired construct expression levels. Following doxycycline addition, protect the cells from light by covering the plate with aluminum foil or by using custom-made dark chambers. All following manipulations to the cells should be performed under dim or red light conditions to avoid undesired optoCDE activation.

3. On the day of the experiment, preheat the microscope to 37 °C for at least 1 h.
4. Using dim light or red-light conditions, exchange the cell culture medium to the preheated imaging medium (Phenol Red-free Opti-MEM or equivalent), supplemented with suitable cell viability reporter dyes.
5. Set up the photostimulation parameters, using the illumination parameters determined in Subheading 3.3. For most of the cell types, the blue light intensity of 5–10 mW/cm^2 is sufficient to induce optoCDE activation and cell death.
6. Set up acquisition. When imaging multiple channels, we recommend acquiring the reporter (red or far-red) channel first to obtain a $t = 0$ for nonilluminated cells.
7. Perform the photostimulation and image acquisition to monitor changes in cell viability and effector activation over time.
8. To quantify the efficiency of cell death induction, divide the number of cells displaying cell death-associated phenotypes (e.g., membrane blebbing, Annexin V staining, CellTox or DRAQ7 influx, increase in caspase reporter signal) by the total number of cells expressing optoCDE construct.

3.5 Single-Cell Targeting in Multicellular Populations Using Optogenetic Cell Death Induction

1. Seed the cells and induce the optoCDE expression as described above.
2. To set up the single-cell photostimulation conditions, activate ROI (region of interest) imaging mode in the imaging software.
3. Select a 5–10 μm ROI corresponding to the cytoplasmic region of the selected cell (but clearly separated from the neighboring cells) and ROI activation parameters, including laser intensity and number of pulses and scans. The optimal set-up typically includes 1–5 pulses of moderate to high-intensity laser power (5–10 mW/cm^2); however, it depends on the biological specimen (including optoCDE expression level and cell type sensitivity to the selected mode of cell death) and the hardware and software parameters and should be determined empirically in a series of trial experiments. Keep in mind that using a high laser power might lead to nonspecific target cell damage, which should be avoided.
4. Acquire a baseline timelapse for 5–10 min.
5. Perform the ROI photoactivation using the parameters defined above.
6. Monitor the target cell and its neighbors for the hallmarks of cell death, such as influx of membrane-impermeable dyes or characteristic morphological changes (nuclear condensation or blebbing).

7. If signs of cellular damage or permeabilization are observed in the neighboring cells, reduce the laser power or ROI area, and repeat the experiment using updated settings. Keep in mind that in some cell types membrane blebbing in bystanders may also result from the loss of adhesion to the dying neighbor and may not necessarily indicate optoCDE activation.
8. In case of insufficient cell death induction, increase the laser power or pulse number and repeat the experiment until efficient cell killing is achieved.

3.6 Using Optogenetics to Monitor Dying Cell Extrusion and Bystander Responses to Cell Death

By enabling ligand-independent cell-autonomous induction of cell death, optogenetic induction of cell death is highly advantageous for studying bystander cell responses. While such studies can be performed using single-cell targeting as described in Subheading 3.5, they can often be more conveniently performed in a co-culture setup, where a small number of "senders" expressing optoCDEs is mixed with the exceeding number of "receivers," which can be either wild-type cells or express reporters for the specific process of interest. Such setup usually enables distinguishing and quantification of the local bystander responses at the dying cell border and more accurate monitoring of dying cell extrusion.

1. The day before the experiment: seed "senders" and "receivers" at a ratio of 1:20–1:100 (depending on the experimental setup and the size of the field of view) into the microscopy-suitable cell vessels. The optimal cell density depends on the experimental setup: for cell extrusion studies, it is essential to achieve confluent monolayers on the day of the experiment, while for studies of chemotaxis and efferocytosis lower cell densities might be used.
2. Induce optoCDE expression using 1–5 μg/mL doxycycline and incubate overnight, protected from light.
3. On the day of the experiment: confirm that the cells have reached the desired confluency (for epithelial cell cultures, the cells should be 90–100% confluent).
4. Before the imaging: exchange the culture medium to the prewarmed imaging medium.
5. (Optional): Prelabel the cocultures with the membrane or actin dyes to mark the plasma membrane and aid cell tracking and segmentation.
6. Preheat the microscope and place the cells in the acquisition chamber.
7. Set up photostimulation and image acquisition parameters, using predetermined conditions. In this type of experiment, the frame acquisition rate is usually determined by the kinetics of the process of interest, for example, for tracking rapid

signaling events, such as calcium waves, we recommend imaging every 5–10 s, while slower processes, such as cell extrusion or mitosis, can be imaged every 3–5 min. Keep in mind that prolonged light exposure even in a sub-cytotoxic range may significantly alter cell signaling and cytoskeletal dynamics (to avoid such nonspecific effects, we recommend performing the imaging with the same parameters of reporter cells alone beforehand to determine the most optimal illumination setup).

8. Perform photostimulation and timelapse imaging.
9. For cell extrusion experiments, a more extended z-stack can be additionally acquired at the end time point to ensure that dying cells are fully expelled from the monolayer.

4 Notes

1. Even low levels of steady-state optoCDE expression and activation might result in negative selection of expressing cells or other nonlethal effects, such as altered genome stability or signaling states. Thus, we strongly recommend using inducible expression systems and limiting light exposure of the cells. Following induction of the optoCDE expression or when using transient expression of noninducible constructs, all cell manipulations should be performed strictly under red light or dim light conditions.
2. When constructing a novel optogenetic tool, we recommend testing both C-terminal and N-terminal Cry2olig fusions and domain configuration, excluding the cases where having an exposed C- or N-terminus is critical for the protein function (such as in cases of N-termini of GSDMD and MLKL). The optimal tool design depends on many factors, such as the knowledge of the target protein structure and functional domain organization, as well as its regulation. While most of our construct designs retain only an effector function domain of the protein of interest, in some cases, additional domains should be retained, as their removal could remove endogenous autoinhibition and result in autoactivation. Moreover, additional modifications may be necessary, such as the removal of the sites susceptible to the endogenous regulation, mutations altering the baseline effector activity level or its duration, or varying linker length to achieve the desired level of proximity while retaining the effector ability to interact with downstream substrates.
3. When assembling the expression vectors, the following considerations should be applied: (1) Insert size: optoCDEs are typically too big to be packed into AAVs, so we typically utilize

lentiviral or retroviral construct delivery; (2) promoter strength and specificity: This should be defined based on the desired expression level, and also considered when using tissue-specific or state-specific promoters; (3) appropriate selection markers: We typically utilize antibiotic selection, as addition of fluorescent proteins diminished the capacity of the tools to be multiplexed with other reporters. Moreover, we recommend using inducible expression systems for tool expression, as a continuous expression of the constructs complicates the cell handling and may lead to low-level dark-state activity, resulting in the reduced viability, altered signaling states, or negative selection of expressing cells.

4. When performing optogenetics experiments, it is essential to ensure that the experimental results are not confounded by nonspecific effects of excessive illumination, which may cause phototoxicity or nonspecific alterations of cell signaling and physiology. To account for this, we recommend performing a separate set of optogenetic stimulation experiments using varying illumination parameters with wild-type cells or cells expressing a photoactuator module alone (the latter also serving as a control for expression-induced toxicity). The experimental design should resemble the one to be used in the following studies, and the cells should be monitored for a minimum of the duration of time planned to be used in future experiments, or longer. Phototoxicity can be assessed using several parameters: changes in cellular morphology, such as cell shrinking or swelling, nuclear condensation, changes in the dynamics of membrane protrusions (such as disappearance of filopodia), or membrane blebbing; activation of stress-related signaling pathways, for example, biochemical assessment of JNK phosphorylation or using JNK reporter cell lines.

5. When performing optogenetic experiments, the following controls should be typically included: nonilluminated cells expressing optoCDEs to monitor for the spontaneous dark-state optoCDE activation; illuminated nonexpressing cells to monitor for phototoxicity and nonlethal illumination-induced effects; cells expressing tagged Cry2olig alone; and cells where optoCDE activation is induced in presence of specific cell death pathway inhibitors. When studying a specific pathway or analyzing the effects of optoCDE activation on the neighboring cells, we also recommend an additional validation step by comparing the optogenetically induced phenotype to the one obtained by conventional RCD triggers.

6. The following parameters can be evaluated: (a) characteristic changes in cell morphology, such as appearance of varying degree of membrane blebbing and production of extracellular vesicles of various sizes (characteristic to all types of cell death),

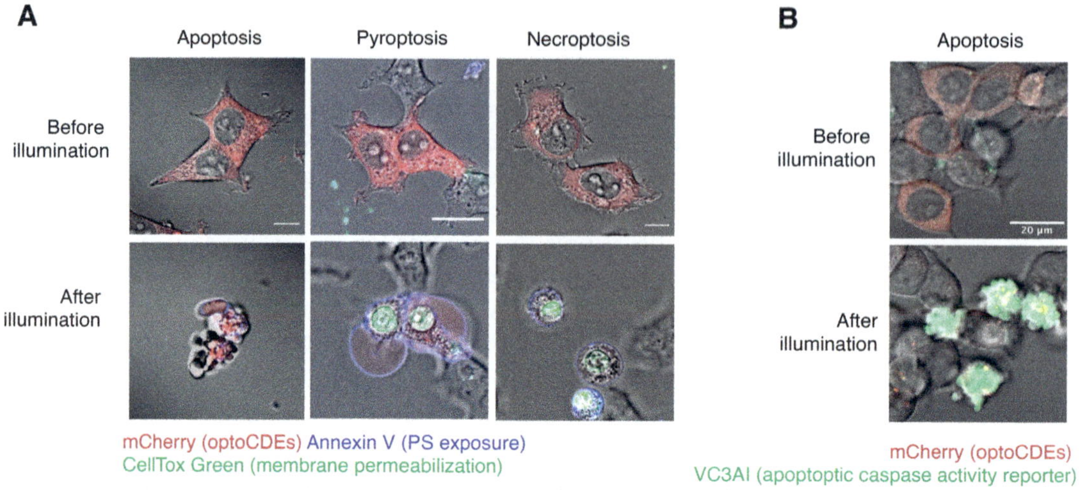

Fig. 3 *Analysis of optoCDE-induced cell death by microscopy.* (**a**) Representative morphological features of HEK293T cells expressing human pyroptotic (opto-Caspase-1), apoptotic (opto-Caspase-8) or necroptotic (opto-RIPK3) effectors before and after stimulation with blue light. Red—mCherry-tagged optoCDEs, green—CellTox Green, blue—Annexin V. (**b**) Representative images of cells expressing genetically encoded caspase-3/7 reporter VC3AI (green) and undergoing light-induced apoptosis

cell shrinking and loss of substrate adhesion (early apoptosis and necroptosis and late pyroptosis), cell swelling and nuclear condensation (pyroptosis and necroptosis), nuclear fragmentation (apoptosis), and membrane rupture and loss of intracellular content (pyroptosis and necroptosis) (Fig. 3).

7. Sublethal induction of cell death. These experiments are performed similarly to the ones mentioned in Subheadings 3.3 and 3.4, but they typically include a single round of low-intensity illumination, after which the cells are left to recover. The optimal light intensity should be determined empirically based on the optimal percentage of "surviving" cells, which are typically detected using the reporters for specific cell death pathways or membrane damage, such as fluorogenic caspase substrates [22, 23] or DRAQ7 influx (Fig. 4). For a more detailed description of such cell survival, we refer the reader to the original publication by Shkarina et al. [14].

Acknowledgments

P.B. is supported by ERC-CoG 770988 (InflamCellDeath) and SNF Project funding (310030B_198005, 310030B_192523). K.S. is a recipient of SNSF Postdoc.Mobility fellowship (P500PB_211096). The protocol is adapted from Shkarina & Broz [24].

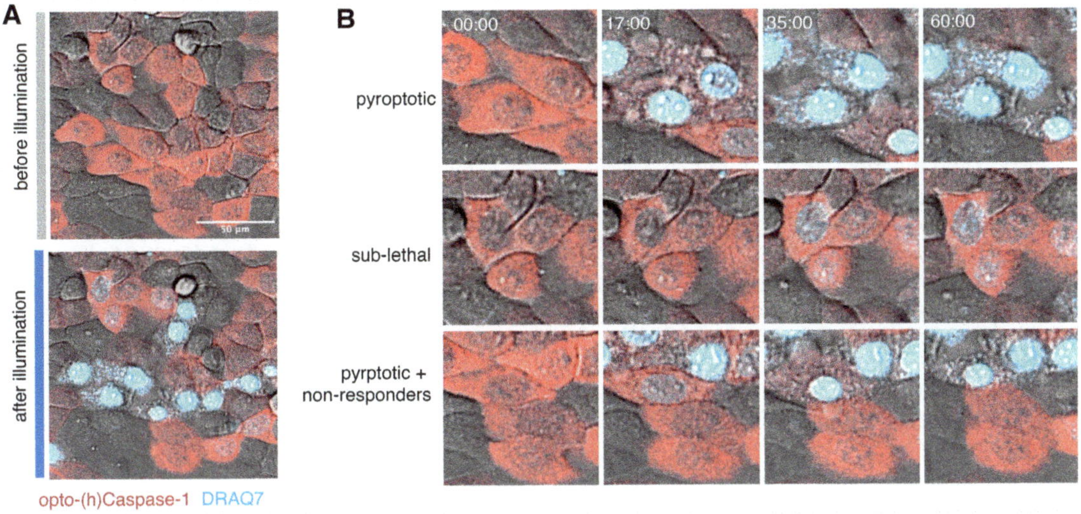

Fig. 4 *Sublethal induction of pyroptosis using transient human opto-Caspase-1 activation in human keratinocytes*. Confluent monolayer of HaCaT cells were transiently stimulated with low-intensity (0.2 mW/cm2/pulse, 3 pulses) blue light at 3 min after the beginning of time-lapse, and the data was acquired for 60 min. (**a**) Representative images of the whole illuminated population before (at 0 min) and after (60 min). DRAQ7 (turquoise) is a membrane-impermeable DNA-binding dye used to visualize pyroptotic membrane permeabilization. (**b**) Close-up images of cells displaying three types of fates: up—pyroptotic, middle—sublethal, bottom—both pyroptotic and nonresponding cells can be observed. Note the strong DRAQ7 signal and loss if cytoplasmic mCherry in pyroptotic cells, and low DRAQ7 positivity in the sublethally activated cells

References

1. Johnson AG, Kranzusch PJ (2022) What bacterial cell death teaches us about life. PLoS Pathog 18:e1010879
2. Ameisen JC (2002) On the origin, evolution, and nature of programmed cell death: a timeline of four billion years. Cell Death Differ 9: 367–393
3. Galluzzi L et al (2018) Molecular mechanisms of cell death: recommendations of the Nomenclature Committee on Cell Death 2018. Cell Death Differ 25:486
4. Green DR (2022) The death receptor pathway of apoptosis. Cold Spring Harb Perspect Biol 14:a041053
5. Green DR (2022) The mitochondrial pathway of apoptosis: part I: MOMP and beyond. Cold Spring Harb Perspect Biol 14:a041038
6. Green DR (2022) Caspase activation and inhibition. Cold Spring Harb Perspect Biol 14: a041020
7. Julien O, Wells JA (2017) Caspases and their substrates. Cell Death Differ 24:1380–1389
8. Broz P, Dixit VM (2016) Inflammasomes: mechanism of assembly, regulation and signalling. Nat Rev Immunol 16:407–420
9. Broz P, Pelegrín P, Shao F (2020) The gasdermins, a protein family executing cell death and inflammation. Nat Rev Immunol 20:143–157
10. Weinlich R, Oberst A, Beere HM, Green DR (2017) Necroptosis in development, inflammation and disease. Nat Rev Mol Cell Biol 18: 127–136
11. Kolb JP, Oguin TH, Oberst A, Martinez J (2017) Programmed cell death and inflammation: winter is coming. Trends Immunol 38: 705–718
12. Shkarina K, Broz P (2024) Selective induction of programmed cell death using synthetic biology tools. Semin Cell Dev Biol 156:74–92
13. Emiliani V et al (2022) Optogenetics for light control of biological systems. Nat Rev Methods Primers 2:55
14. Shkarina K et al (2022) Optogenetic activators of apoptosis, necroptosis, and pyroptosis. J Cell Biol 221:e202109038

15. Hasel De Carvalho E et al (2023) The Opto-inflammasome in zebrafish as a tool to study cell and tissue responses to speck formation and cell death. elife 12:e86373
16. Taslimi A et al (2014) An optimized optogenetic clustering tool for probing protein interaction and function. Nat Commun 5:4925
17. Gerhardt KP et al (2016) An open-hardware platform for optogenetics and photobiology. Sci Rep 6:35363
18. Repina NA et al (2020) Engineered illumination devices for optogenetic control of cellular signaling dynamics. Cell Rep 31:107737
19. Höhener TC et al (2022) LITOS: a versatile LED illumination tool for optogenetic stimulation. Sci Rep 12:13139
20. Bugaj LJ, Lim WA (2019) High-throughput multicolor optogenetics in microwell plates. Nat Protoc 14:2205–2228
21. Sweeney K, Moreno Morales N, Burmeister Z, Nimunkar AJ, McClean MN (2019) Easy calibration of the Light Plate Apparatus for optogenetic experiments. MethodsX 6:1480–1488
22. Tang HL, Tang HM, Fung MC, Hardwick JM (2015) In vivo CaspaseTracker biosensor system for detecting anastasis and non-apoptotic caspase activity. Sci Rep 5:9015
23. Nano M, Mondo JA, Harwood J, Balasanyan V, Montell DJ (2023) Cell survival following direct executioner-caspase activation. Proc Natl Acad Sci USA 120:e2216531120
24. Shkarina K, Broz P (2023) Optogenetic induction of pyroptosis, necroptosis, and apoptosis in mammalian cell lines. Bio Protoc 13(14): e4762

Chapter 6

Modulating Optogenetic YAP In Vitro and In Vivo

Pearlyn Jia Ying Toh

Abstract

YAP is a central regulator of the Hippo-YAP signaling axis, an evolutionarily conserved pathway that modulates organ growth and regeneration. Dysregulation of YAP signaling leads to uncontrolled proliferation, promoting epithelial-to-mesenchymal transition and invasion in cancer metastasis. Exogenous manipulation of YAP activity at the second-to-minute timescale is an important step in studying the signaling pathway. We present an optogenetic system to control the subcellular localization of YAP and therefore its activity as a transcriptional co-activator. We used the LOV2-Jα interacting domains to photocage a nuclear localization signal (NLS) attached to YAP. Under 488 nm light, the Jα helix unfolds and the interaction with LOV2 is disrupted, thereby exposing the NLS and allowing for the entire optogenetic construct to be shuttled into the nucleus. This nuclear translocation is reversible and tuneable and demonstrates functional activity after nuclear localization both in vitro and in vivo.

Key words Optogenetics, Hippo-YAP, Spatiotemporal dynamics, Zebrafish

1 Introduction

YAP is a mechanotransducer that changes its subcellular localization in response to different external mechanical stimuli, effecting changes within the cell that range from proliferation, migration, and changes in cytoskeletal structure to survival [1, 2]. Activated YAP translocates into the nucleus to exert its transcriptional coactivator activity by binding to transcription factors, thereby inducing transcription of target genes [3]. In contrast, prolonged activation of YAP can result in uncontrolled proliferation and cancer. Therefore, there is a need for a tightly regulated exogenous manipulation of YAP to address basic research questions in the field.

Optogenetics has been established to control well-defined events within biological systems, especially in the field of neurobiology, where it enables specific activation of targeted neurons [4]. However, molecular biologists soon realized its potential in controlling intercellular signaling pathways, having an advantage of tight spatiotemporal control as compared to conventional methods

Armin Baumschlager (ed.), *Optogenetics: Methods and Protocols*, Methods in Molecular Biology, vol. 2840,
https://doi.org/10.1007/978-1-0716-4047-0_6,

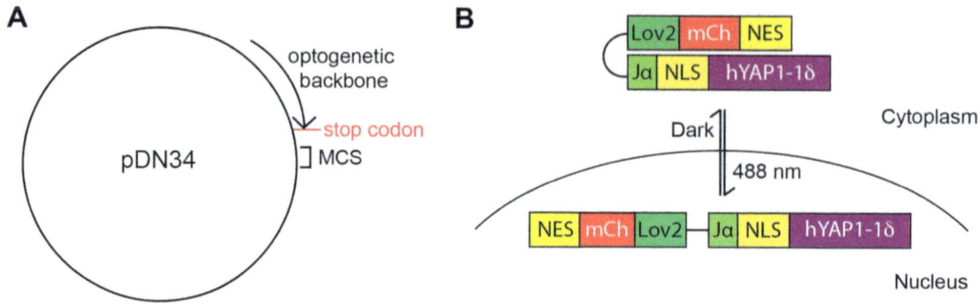

Fig. 1 Schematic showing details of optogenetic YAP containing light-sensitive domains LOV2 and Jα, mCherry fluorophore (mCh) for imaging, nuclear export signal (NES), nuclear localization signal (NLS), and hYAP1-1δ. The construct is cytoplasmic in the dark and upon 488 nm activation, the interaction between LOV2 and Jα is disrupted and the construct translocates into the nucleus

of biochemical perturbations. Therefore, we wanted to develop a tool that allows for specific, reversible, independent, and fast activation at a timescale of seconds to minutes, to spatiotemporally activate YAP.

Here, we outline our optogenetic method to modulate the subcellular localization of YAP in vitro in a range of different cell lines as well as in vivo in the zebrafish embryo [5]. The construct is adapted from a previously developed tool, biLINuS, that exploits the light-mediated interaction between the LOV2 domain and the Jα helix [6]. We chose the LOV photosensitive protein over other optogenetic systems such as PhyB or CRY2 due to the simplicity of the system. LOV systems only require a single protein to be introduced into cells and an activating 488 nm wavelength that is widely available on conventional microscopes. The crux of the optogenetic tool is the spontaneous photocaging of a nuclear localization signal (NLS) within the LOV2-Jα interacting domain in the dark. In the presence of 488 nm light, the Jα helix undergoes a conformational change and unfolds from the LOV2 interaction, thereby exposing the NLS and allowing it to be recognized and bound by endogenous importins, inducing nuclear translocation of the whole construct (Fig. 1). By tagging YAP to the biLINuS construct, we showed that optogenetic YAP responds to 488 nm activation and is functional after light-induced nuclear localization by measuring downstream transcriptional activity of YAP [5]. The optogenetic construct is also responsive to different activating laser intensities and can undergo consecutive rounds of activation and recovery without decreasing sensitivity.

This chapter describes protocols to introduce optogenetic YAP into both tissue culture cells and zebrafish embryos, illumination protocols for short- and long-term activation, as well as downstream functional analyses of activated optogenetic YAP. Although this protocol is optimized in a select number of cell lines, it can be easily adapted to other cell lines of interest.

2 Materials

2.1 Plasmid Transformation into Immortalized Cell Lines

1. pDN34-BiNLS-hYAP1-1δ plasmid (Addgene #169471).
2. HEK293T cells or MKN28 cells.
3. HEK293T media: DMEM, 10% fetal bovine serum (FBS), 1% penicillin/streptomycin. MKN28 media: RPMI, 10% FBS, 1% penicillin/streptomycin.
4. 35 mm glass-bottom dish with 12 mm glass coverslip.
5. Lipofectamine 2000 (for HEK293T) or K2 Transfection System (for MKN28).
6. Opti-MEM.

2.2 Synthesis of Optogenetic fYap mRNA for Zebrafish Microinjection

1. pCS2-BiNLS-fYap plasmid (Addgene #176623).
2. NotI restriction enzyme.
3. mMESSAGE mMACHINE SP6 Transcription Kit.
4. QIAquick Cleanup Kit.
5. RNA Clean & Concentrator.
6. Nuclease-free water (NFW).
7. PCR tube strips.
8. Agarose gel electrophoresis system.
9. Spectrophotometer.

2.3 Zebrafish Microinjection

1. Wildtype zebrafish line, such as from the AB strain.
2. Crossing tank for zebrafish spawning.
3. 60X embryo media: 34.8 g NaCl, 1.6 g KCl, 5.8 g $CaCl_2 \cdot 2H_2O$, 9.78 g $MgCl_2 \cdot 6H_2O$, dissolved in 2 L H_2O and adjusted to pH 7.2 with NaOH and autoclaved.
4. 0.5% phenol red solution.
5. Fluorescence stereomicroscope.
6. 1% low melting agarose (LMA).
7. Micro dissecting forceps.
8. Glass Pasteur pipettes and glass petri dish.

2.4 Imaging Equipment for Activation and Imaging of Optogenetic YAP Using Confocal Microscopy

1. Inverted spinning disk confocal microscope.
2. 60x high numerical aperture (>1.2 NA) water immersion objective.
3. Any dual or quad-band filter adapted to GFP and mCherry spectra.
4. Environmental chamber with temperature and CO_2 controls.
5. 488 nm and 561 nm light source.
6. sCMOS camera.

2.5 Long-Term Activation of Optogenetic YAP in Tissue Culture Cells and Zebrafish Embryos

1. CO_2 incubator with temperature control.
2. 5W white LED light strip with temporal control pulsing at 1 s every 30 s.
3. Aluminium foil.
4. TRIzol.
5. Pestle for Eppendorf tubes.
6. RNeasy Plus Mini Kit and QIAshredder for tissue culture cells and Direct-zol RNA kit for zebrafish embryos.
7. High Capacity RNA-to-cDNA kit.

2.6 RT-qPCR

1. PowerUp SYBR Green Master Mix.
2. Thermal cycler for real-time qPCR.
3. Primers listed in Table 1 were diluted to 10 μM forward and reverse primer mix.

Table 1
Primers used for RT-qPCR

Target gene	Forward primer	Reverse primer
EIF1B	TGATGCAACT AAGGGTGACG	CAGTGTCTTT CTGCCGTTCC
ANKRD1	ATGCTTGGAA GGACATTTGG	TGGCTGTGGA TTCAAGCATA
CTGF	TGCATCCGTA CTCCCAAAAT	ATGTCTTCAT GCTGGTGCAG
CYR61	TCCCTGTTTT TGGAATGGAG	GAGCACTGGG ACCATGAAGT
EPO	GCATGTGGATA AAGCCGTCAGTG	GAGTTTGCGGA AAGTGTCAGCAG
MYC	CCCGCTTCTCT GAAAGGCTCTC	CTCTGCTGCTGC TGCTGCTGGTAG
rpl13	TAAGGACGGA GTGAACAACCA	CTTACGTCTGC GGATCTTTCTG
arhgap27	CACAATCTCT GCAGCCAGGA	CCGCCAGATT TCCACTGACT
ctgfa	GTGTTCACCTGGT GTAAGCCTAGTTC	ACCGTCCAGACA CGTGCACTGGTAT
cyr61	ACGCGAAGA TGTTTGCTTGG	CCGGCCCTCT GATTTAGCTC
vgll2b	GACAAAGAGG TCGGAGAGGC	CGCCGACTGC TTGAGGATAA
vgll4l	GTGCGTTGAAG ATTCAGGAATG	CACTCTGTCGC TGGTCATGTGC

2.7 Cell Proliferation Assay and Anchorage-Independent Growth Assay

1. 96-well plates.
2. CyQuant NF cell proliferation assay and 96-well cell transformation assay.
3. Standard fluorescence microplate reader.

3 Methods

3.1 Plasmid Transfection into HEK293T Cell Line

This subheading describes transfection for HEK293T cells, but other tested cell lines that follow the same protocol are HFF and H9c2 cells.

1. Seed cells 24 h before transfection onto a 35 mm glass-bottom dish (for confocal imaging) or 6-well plates (for long-term activation) so that cells are around 80% confluent at transfection (*see* **Note 1**).
2. Prepare in separate Eppendorf tubes transfection mixtures as follows, pipetting up and down to mix. Each mixture is only sufficient for one 35 mm glass-bottom dish or one well of a 6-well plate.

   A: 150 μL opti-MEM + 6 μL Lipofectamine 2000.

   B: 175 μL opti-MEM + 3.5 μg pDN34-BiNLS-hYAP1-1δ plasmid.
3. Combine 150 μL from each A and B mixture into another Eppendorf tube, pipetting up and down to mix. Let it stand at room temperature for 5 min, then add all to cells.
4. Incubate cells for at least 24 h before imaging or using for downstream functional assays.

3.2 Plasmid Transfection into MKN28 Cell Line

1. Seed cells 24 h before transfection onto 35 mm glass-bottom dish (for confocal imaging) or 6-well plates (for long-term activation). Cells should be around 80% confluent at transfection (*see* **Note 1**).
2. Add 20 μL K2 multiplier to each dish or well and incubate for 2 h at 37 °C.
3. Prepare in separate Eppendorf tubes transfection mixtures as follows, gently pipetting up and down to mix. Each mixture is only sufficient for one 35 mm glass-bottom dish or one well of a 6-well plate.

   A: 130 μL opti-MEM + 9.6 μL K2 transfection reagent.

   B: 130 μL opti-MEM + 2.4 μg pDN34-BiNLS-hYAP1-1δ plasmid.

4. Combine both A and B solutions together, mix by gentle pipetting, and let stand at room temperature for 15 min before adding to cells.
5. Incubate cells for at least 24 h before imaging or using for downstream functional assays.

3.3 Synthesis of Optogenetic fYap mRNA for Zebrafish Microinjection

1. Perform standard restriction enzyme digestion of 2 μg pCS2-BiNLS-fYap plasmid using NotI restriction enzyme at 37 °C for 2 h to ensure all plasmid DNA material is linearised (*see* **Note 2**).
2. Using the QIAquick Cleanup Kit, add 5 volumes of PB binding buffer to the linearised plasmid solution, and spin down in a QIAquick spin column at 12,000× *g* for 30 s.
3. Discard the flowthrough, add 750 μL PE wash buffer from QIAquick Cleanup Kit, and spin down at 12,000× *g* for 30 s.
4. Add 20 μL NFW directly onto the membrane and elute the DNA by spinning down at 12,000× *g* for 30 s (*see* **Note 3**).
5. Prepare the synthesis of capped mRNA with the purified linearized plasmid from the previous step using the mMESSAGE mMACHINE SP6 Transcription Kit as follows: 10 μL of 2X NTP/CAP, 2 μL of 10X reaction buffer, 2 μL of enzyme mix, and 6 μL of purified linearized plasmid from **Step 4**.
6. Mix thoroughly and gently spin down, incubating at 37 °C for 2 h.
7. Add 1 μL of Turbo DNase to the mixture and further incubate at 37 °C for another 15 min.
8. Run 1 μL of sample through a 1% agarose gel electrophoresis to verify the size of synthesized mRNA. The gel should only contain one band around 6500 bp.
9. Purify capped mRNA using RNA Clean & Concentrator according to the manufacturer's instructions and elute with 20 μL NFW.
10. Measure the concentration of mRNA using a spectrophotometer and dilute it to a working concentration of 200 ng/μL. Aliquot diluted mRNA in PCR tube strips at 1 μL per tube and store at −80 °C prior to microinjections.

3.4 Zebrafish Microinjection and Mounting for Imaging

1. Set up a female and male pair of AB strain of *Danio rerio* in a crossing tank overnight, about 2 h before the end of the light cycle, separating the two sexes using a divider (*see* **Note 4**).
2. In the morning, remove the divider and allow the fish to mate for 15–30 min. Collect embryos immediately from zebrafish pairs that were successfully bred.

3. Prepare the microinjection set up and injection mixture as follows: 2 μL of phenol red, 1 μL of mRNA (from Subheading 3.3), and 7 μL of NFW.
4. Inject approximately 1 nL of injection mixture directly into the cytoplasm of the one-cell stage zebrafish embryos (*see* **Note 5**). Set injected embryos back into clean 1X embryo media diluted from 60X stock solution with water. Incubate at 28.5 °C, with around 50 embryos in a 10 cm petri dish to prevent overcrowding. Set aside about one-fifth of uninjected embryos as controls. For long-term activation, skip to Subheading 3.7.
5. Check embryos 3 h after injection to remove any dead embryos and change to fresh embryo media.
6. Around 6 h post fertilization (hpf), or when embryos reach the shield stage, screen embryos using a fluorescence stereomicroscope for positive mCherry signal present in the embryo (*see* **Note 6**). Set aside positive embryos in a glass petri dish filled with fresh embryo media.
7. Prepare 1% low melting agarose (LMA) dissolved in embryo media in an Eppendorf tube and keep it at 42 °C in a thermal bath (*see* **Note 7**).
8. Dechorionate positive embryos in the glass petri dish carefully using sharp, clean forceps. Use glass Pasteur pipettes to transfer dechorionated embryos into fresh embryo media as they tend to get stuck on plastic surfaces, causing them to burst.
9. Collect 2–3 dechorionated embryos and put them into the 1% LMA Eppendorf tube. Aspirate a single drop of LMA containing the embryos onto the glass surface of the imaging dish, working quickly to orient the embryos with the animal pole facing downwards as close to the surface of the glass-bottom dish. Allow the LMA to solidify before filling the rest of the dish dropwise with embryo media to prevent the LMA from drying out (*see* **Note 8**).
10. Cover the imaging dish with aluminum foil to reduce light exposure before imaging.

3.5 Activation and Imaging Protocol for Cells and Embryos on a Confocal Microscope

1. Turn on the lasers at least 1 h before imaging to allow the lasers to warm up and stabilize (*see* **Note 9**). For live cells, the environment chamber needs to be stable at 37 °C to reduce drift in the z-axis. For zebrafish embryos, imaging should be performed at 25 °C (*see* **Note 10**).
2. Wrap glass-bottom dish containing samples with aluminum foil prior to transport and imaging to prevent excess exposure to ambient light.

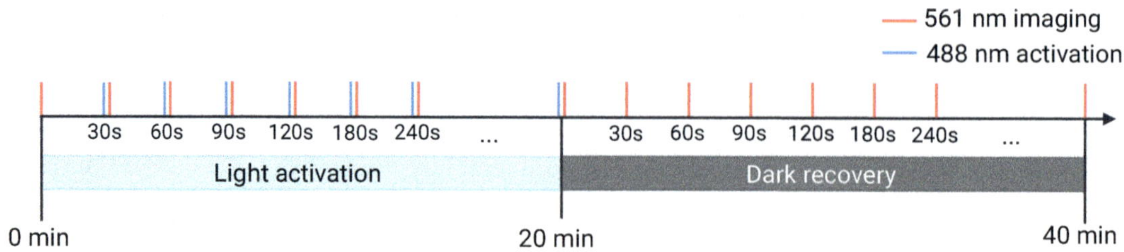

Fig. 2 Activation protocol of optogenetic YAP with a 1 s pulse of 488 nm light every 30 s over a period of 20 min. Imaging is taken at the same interval with 561 nm laser line over 40 min to track the subcellular localization of YAP

3. Mount dish containing samples onto the microscope and observe samples using brightfield. Favor the use of red LED light to prevent unwanted activation of the construct.
4. Image with a 60x high numerical aperture water immersion objective with 488 nm and 561 nm laser lines (*see* **Note 11**).
5. Before light activation with 488 nm laser, image samples with 561 nm laser to obtain the basal optoYAP distribution in the dark.
6. Pulsatile activation of optoYAP is as follows: 1 s pulse of 0.15 mW 488 nm laser, followed by a 30 s dark phase. This pulsatile activation continues for 20 min then immediately followed by a recovery period of a further 20 min incubation in the dark (Fig. 2). During the entire 40 min of light activation and recovery, imaging of the sample is performed concurrently with the 561 nm laser every 30 s (*see* **Note 12**).

3.6 Long-Term Activation of Tissue Culture Cells for RT-qPCR

1. Grow sufficient cells for two 6-well plates and seed them into two 6-well plates. Depending on the cell line, follow either Subheading 3.1 or 3.2 for transfection. Transfect only three wells of each 6-well plate.
2. Wrap one of the plates with aluminum foil to shield from light exposure and keep both plates in an incubator fitted with a LED light strip that pulsed at 1 s every 30 s. This gives three wells each for four different conditions—(i) transfected + dark, (ii) transfected + light, (iii) untransfected + dark, and (iv) untransfected + light (Fig. 3).
3. Subject cells to pulsatile light activation in the incubator for 48 h with minimal disruption.
4. Extract total RNA from cells using RNeasy Plus Mini Kit and QIAshredder and elute RNA with 40 μL NFW. Store RNA at −80 °C.
5. Synthesize cDNA using 1 μg of RNA with a commercially available cDNA synthesis kit using oligo d(T) primers and store cDNA at −20 °C.

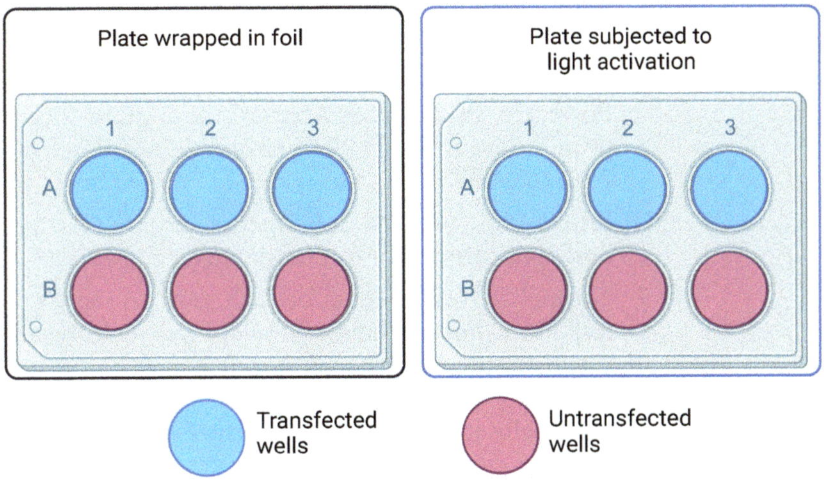

Fig. 3 Illustration of 6-well plates containing transfected (in blue) and untransfected (in pink) cells, with one plate wrapped in aluminum foil to prevent exposure to activating light

3.7 Long-Term Activation of Zebrafish Embryos for RT-qPCR

1. Separate injected and uninjected zebrafish embryos into 50 embryos per 10 cm petri dish filled with embryo water to prevent overcrowding.
2. Check cells around 6 hpf to remove dead embryos and screen for positive embryos containing mCherry signal.
3. At the tailbud stage around 10 hpf, split the surviving embryos into four different conditions—(i) injected + dark, (ii) injected + light, (iii) uninjected + dark, (iv) uninjected + light. Wrap dishes for conditions (i) and (iii) with aluminum foil and place all samples into an incubator at 28.5 °C fitted with a LED light that pulses at 1 s every 30 s.
4. Incubate samples overnight (around 14 h) under pulsatile light activation.
5. Remove dead embryos and pool five embryos into a single Eppendorf for each condition for all remaining embryos, having at least five replicates for each condition.
6. Remove as much embryo media as possible and add 100 μL of TRIzol into each Eppendorf. Homogenise with a clean pestle.
7. Extract RNA using the Direct-zol RNA kit, including the DNase I treatment, and elute in 30 μL NFW. Measure the concentration of RNA using a spectrophotometer.
8. Synthesize cDNA using 1 μg of purified RNA with High-Capacity RNA-to-cDNA kit and store cDNA at −20 °C.

3.8 RT-qPCR for Tissue Culture Cells and Zebrafish Embryos

1. Perform real-time qPCR in triplicates using PowerUp SYBR Green Master Mix for 40 cycles in a real-time PCR thermal cycler, with each well containing the mixture as follows: 5 μL of PowerUp SYBR Green Master Mix, 2 μL of forward + reverse primer mix, 1 μL of cDNA (from Subheading 3.6 or 3.7), and 2 μL of NFW.
2. Primer sequences for target genes and housekeeping genes for both cell culture and zebrafish are listed in Table 1.
3. Normalize the Ct value for each gene to the Ct value of their respective housekeeping gene for cell culture and zebrafish. Calculate the relative fold change using the ΔΔCt method.

3.9 Cell Proliferation Assay

1. Seed cells into two 96-well plates at 1000 cells per well in 100 μL of media and fill the remaining empty wells with 100 μL PBS (Fig. 4).
2. Keep one plate under pulsatile light activation and the other wrapped in aluminum foil for 7 days, changing media every 2 days. Use the same incubator with LED set up in Subheading 3.6.

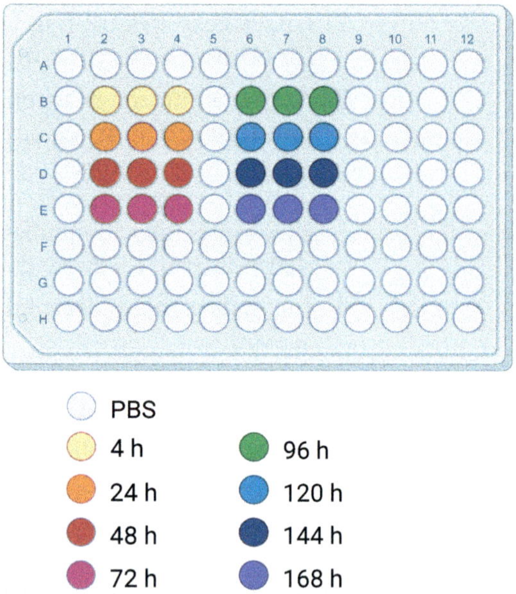

Fig. 4 Illustration of cell seeding in 96-well plate for long-term cell proliferation assay, where each timepoint indicates a terminal measurement of cell proliferation using the CyQuant cell proliferation assay. Blank wells without cells should be filled with PBS to prevent media from drying out

3. 4 h after seeding, remove media and add 100 μL of 1X dye binding solution into each well for 4 h and incubate for 30 min. Measure fluorescence intensity using a plate reader at 485 nm excitation and 530 nm emission.
4. Repeat intensity measurements at every 24 h timepoint until 168 h.

3.10 Anchorage-Independent Growth Assay

1. Prepare 1.2% agar solution and 2X DMEM/20% FBS reagent following Cell Transformation Assay kit instructions prior to use.
2. Melt 1.2% agar solution in a microwave and let cool in a 37 °C water bath. Warm 2X DMEM/20% FBS solution in a 37 °C water bath.
3. Add equal volumes of 1.2% agar solution and 2X DMEM/20% FBS solution in another tube and mix by inversion (*see* **Note 13**). Transfer 50 μL of the mixture into each well of a 96-well plate, avoid introducing air bubbles into the mixture and avoid the wells at the border of the plate (*see* **Note 14**). Gently knock the plate to allow the agar mixture to cover the wells entirely.
4. Transfer plate to 4 °C to allow agar to solidify for 30 min, after which store at 37 °C for at least 15 min before adding cells.
5. Harvest and resuspend cells in 2×10^5 cells/mL, keeping cell suspension in a 37 °C water bath.
6. Mix equal volumes of 1.2% agar solution, 2X DMEM/20% FBS solution, and cell suspension in another tube and mix by inversion (*see* **Note 13**). Transfer 75 μL of the mixture into each well of the 96-well plate that already contains the solidified base layer (*see* **Note 15**). Set aside some control wells that do not contain any cells, replacing the cell suspension with culture media.
7. Transfer plate to 4 °C for 15 min to allow cell agar layer to solidify.
8. Add 100 μL of appropriate culture media to the wells and incubate at 37 °C for 7 days under light activation or wrapped in aluminum foil under dark condition. Use the same LED set up in Subheading 3.6. Carefully change with fresh culture media every 2 days and avoid disrupting the agar (*see* **Note 16**).
9. Observe cell colony formation under 10x low numerical aperture air objective with a brightfield microscope. Colonies formed appear as dark rounded spheres.
10. To measure the total amount of DNA present in each well after 7 days, prepare sufficient amount of CyQuant working solution immediately before diluting the CyQuant GR dye 400X with PBS.

11. Remove culture medium by placing it upside down onto a paper towel and tap gently.
12. Add 50 μL of agar solubilization solution to each well of the 96-well plate and incubate at 37 °C for 1 h.
13. Pipette each well thoroughly to ensure complete agar solubilization.
14. Add 25 μL of 8X lysis buffer to each well, pipetting well to get a homogeneous mixture, and incubate at room temperature for 15 min.
15. Transfer 10 μL of the mixture into another 96-well plate and add 90 μL of CyQuant working solution to each well. Incubate at room temperature for 10 min.
16. Assess the amount of DNA present by using a plate reader to read fluorescence at 485 nm excitation and 538 nm emission wavelength.

4 Notes

1. Cells should not be too confluent during transfection, which reduces the transfection efficiency.
2. Good quality mRNA is essential for microinjections as impurities present may affect the developmental process or expression of the optogenetic protein in the larvae. To reduce contaminants during the preparation of mRNA, always use filtered pipette tips that are nuclease-free.
3. NFW added on the membrane of the columns can be left for 1–2 min to increase the product yield.
4. Successful zebrafish breeding is dependent on several environmental factors, such as water quality or feeding frequency, as well as other stressors on the fish. To improve successful breeding, fish should have at least 1 week recovery period between breeding. Additionally, the ratio can be increased to one male to two females in a single breeding tank during the set up the day before.
5. The amount of mRNA injected into the one-cell stage embryo should be around 20 pg, but the injection volume depends on the size of the injection droplet. However, we have noticed that the embryo can tolerate injections of up to 50 pg mRNA. Additionally, to reduce mosaic expression of the protein, do not inject into embryos that have already developed into the 2-cell stage or into the yolk.
6. It is possible during microinjection that the mRNA flows into the yolk, producing a strong mCherry signal seen in the yolk during screening. Ensure that the embryos selected are not false positives.

7. The temperature of the LMA should not exceed 42 °C as it will cause the embryos to burst when mounting. Similarly, LMA should not be left at room temperature prior to mounting as it will solidify. Extra tubes of LMA can be prepared in advance and stored at room temperature for up to 2 weeks, thus only requiring heating when preparing for mounting.
8. Reduce perturbation of the solidified LMA by doing a dropwise addition of the embryo media on the walls of the dish, allowing the media to gently cover the LMA. Solidified LMA may float up from the glass-bottom surface if media is added directly to it or during transport to the microscope. Freshly prepared LMA helps attachment to the glass-bottom dish.
9. 10 mins if using diode laser.
10. Avoid imaging at room temperature as fluctuation in the temperature of the room will cause drifting of the stage.
11. The laser power of the 488 nm line should be calibrated and measured to be 0.15 mW at the focal plane.
12. Z-stack imaging should be performed during the activation protocol for the zebrafish embryos, starting at the most superficial enveloping layer to the deep cells below.
13. Do not mix by pipetting to reduce the addition of air bubbles into the mixture.
14. Add 100 μL PBS to all wells along the four borders of the 96-well plate to prevent the agar from drying out.
15. Gently add cell agar solution into the wells to prevent excess disruption to the base agar layer. The final concentration of agar solution containing cells amounts to 0.4%, which corresponds to Young's modulus of <2 kPa [7].
16. Turn off the light in the tissue culture hood, as well as in the room or working area to prevent excessive light interference.

References

1. Panciera T, Azzolin L, Cordenonsi M et al (2017) Mechanobiology of YAP and TAZ in physiology and disease. Nat Rev Mol Cell Biol 18:758–770
2. Totaro A, Panciera T, Piccolo S (2018) YAP/-TAZ upstream signals and downstream responses. Nat Cell Biol 20(8):888–899
3. Zhao B, Ye X, Yu J et al (2008) TEAD mediates YAP-dependent gene induction and growth control. Genes Dev 22:1962–1971
4. Mahmoudi P, Veladi H, Pakdel FG (2017) Optogenetics, tools and applications in neurobiology. J Med Signals Sens 7:71–79
5. Toh PJY, Lai JKH, Hermann A et al (2022) Optogenetic control of YAP cellular localisation and function. EMBO Rep 23:e54401
6. Niopek D, Benzinger D, Roensch J et al (2014) Engineering light-inducible nuclear localization signals for precise spatiotemporal control of protein dynamics in living cells. Nat Commun 5:1–11
7. Li C, Huang Z, Wang RK (2011) Elastic properties of soft tissue-mimicking phantoms assessed by combined use of laser ultrasonics and low coherence interferometry. Opt Express 19:10153–10163

Chapter 7

Design and Use of AsLOV2-Based Optogenetic Tools for Actin Imaging

Kourtney L. Kroll, Tobin R. Sosnick, and Ronald S. Rock

Abstract

We present protocols for using an optogenetic tool called LILAC for actin imaging. LILAC is a light-controlled version of Lifeact that uses the Avena sativa LOV2 (AsLOV2) domain. By significantly reducing Lifeact's affinity for the cytoskeleton in the dark, LILAC reduces concentration-dependent negative side effects while enabling new image processing methods. We discuss the considerations for using this probe of live-cell actin dynamics, including fluorescent protein selection, cell maintenance, microscopy protocols, and image processing. Our work highlights the potential of AsLOV2-based optogenetics for novel imaging and control tools in cell biology.

Key words Actin, Lifeact, Fluorescence microscopy, LOV2, Image processing, Nonperturbative imaging probes

1 Introduction

The actin cytoskeleton is critical for myriad cellular processes, including cell transport, division, movement, and more [1]. Therefore, visualization of the cytoskeleton with microscopy is important for the field of cellular biology. Lifeact is an actin-binding peptide widely used for live-cell actin imaging [2]. However, Lifeact stabilizes actin filaments when bound and can have concentration-dependent negative side effects on cells [3–6]. To overcome these issues, we designed LILAC (Light-induced localization to actin), an optogenetic version of Lifeact [7]. By reducing Lifeact's affinity for the cytoskeleton in the dark, we aimed to reduce these negative effects. Moreover, the ability to switch Lifeact on and off with light allows for new image processing methods for enhanced actin imaging. In this chapter, we discuss features to consider when swapping

Armin Baumschlager (ed.), *Optogenetics: Methods and Protocols*, Methods in Molecular Biology, vol. 2840,
https://doi.org/10.1007/978-1-0716-4047-0_7,

the fluorescent protein at the LILAC N-terminus, S2 cell maintenance and transfection protocols when working with LILAC, and microscopy and imaging processing protocols to extract the most information from an imaging session.

2 Materials

2.1 S2 Cell Maintenance

1. T25 or T75 tissue culture flasks.
2. Schneider's insect media (Sigma Aldrich).
3. 10× Insect medium supplement (Sigma Aldrich).
4. Fetal bovine serum (optional).
5. 10,000 U/mL Penicillin/Streptomycin.
6. Kimwipes.
7. 70% ethanol.
8. Autoclaved 1.5 mL microcentrifuge tubes.
9. Hemocytometer.
10. Phase contrast microscope.
11. Tissue culture hood.

2.2 S2 Cell Transfection

1. DDAB: 250 μg/mL Didodecyldimethylammonium bromide in water (*see* **Note 1**). Filter to sterilize using a 0.4-micron filter.
2. Plasmid DNA for LILAC expression in insect cells.
3. 100 mL beaker.
4. 50 mL conical tube.
5. Vacuum filter/receiver system.
6. 1.5 mL microcentrifuge tubes.
7. 0.5 M $CuSO_4$.
8. Sonicator.
9. Hemocytometer.
10. Phase contrast microscope.
11. Tissue culture hood.

2.3 Live-Cell Microscopy

1. ConA: 0.5 mg/mL Concanavalin A in water.
2. 35 mm glass bottom dishes.
3. TIRF microscope with mCherry/GFP channels.

3 Methods

3.1 Assay Development and Fluorescent Protein Selection

When working with light-sensitive proteins, the selection of fluorescent proteins (FPs) is crucial for visualization of both the optogenetic tool and any other proteins that need to be visualized [8]. LOV2 is excited by blue light, complicating the use of FPs that are either blue or excited by blue light, such as green fluorescent protein (GFP) and cyan fluorescent protein (CFP) [9]. The easiest option is to use red-shifted FPs, such as mCherry, or near-infrared FPs like mIFP663 [10, 11]. By using red-shifted labels, one can ensure that the excitation and emission spectra of the FPs would not excite LOV2, allowing for visualization of these FPs while LOV2 still is in its dark state.

On the other hand, blue-shifted FPs including GFP and CFP can be used in conjunction with LOV2-based tools, if they are used strategically. For example, if only the photo-excited image is necessary, GFP can be directly attached to the LOV2-based protein and both can be excited simultaneously. Alternatively, if an RFP is attached to LOV2, it can be visualized first, in the dark. Then a second GFP-tagged protein can be imaged, simultaneously exciting LOV2 and promoting its binding to actin. Then the RFP-LOV2 can be imaged immediately thereafter. If the imaging system permits, simultaneous excitation of GFP and RFP will maintain the RFP-LOV2 in the lit state and on actin.

3.2 S2 Cell Maintenance

Drosophila melanogaster Schneider 2, or S2 cells are a commonly used insect cell line derived from early stage fly embryos. We use S2 cells for a number of reasons as they are popular, have a characteristic and prominent actin ring after spreading, and are simple to transfect and maintain. This maintenance protocol is based on that of Bunch et al. [12].

1. Prepare Schneider's insect media supplemented 10% insect media supplement (*see* **Note 2**). Prewarm the media to room temperature.
2. Resuspend the cells (*see* **Note 3**).
3. Dilute the cells 1:10 by adding 100 μL of cells to a microcentrifuge tube with 900 μL of media.
4. Remove the tube from the tissue culture hood and mount 10 μL of the dilution in a hemocytometer.
5. Count the number of cells in one section of the hemocytometer under the microscope. Multiply this count by 10^5 to obtain the number of cells per mL (*see* **Note 4**).
6. Split cells to 2×10^6–4×10^6 cells/mL by adding a small amount of cells to a new flask with fresh media (*see* **Note 5**).

7. Maintain S2 cells in the dark at room temperature without CO_2. Cells can either be shaken in suspension flasks or grown in a monolayer in tissue culture flasks, as they are semi-adherent. Cells can be stored up to 28 °C for faster growth.

3.3 S2 Cell Transfection

Here, we describe transfection (*see* **Note 6**) of mCherry-LILAC in the pMT vector. This vector contains the metallothionein promoter from Drosophila. This promoter allows for the concentration-dependent induction of expression with the addition of copper to the cell media with precise timing. Of note, each cell will uptake a differing plasmid copy number. This causes cell-to-cell variation in expression level given a constant copper sulfate concentration, allowing for analysis of concentration-dependent effects within a single transfection. This transfection protocol is based on the protocol developed by Han et al. [13].

1. Prepare a DDAB:media solution by mixing media and DDAB 2:1. For a 6-well plate transfection, each well will need 145 μL of the mixture (*see* **Note 7**). Incubate for 5 min.
2. While waiting for the media:DDAB mixture, count the cells (*see* Subheading 3.2, **Steps 3–5**).
3. Prepare a mixture of mCherry-LILAC plasmid DNA with DDAB:media. For a 6-well plate, add 100–500 μg of DNA to a labeled 1.7 mL tube for each well (*see* **Note 8**). Gently add 145 μL of the media:DDAB mixture to each tube. Incubate for 15 min.
4. While incubating the DNA:DDAB:media mixture, seed the cells in the wells at 1.35×10^6 cells/mL (*see* **Note 9**).
5. Add 145 μL of DNA and media:DDAB mixture to each cell dropwise and slowly while gently shaking the 6-well plate in a circular motion. Store cells in the dark.
6. If using the pMT promoter, the next day induce expression by adding between 0.5 and 2 μL of 0.5 M $CuSO_4$ (*see* **Note 10**). Give the cells at least 12 h to express protein before imaging, especially if using a fluorescent protein which takes time to fold (*see* **Note 11**).

3.4 Live-Cell TIRFM

Here, we image mCherry-LILAC by exciting either just the mCherry tag (dark) or by exciting both the mCherry and LOV2 (lit) (*see* **Note 12**). This fluorophore/LOV2 design allows us to determine if there is a difference in actin labeling in the LOV2 dark state versus the lit state. Our imaging protocol is based on the protocol described by Buster et al. [14].

1. To prepare dishes for imaging, add 15–50 μL of ConA to the dish in the tissue culture hood to cover the glass imaging area. Spread the solution gently using the pipette tip, and leave the

dishes in the hood for 15–20 min until the water evaporates. Put the dishes in a sterile container and store them for later use.

2. Harvest cells for imaging 12–24 h postinduction. Use the pipette to dislodge cells and add 100–300 μL of cells to the ConA-coated glass-bottom dish (*see* **Note 13**).
3. Ensure the microscope has the appropriate channels for imaging the LILAC construct. For our standard variant, mCherry-LILAC, an mCherry filter cube (green excitation, red emission filter) is necessary. It is also necessary to excite the LOV2 domain of LILAC. This may be achieved using the GFP channel of the microscope (*see* **Note 14**).
4. Load the imaging dish on the microscope. When finding focus, use the mCherry channel, so that it is not necessary to wait ~2 min for LILAC to return to the dark state for a preexcitation image. Try to focus on the glass surface, as the actin-rich filopodia and lamellipodia will be found there (*see* **Note 15**).
5. After locating a cell of interest, first image the cell with the mCherry signal. This gives the preexcitation image of LILAC in the cell body.
6. Next, photoexcite LOV2 of LILAC with the GFP channel for ~1 s, followed by the mCherry channel (*see* **Note 16**). If desired, continue imaging with the mCherry signal for ~2 min until LILAC recovers to the dark state.

3.5 Image Processing

Images taken using LILAC can be processed and analyzed in a variety of ways to gain clearer visualization of actin, insight into LILAC dynamics, and more. All image processing is done in FIJI, an open-source and widely used software. The methods of image acquisition and desired information determine which image processing methods are most useful with LILAC.

3.5.1 Image Subtraction to Enhance Actin Contrast

If images of LILAC have been acquired both pre- and postexcitation, image subtraction can be used to highlight the actin structures labeled by LILAC with blue light (Fig. 1).

1. In FIJI, make a substack of the preexcitation image.
2. Using the Image Calculator plug-in, select the postexcitation stack and subtract the preexcitation substack (*see* **Note 17**).

3.5.2 Actin Labeling Ratio

To quantify the labeling of actin for each image, use the actin labeling ratio (ALR). This is the ratio of the average pixel intensity of the outer lamellipodial actin ring and the average pixel intensity of the cell interior (Fig. 2a, Eq. 1). The higher the ALR, the brighter the lamellipodial actin is in relation to the interior of the cell.

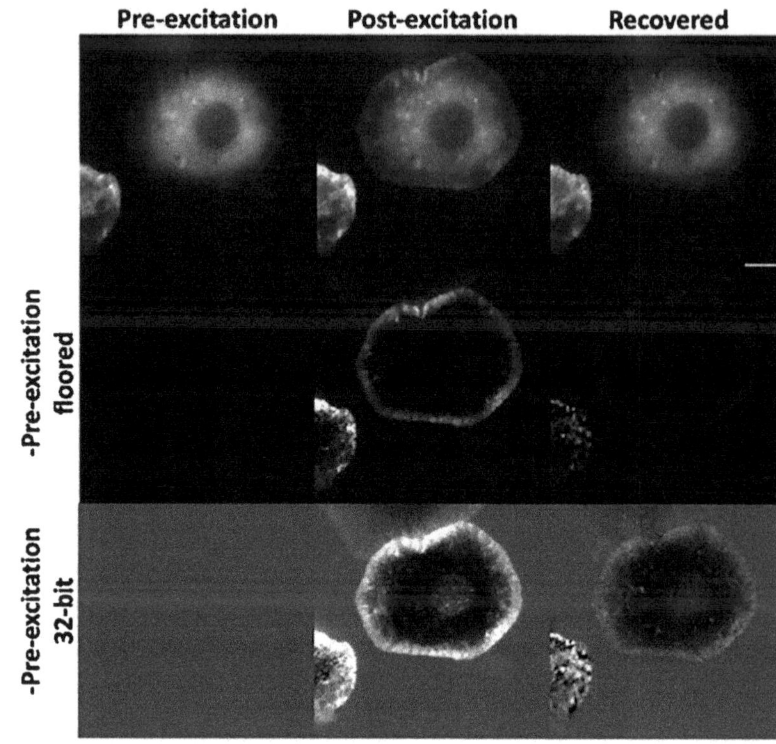

Fig. 1 Image subtraction highlights actin structures in live cells. TIRF images of S2 cells expressing LILAC at various stages of excitation in their raw form (top row). The preexcitation image can be subtracted by either flooring negative values (center row) or by outputting a 32-bit float (bottom row), showing where LILAC has been depleted. Scale bar: 5 μm

$$\mathrm{ALR} = \frac{\mathrm{meanpixelintensity}_{\mathrm{outer}}}{\mathrm{meanpixelintensity}_{\mathrm{inner}}} \tag{1}$$

1. Using the FIJI freehand selection tool, trace the outer edge of the cell and save the region of interest (ROI) using the [t] shortcut.
2. Trace the inner boundary of the lamellipodia and save the ROI.
3. Select both ROIs and use the XOR function to create an ROI that contains only the lamellipodial ring.
4. Use multi-measure to record the average pixel intensity for each frame, as well as whatever shape parameters are of interest for both the outer and inner regions.

3.5.3 Switching Ratio

As a proxy for the ratio of LILAC-actin binding in the light versus the dark, use the actin switching ratio (ASR, Fig. 2a). The switching is defined as the ALR postexcitation and preexcitation, such that a switching ratio over 1 indicates enriched actin labeling postexcitation (Eq. 2).

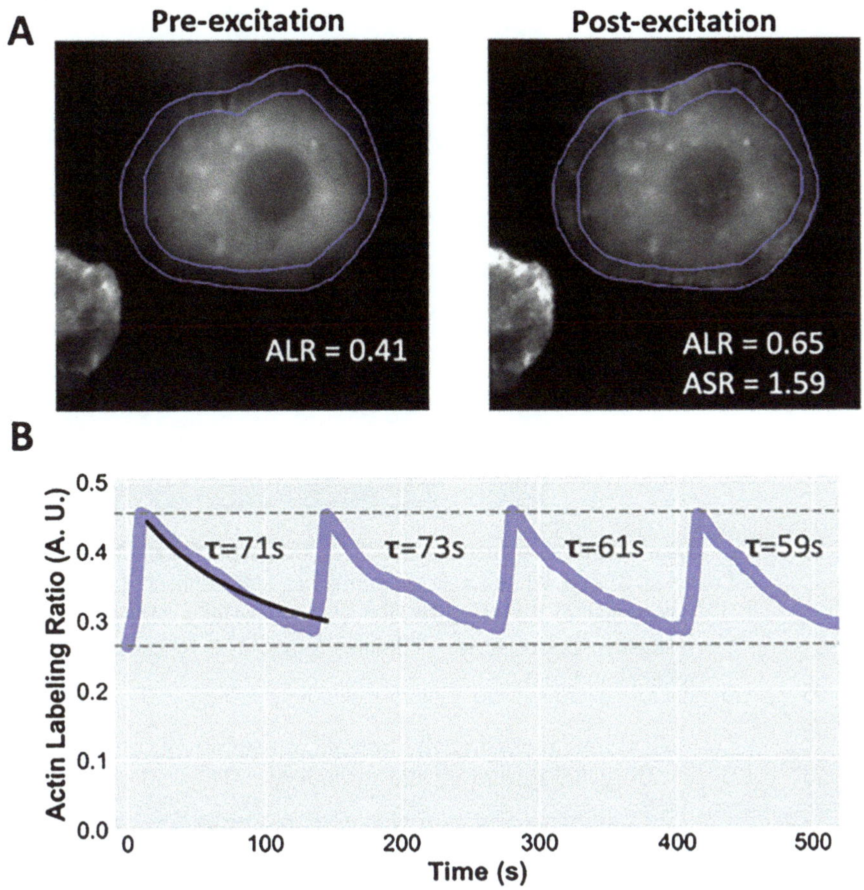

Fig. 2 Actin labeling ratio (ALR) is used to measure actin binding in S2 cells. (**a**) TIRF images of S2 cells expressing LILAC pre- and postexcitation. Blue highlights the traced regions for calculating the ALR. (**b**) ALR over time for a repeatedly excited cell. Recovery time constants were calculated using an exponential decay fit. Dashed lines indicate the upper and lower bounds of the ALR

$$\mathrm{ASR} = \frac{\mathrm{ALR}_{\mathrm{post}}}{\mathrm{ALR}_{\mathrm{pre}}} \tag{2}$$

3.5.4 Recovery Time Constants

It is important to know the recovery time of LILAC's photocycle when determining how often to excite cells for optimal imaging results. For example, if cells are recovering to the dark state quickly, it is important to minimize the time between blue light excitation and capturing the postexcitation image. Conversely, if cells are recovering slowly, if multiple excitations are desired with both pre- and postexcitation images, enough time must pass such that LILAC fully recovers to the dark state.

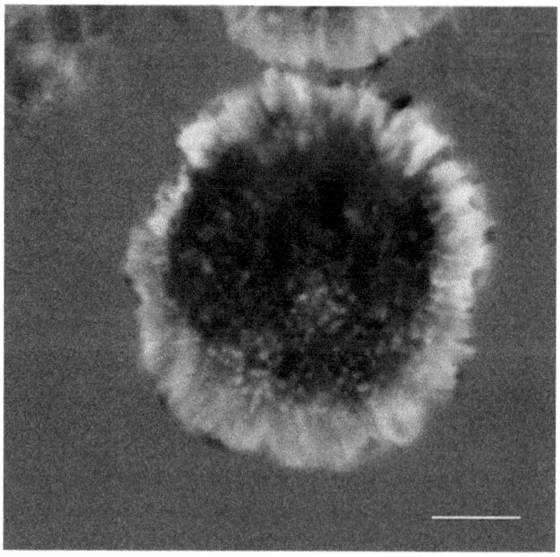

Fig. 3 Correlative imaging with LILAC. OLID (optical lock-in detection) [15] image of an S2 cell. White pixels are correlated, black anticorrelated, and mid-gray are uncorrelated. We used the first 150 frames of a movie (0.5 s/frame) where excitation pulses arrive in frames 2 and 102 to generate the OLID image. Scale bar: 5 μm

1. When acquiring the images, pulse blue light, followed by exciting/imaging just the mCherry.
2. Calculate and plot the ALR of the cell for each frame. Fit the data to an exponential decay (Fig. 2b).

3.5.5 Correlative Imaging

Another advantage of imaging actin with LILAC is the ability to use correlative imaging (Fig. 3). Correlative imaging is the method of perturbing the fluorescence signal in a controlled manner, and then applying lock-in detection analysis to amplify the signal of interest [15]. This highlights pixels that are responding temporally to the perturbation, allowing for increased visualization of dynamic, but dim pixels. In this case, we modulate the fluorescence using activation of LILAC with blue light.

1. Identify a rectangular region in the lamellipodia that responds to blue light excitation in FIJI. This response function should look similar to the ALR decay.
2. Sum the intensity of this area for each frame to obtain the response function.
3. Calculate the Pearson correlation coefficient between the response function and the intensity of each pixel along the time axis.

4 Notes

1. DDAB is difficult to solubilize in water. This can be solved by heating up the solution at 60 °C until it is no longer cloudy. Then, sonicate for at least 30 min to ensure it is in solution. DDAB solutions are stable for approximately 1 year when stored at 4 °C.
2. S2 cells and medium must be handled in a tissue culture. S2 cells prefer Schneider's insect media supplemented with 10% fetal bovine serum (FBS) or 10% insect media supplement. Because media supplement is more consistent and cheaper it is preferred, but notably, cells are often shipped/frozen in FBS and must slowly be weaned to the media supplement. To reduce contamination, antibiotics such as penicillin (50 units/mL) and streptomycin (50 μg/mL) are often also added, although we recommend avoiding antibiotics because they can mask persistent bacterial contamination.
3. At around 5×10^6 cells/mL, cells are confluent and typically float and form clumps, which does not have negative effects on the cells. When working with S2 cells, pipette up and down, ensuring that cells are dislodged from the surface while pipetting. The bottom of the flask should look clearer where cells have been dislodged, and the clumps will be dispersed. Cells can begin to reattach within 1 min, so be sure to resuspend them each time they are pipetted.
4. Normally, the number of cells is multiplied by 10^4 to get the density of cells/mL. However, we have diluted the cells 1:10, in which case multiply by 10^5. For example, if 150 cells are counted, the density is 15×10^6 cells/mL. Consult the instructions for your hemocytometer for the definition of an imaging section.
5. Cells double every 24–36 h and should be split every 3–6 days. They should never be split below 0.5×10^6 cells/mL and should not grow to higher than 15×10^6–20×10^6 cells/mL.
6. Transfection is the process of inserting foreign DNA into eukaryotic cells. It allows for the expression of labeled proteins, proteins from other cell types, and for expression level control. There are a variety of transfection methods including physical methods like electroporation and microinjection, as well as chemical methods such as calcium phosphate precipitation and lipofection. Lipofection using cationic lipids such as DDAB is advantageous because it results in high transfection efficiency and low cell toxicity.

7. Make 7 reactions to have extra on hand. That means $7 \times 145 = 1015$ μL total. Therefore, 677 μL media should be combined with 338 μL of DDAB. If using a different diameter well, scale the volume used here by the well surface area divided by the area of a well in the 6-well plate (9.5 cm^2).
8. The optimum amount of DNA can vary, depending on the batch of DDAB stock solution. It is useful to test a variety of DNA concentrations to find the optimum transfection efficiency and expression conditions.
9. Transfections should be done 2–4 days after cells are split while they are quickly dividing. Calculate the volumes of cells and media needed to dilute to 1.35×10^6 cells/mL. For example, if using all six 2 mL wells, make 14 mL of cells at the desired concentration. If the cells are at 5.4×10^6 cells/mL, add 3.5 mL cells and 10.5 mL of media to a 15 mL tube. Then add 2 mL of the diluted cells to each well.
10. The copper will induce the cells to express the protein of interest, and the more copper is added, the more they will express on average. Test various concentrations to find the optimal conditions.
11. If the cells express high concentrations of exogenous proteins for an extended period of time, they often begin to localize the exogenous protein to lysosomes. This can be seen by imaging the cells with fluorescence microscopy.
12. There are a number of different fluorescence microscopy methods, most of which illuminate a large volume of the sample. This has two disadvantages: increased photobleaching of the sample, and the illumination of the sample outside of the focal plane resulting in increased background photons. To avoid this, we image with total internal reflection fluorescence microscopy (TIRFM). With TIRFM, the illumination is on an angle such that it induces total internal reflection, and an evanescent wave is created. This wave excites only a thin layer of the sample (ca. 200 nm) at the surface of the glass. This is particularly useful for imaging S2 cells, as the main actin structure—the lamellipodium—is thin and near the surface of the glass. TIRF is also highly favorable for single-molecule imaging of actin filaments in vitro, as filaments are specifically visualized as they come into contact with the glass surface.
13. S2 cells spread on ConA-coated glass. Cells take around 30 min to spread. To see filopodia in landing cells, image cells immediately to watch them spread.
14. Ideally, simultaneous excitation of mCherry-LILAC with green and blue light can be achieved, while observing with the red emission filter. If this is not an option, the green fluorescent

protein (GFP) channel (blue excitation, green emission) can be used to excite LILAC (but not visualize it), followed by switching to the mCherry channel for visualization. If available channels are limited, an eGFP may be substituted for mCherry on the LILAC construct. In this case, the GFP-LILAC will bind when excited. The tradeoff is that it will be more difficult to capture a preexcitation image for the image subtraction methods using this probe.

15. Because S2 cells grow at room temperature, stage heating and CO_2 are not needed for live imaging.
16. LILAC recruits to actin structures in under 1 s, and in normal media, recovers to the dark state with a time constant of about 1 min. LILAC may be continually excited for continuous imaging of actin over brief periods, but be sure to monitor for signs of cellular defects with extended illumination times.
17. If the 32-bit float box is not selected, negative values will be floored to zero. This means that areas such as the center of the cell, where LILAC is depleted, postexcitation will all be floored to zero. The contrast will likely need to be adjusted in this image. If the 32-bit float box is selected, enriched, positive values will be white, unchanged values will be zero, and depleted, negative values will be black. This allows for visualization of areas where LILAC is depleted upon excitation. It is suggested to try both and see which images are preferred.

Acknowledgments

This work was supported by the David Grier Prize for Innovative Research in Biophysical Sciences and an American Heart Association Predoctoral Fellowship 915957 (to K.L.K.), NIH grants EB009412 and GM055694 (to T.R.S.), and NIH grant GM149073 (to R.S.R.).

References

1. Blanchoin L, Boujemaa-Paterski R, Sykes C, Plastino J (2014) Actin dynamics, architecture, and mechanics in cell motility. Physiol Rev 94: 235–263. https://doi.org/10.1152/physrev.00018.2013
2. Riedl J, Crevenna AH, Kessenbrock K et al (2008) Lifeact: a versatile marker to visualize F-actin. Nat Methods 5:605–607. https://doi.org/10.1038/nmeth.1220
3. Spracklen AJ, Fagan TN, Lovander KE, Tootle TL (2014) The pros and cons of common actin labeling tools for visualizing actin dynamics during Drosophila oogenesis. Dev Biol 393: 209–226. https://doi.org/10.1016/j.ydbio.2014.06.022
4. Courtemanche N, Pollard TD, Chen Q (2016) Avoiding artefacts when counting polymerized actin in live cells with LifeAct fused to fluorescent proteins. Nat Cell Biol 18:676–683. https://doi.org/10.1038/ncb3351
5. Flores LR, Keeling MC, Zhang X et al (2019) Lifeact-TagGFP2 alters F-actin organization, cellular morphology and biophysical behaviour. Sci Rep 9:3241. https://doi.org/10.1038/s41598-019-40092-w

6. Xu R, Du S (2021) Overexpression of Lifeact-GFP disrupts F-actin organization in cardiomyocytes and impairs cardiac function. Front Cell Dev Biol 9:746818. https://doi.org/10.3389/fcell.2021.746818
7. Kroll KL, French AR, Sosnick TR, Rock RS (2023) LILAC: enhanced actin imaging with an optogenetic lifeact. Nat Methods 20:214–217. https://doi.org/10.1038/s41592-022-01761-3
8. Rodriguez EA, Campbell RE, Lin JY et al (2017) The growing and glowing toolbox of fluorescent and photoactive proteins. Trends Biochem Sci 42:111–129. https://doi.org/10.1016/j.tibs.2016.09.010
9. Chalfie M, Tu Y, Euskirchen G et al (1994) Green fluorescent protein as a marker for gene expression. Science 263:802–805. https://doi.org/10.1126/science.8303295
10. Shaner NC, Campbell RE, Steinbach PA et al (2004) Improved monomeric red, orange and yellow fluorescent proteins derived from discosoma sp. red fluorescent protein. Nat Biotechnol 22:1567–1572. https://doi.org/10.1038/nbt1037
11. Liu F, Hu H, Deng M et al (2022) A bright monomeric near-infrared fluorescent protein with an excitation peak at 633 nm for labeling cellular protein and reporting protein-protein interaction. ACS Sens 7:1855–1866. https://doi.org/10.1021/acssensors.2c00286
12. Bunch TA, Grinblat Y, Goldstein LSB (1988) Characterization and use of the Drosophila metallothionein promoter in cultured Drosophila melanogaster cells. Nucleic Acids Res 16:1043–1061. https://doi.org/10.1093/nar/16.3.1043
13. Han K (1996) An efficient DDAB-mediated transfection of Drosophila S2 cells. Nucleic Acids Res 24:4362–4363. https://doi.org/10.1093/nar/24.21.4362
14. Buster DW, Nye J, Klebba JE, Rogers GC (2010) Preparation of Drosophila S2 cells for light microscopy. J Vis Exp. https://doi.org/10.3791/1982
15. Marriott G, Mao S, Sakata T et al (2008) Optical lock-in detection imaging microscopy for contrast-enhanced imaging in living cells. Proc Natl Acad Sci 105:17789–17794. https://doi.org/10.1073/pnas.0808882105

Chapter 8

Optogenetic Engineered Living Materials

Varun Sai Tadimarri and Shrikrishnan Sankaran

Abstract

Engineered living materials (ELM) is a new frontier in materials research that uses living microorganisms to augment nonliving materials with lifelike capabilities, such as responding to external stimuli. This is achieved by genetically programming the microorganisms in an ELM with stimulus-sensing modules. A popular stimulus to remotely control various ELM functions is light, which has been realized thanks to optogenetics. This chapter describes methods to create a simple ELM capable of sensing light and responding to it by producing and releasing a drug. The material component of the ELM will be a Pluronic F127-based hydrogel and the living component will be optogenetically engineered bacteria. Such ELMs are being developed to create smart therapeutic solutions for challenging chronic diseases and variations in the design to improve performance and safety will be mentioned.

Key words Engineered living materials, Agarose, *E. coli*, Blue light, Drug delivery, Living therapeutics

1 Introduction

Engineered living materials (ELMs) represent a promising approach to develop innovative materials that exhibit dynamic and adaptive behaviors resembling those of living organisms. ELMs achieve this by integrating living microorganisms with supportive materials. By doing so, smart functions can be programmed in ELMs by engineering microorganisms with genetic circuits. For instance, remote control over the activity of the ELM can be realized by engineering microorganisms to respond to external stimuli. A popular stimulus for remotely controlling ELM functions is light since it is inexpensive, abundant, and can be finely tuned in terms of intensity, wavelength, duration, and localization. Most importantly for ELMs, advances in optogenetics allows precise and reversible control of biological systems with light. By incorporating optogenetically engineered cells into ELMs, it is possible to remotely control various functions of the materials, such as patterned coloration, [1] or drug release. [2, 3].

Armin Baumschlager (ed.), *Optogenetics: Methods and Protocols*, Methods in Molecular Biology, vol. 2840,
https://doi.org/10.1007/978-1-0716-4047-0_8,

Creating ELMs that respond to light requires careful selection of both the nonliving and living components of the material. Since ELMs contain genetically modified organisms, it is desirable that they remain encapsulated within the material component and cannot escape. This is particularly important for biomedical applications such as sensing or drug delivery in the body. Thus, it is desirable to use a material component that is biocompatible but cannot be degraded within the body. In this perspective, chemically cross-linkable versions of the polymer Pluronic F127 (e.g., Pluronic F127 diacrylate) have gained popularity in recent times due to its biocompatibility and usability in versatile fabrication techniques like mold casting [4] and 3D printing. [5] This polymer forms micelles in solution, which undergo noncovalent association into a physically assembled gel at temperatures above 14 °C and concentrations above 15% w/v. This is highly suitable for fabrication by mold casting and 3D printing. Once the desired shape is made, it can be stabilized through chemical cross-linking. Such stabilized gels can confine the growth of microorganisms within them and when fabricated in a core-shell format, they can prevent the escape/outgrowth of those microorganisms to the surroundings. [4, 6] Regarding the living component, bacteria are a popular choice due to their robustness, ease of genetic manipulation, and ability to secrete bioactive molecules. In particular, *Escherichia coli* is most commonly used due to the availability of a large genetic toolbox including optogenetic modules. For applications in the body, biocompatible variants of *E. coli*, such as ClearColi [2, 3] and *E. coli* Nissl1917 [7] are most commonly explored. By designing and programming these bacteria to produce a specific drug upon light exposure, ELMs can provide a highly controlled and localized therapeutic effect for treating chronic diseases, such as chronic wounds or cancer.

In this chapter, a methodology is described to create ELM constructs capable of producing and releasing drugs in a light-regulated manner. [8] The material component used is Pluronic F127 diacrylate, and the bacteria used is ClearColi BL21(DE3). Light regulation is encoded using the optogenetic plasmid, pDawn, [9] which enables induction of gene expression in response to blue light irradiation. While several protein drugs can be encoded for production and secretion, these methods are based on the experience gained from engineering a protein drug that promotes blood vessel regeneration. [8].

2 Materials

2.1 Hydrogel Constructs

1. Pluronic F127 diacrylate polymer—The polymer can be synthesized as described here [10] or can be purchased from Sigma Aldrich.

2. Polydimethylsiloxane (PDMS) kit (SYLGARD™ 184 silicone elastomer base, Dow chemicals, USA).
3. 13 mm (about 0.51 in) Borosilicate glass cover slips (VWR chemicals, USA).
4. Irgacure 2959 (2-Hydroxy-4′-(2-hydroxyethoxy)-2-methylpropiophenone).
5. UV irradiation lamp (Gel doc EZ system from cell Biosciences, Bio-Rad) with irradiation in range of 302–365 nm and intensity of 6 mW/cm^2.
6. 3-(Trimethoxysilyl) propyl acrylate (3-APS silane, Sigma-Aldrich).

2.2 Bacteria and Nutrient Broth

For Bacterial culture and polymer dissolution: LB Miller medium—10 g/L Tryptone, 5 g/L Yeast Extract, and 10 g/L NaCl dissolved in Ultrapure water and autoclaved at 121 °C for 15 min. Addition of antibiotic depending on the selection marker: Kanamycin (50 μg/mL final concentration in culture); Ampicillin (50 μg/mL final concentration in culture).

For transformation by electroporation : Expression recovery medium provided with the ClearColi BL21(DE3) competent cell kit or SOC medium—20 g/L tryptone, 5 g/L yeast extract, 10 mL/L of 1 M NaCl, and 2.4 mL/L of 1 M KCl in Ultrapure water with pH adjusted to 7.0 and autoclaved.

- ClearColi Bl21(DE3) competent cells.

  ClearColi are genetically modified *E. coli* strains in which the production pathway of the endotoxic lipopolysaccharide has been disrupted through seven mutations, thereby improving its immune compatibility.
- Escherichia Coli Nissle 1917.

  Escherichia Coli Nissle 1917 is a gram-negative and common probiotic organism found in the lower intestine of warm-blooded animals, including human beings. This organism has been extensively used to study its probiotic properties assumed to contribute to the treatment of inflammatory bowel diseases.

2.3 Cloning

2.3.1 pDawn

pDawn is an engineered single plasmid encoding a two-component system for light-regulated gene expression. pDawn enables blue light-responsive gene expression in bacteria, especially *E. coli*. pDawn contains a blue-light sensing protein YF1, a histidine kinase by function. In the absence of light, YF1 phosphorylates FixJ which in turn binds to the FixK2 promoter to induce transcription of phase repressor cl. Repressor cl inhibits transcription from phage promoter pR, preventing the expression of a reporter gene (present in multiple cloning sites). In the presence of light, YF1 remains inactive, hence preventing synthesis of repressor cl and therefore allowing transcription of a reporter gene.

2.3.2 Cloning Kits and Enzymes

1. Synthetic oligonucleotides including gene of interest and primers.
2. Q5 High-Fidelity 2X Master mix.
3. 2X HiFi assembly master mix.
4. DH5 alpha-competent cells.
5. PCR product clean-up kit.
6. Plasmid isolation kit.

2.3.3 Transformation

Transformation of ClearColi cells is performed according to the electroporation protocol explained in Subheading 3.

2.3.4 Plasmid Extraction

Plasmid extraction is carried out with the QIAprep Spin Miniprep kit by following the protocol provided along with the kit by Qiagen. The protocol is explained in Subheading 3.

2.3.5 Sequencing

Plasmids are sent to Eurofins Genomics Germany Gmbh, where sequencing is done using the Sanger sequencing technique. At least one microgram of the plasmid has to be sent for reliable sequencing results.

2.4 Light Irradiation

OptoWELL (Opto biolabs, Freiburg, Germany) device was used in our experiments for light irradiation setup.

OptoWELL is a smart LED instrument that allows users to perform light-dependent experiments in a 24-well plate format, with customizable wavelengths between 365 and 950 nm and a wide range of intensities that can be individually programmed for each well.

3 Methods

3.1 Bacterial Cloning

3.1.1 Gibson Assembly of Gene Encoding Protein of Interest in pDawn

1. Design the insert gene encoding the protein of interest using a DNA sequence editor such as A Plasmid Editor (APE) or Snapgene. For secretion of the protein of interest, it needs to be fused to a carrier protein, like YebF at its N-terminal. The DNA sequence of this carrier protein needs to be included in the insert gene.
2. Gibson assembly is a molecular cloning technique that enables the quick and error-free joining of DNA fragments with 15–30 base pair overlaps in a one tube reaction. This is performed using NEB's Hifi Assembly kit. The components needed for setting up HiFi assembly are vector backbone, insert DNA (synthetic oligonucleotide or amplified from genomic/plasmid DNA), and HiFi DNA assembly master mix (2x).
3. Use the NEBuilder assembly tool (https://nebuilder.neb.com) to generate the primers required for the assembly of the

insert gene and the pDawn plasmid using the NEB Hifi assembly kit. The insert gene needs to be localized in pDawn downstream of the RBS of the pR promoter, where the plasmid contains a multiple cloning site (MCS).

4. Order the primers from a custom synthesized oligonucleotide service provider like Eurofins Genomics or Integrated DNA technologies (IDT).
5. Vector backbone is generated by setting up PCR to amplify the plasmid with a set of primers which yields amplified DNA with two overhanging ends. Insert DNA can also be amplified from an existing plasmid or ordered as a synthetic oligonucleotide, but with at least 20 bp overhangs, which should be exactly the same as overhangs of vector backbone. The amplification of both vector backbone and the insert is confirmed by agarose gel electrophoresis of the PCR-amplified products.
6. Purification of the PCR products containing the amplicon is done to remove the primers, salts, enzymes, and short/failed PCR products. Also, Dpn 1 digestion can be done to destroy the template or parental DNA to increase the efficiency of HiFi assembly.
7. The HiFi assembly of purified insert and vector backbone is set up by following the protocol provided by the manufacturer (https://international.neb.com/protocols) to obtain the HiFi product, which is transformed into ClearColi BL21(DE3) cells using the protocol explained in detail below.

3.1.2 Transformation in ClearColi BL21(DE3) Competent Cells

1. Thaw the electrocompetent ClearColi BL21(DE3) cells by placing them on ice for 10 min. Also, place the 1 mm (about 0.04 in) electroporation cuvette on ice.
2. Transfer 50–60 μL of competent cells to a 1.5 mL microcentrifuge tube and add 50–100 ng equivalent of plasmid DNA to the cell mixture. Carefully flick the tube a few times to ensure uniform mixing.
3. Place the mixture on ice for 30 min. Transfer the mixture to the chilled 1 mm cuvette, place the cuvette in the provided module of the Bio-Rad electroporator device set at 1.8 kV, 200 ohms, and 25 μF, and electroporate with a single pulse.
4. Immediately add 900 μL of prewarmed (37 °C) Expression Recovery or SOC medium to the cuvette, gently mix a couple of times by pipetting, and transfer it to a 1.5 mL microcentrifuge tube.
5. Incubate the tube with shaking at 250 rpm, 37 °C for 60 min.

Centrifuge the mixture tube at 1400 rcf for 4 min and spread 100–200 μL cells onto a prewarmed (37 °C) LB NaCl agar + kanamycin plate. Incubate the plate at 30 °C for 18 h or at 37 °C for 12 h.

3.1.3 Plasmid Extraction and Sequencing

This protocol is based on the QIAprep Spin Miniprep Kit by Qiagen® used for the experiments mentioned here.

1. Inoculate the bacteria from a glycerol stock or the colony from an agar plate into 4–6 mL of LB medium supplemented with kanamycin (50 μg/mL final concentration in culture), and incubate it overnight in a shaker at 250 rpm, 37 °C.
2. Resuspend the pellet in 250 μL of buffer P1 and transfer it to a 1.5 mL microcentrifuge tube. Add 250 μL of buffer P2 and mix thoroughly by inverting the tube 4–6 times until the solution becomes clear. Do not allow the reaction to proceed for more than 5 min.
3. Add 350 μL of buffer N3 and mix immediately by inverting the tube a few times.
4. Centrifuge the resulting solution in a table-top centrifuge for 10 min at 15,000 rcf. Transfer 800 μL of supernatant from the previous step to a QIAprep 2.0 spin column followed by centrifugation at 15,000 rcf for 30–60 s and discard the flow through.
5. Wash the spin column by adding 800 μL of buffer PB followed by centrifugation at 15,000 rcf for 30–60 s and discard the flow-through. Repeat the step with 750 μL of buffer PE.
6. Discard the flowthrough and centrifuge the spin column for a minute at 15,000 rcf to discard the residual buffer.
7. Place the QIAprep spin column in a clean 1.5 mL microcentrifuge tube, add 50 μL of elution buffer (10 mM Tris-HCl, pH 8.5), let it incubate for 2 min at room temperature, and centrifuge for 2 min at 15,000 rcf to elute the DNA.
8. Measure the concentration of eluted DNA (Plasmid) on a nanodrop device and send approximately 1 μg of DNA for Sanger sequencing.

3.1.4 Preparation of Glycerol Stock

From the LB Miller agar + Kanamycin plate with bacterial colonies, inoculate the colony harboring the plasmid with the correct sequence alignment (result from Sanger sequencing) in LB Miller media supplemented with kanamycin (50 μg/mL final concentration in culture). Incubate the culture overnight at 34 °C and 300 rpm so that the culture does not overgrow beyond log phase (OD_{600nm} 0.2–0.8) by the next morning. Add 750 μL of the overnight culture to 250 μL of 60% autoclaved glycerol in a 1.5 mL Eppendorf tube or a cryovial and gently mix. Freeze the glycerol stock at −80 °C, which can be used for many years.

3.2 Bacterial Culturing and ELM Fabrication

3.2.1 Inoculation from Glycerol Stock and Overnight Culture Growth

Pick/recover bacteria from glycerol stock (stored at −80 °C) with a sterile loop or sterile pipette tip and suspend it in a test tube with 5 mL of LB Miller medium supplemented with kanamycin (50 μg/mL final concentration in culture). Incubate with shaking at conditions that ensure that the overnight culture does not grow beyond the active logarithmic phase and is ideally at an OD_{600nm} between 0.4 and 0.8. The growth phase of bacterial culture has a direct impact on the health of the bacterial population, and in turn, the concentration of drug released from the encapsulated bacteria. The bacteria grow fastest at temperatures close to 37 °C, and their growth can be slowed by incubation at lower temperatures. It is also possible to modulate the growth rate by adjusting the shaking speed. However, it must be ensured that the shaking speed is high enough for turbulent mixing of the culture to allow growth under aerobic conditions, which has been found to improve light responsiveness and protein expression. Thus, temperature, shaking speed, and incubation times must be optimized to achieve the desired growth rate. For example, inoculated culture can be incubated at 34 °C and 200 rpm overnight to ensure the bacterial population will be in their log phase in the morning.

3.2.2 Spin Down and Concentrate to Desired OD

Measure the OD600 of overnight culture using a Nanodrop or UV spectrophotometer and calculate the volume of culture needed to prepare 1 mL of the culture with OD600 of 1. Based on the calculated volume, spin down the specific volume of the culture at 1400 rcf for 5 min. Discard the supernatant and resuspend the cell pellet in 1 mL of LB miller medium to make up the OD_{600nm} to 1.

3.2.3 Bilayer Bacterial Thick Film Fabrication

Preparation of Polydimethylsiloxane (PDMS) Molds

1. Add around 50 mL of 99% ethanol to a 100 mL glass beaker and sonicate it for 5 min. Discard the ethanol and use this beaker for mixing the components needed to fabricate the PDMS molds (*see* **Note 1**).
2. Mix 10 g of SYLGARD™ 184 silicone elastomer base and 1 g of SYLGARD™ 184 silicone elastomer curing agent (Crosslinker) in the sonicated beaker (*see* **Note 2**).
3. Place the beaker with contents in a vacuum-desiccator (DN 150 Duran) for 10 min to remove the air from the mixture. Take out the beaker from the desiccator after making sure there are no bubbles left.
4. Calculate the volume of PDMS mixture needed for the respective glass flask to be used to obtain molds of desired heights and pour the mixture into it. Use the formula of volume of a cylinder ($\pi r^2 h$) to calculate the amount of PDMS mixture needed per mold, for example, we have used a glass flask of 7 cm base diameter, 2 cm height, and 2 mm thickness to cast PDMS molds of desired dimensions.

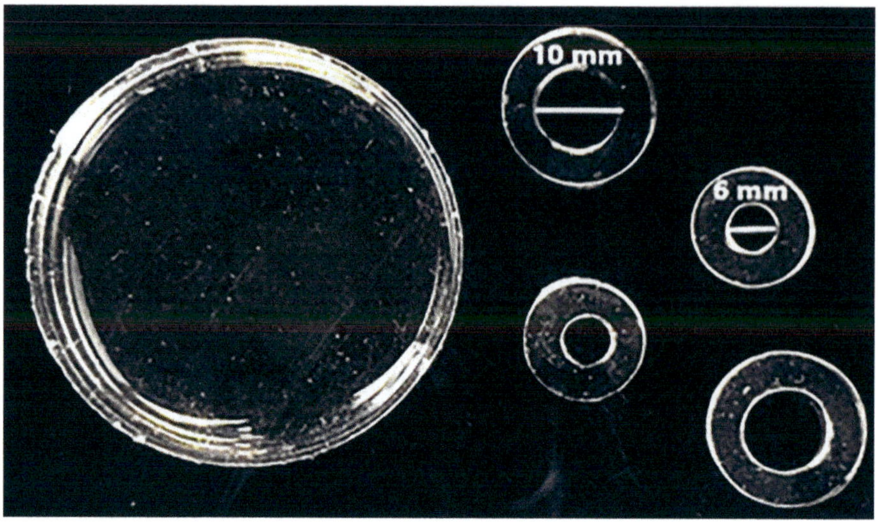

Fig. 1 Solidified PDMS mold along with 10 mm and 6 mm sized molds used for ELM fabrication

5. Place the glass flask with the PDMS mixture in a hot air oven at 95 °C for 2 h for polymerization. Take them out after 2 h and leave them at room temperature overnight.
6. Solidified PDMS molds are easily peeled out from the glass flask and holes of desired diameter are punched out of the solidified PDMS molds using a Boehm hollow punch set. The punch set consists of hollow metal punches ranging from 3 to 20 mm. In our ELM fabrication, desired dimensions of PDMS molds are 6 mm and 10 mm each for the core and shell, respectively (as seen in Fig. 1).

Silanization of Coverslips

1. Arrange the required number of 13 mm coverslips made of borosilicate glass in a Teflon coverslip holder (e.g., Invitrogen Coverslip Mini-Rack).
2. Wash the coverslips with 99% ethanol followed by Milli-Q water and with 99% ethanol again.
3. Suspend the whole setup in a 100 mL beaker filled up to half with 99% ethanol and sonicate it for 10 min.
4. Transfer the setup to a beaker containing 20 mL of 95% v/v ethanol, 4% v/v Milli-Q water, and 1% v/v of 3-(trimethoxy silyl) propyl acrylate. Make sure the coverslip setup is completely immersed in this solution.
5. Place the beaker on a rotary shaker for at least 6 h at 100 rpm. Silanized coverslips should not be allowed to be exposed to air or any other gases. They can be stored by suspending ethanol or a case protected from exposure to air (*see* **Note 3**).

Preparation of Pluronic Diacrylate Polymer Mix

Initially, prepare 0.2% w/v solution of irgacure 2959 photoinitiator in LB Miller medium or any suitable medium which supports bacterial growth. Dissolution of irgacure 2959 can be accelerated by sonication for 10 min followed by vortexing for 5–10 min. Then transfer the solution to a 10 mL amber glass bottle with a screw top. Weigh the Pluronic diacrylate to prepare a 30% w/v solution and add it to the amber glass bottle with the irgacure solution. Set the glass bottle in a rotary shaker and leave it overnight on rotation at 4 °C to ensure complete dissolution. Store the polymer solution at 4 °C until further use (*see* **Note 4**).

Mixing Bacterial Culture with Pluronic Diacrylate Solution

Collect overnight grown log-phase bacterial culture and make it up to an OD_{600nm} of 1 with fresh nutrient broth (LB Miller or modified cell culture medium). Prepare the bacteria–Pluronic diacrylate mixture needed for the fabrication of the core com of ELM by mixing them in the ratio of 1:9 by volume. For example, approximately 1 mL of the core mixture can be prepared by mixing and vortexing 100 μL of bacterial culture with OD_{600nm} of 1 with 900 μL of 30% Pluronic diacrylate solution, such that the final bacterial density in the mixture is equivalent to OD_{600nm} of 0.1.

Fabrication of Core and Shell of ELM

Arrange the Pluronic diacrylate solution (on ice), core mixture (on ice), 3-APS coated coverslips, sterilized PDMS molds and a Petri dish lid covered with parafilm in sterile hood (as seen in Fig. 2).

Core fabrication: Place the silanized coverslip on a flat sheet of parafilm, then place the cylindrical PDMS mold of 6 mm diameter and 0.8 mm thickness on the coverslip with conformal contact. Pipette 22–25 μL of core solution (polymer-bacterial mixture) to make a core of 6 mm diameter and thickness of 0.7 mm. Leave it undisturbed for 2 min, then transfer the PDMS + coverslip setup to a UV irradiation device like a Gel-Doc system (e.g., Biozym scientific GmbH, FluorChemQ). Initiate the cross-linking (photopolymerization) by irradiating it with ultraviolet (UV) transillumination (365 nm) at 4 mW/cm^2 for 60 s and incubate in the dark at room temperature for 10 min to allow cross-linking to go to completion. Transfer the polymerized core along with the setup back to the sterile hood and carefully remove the PDMS mold without disturbing the polymerized core.

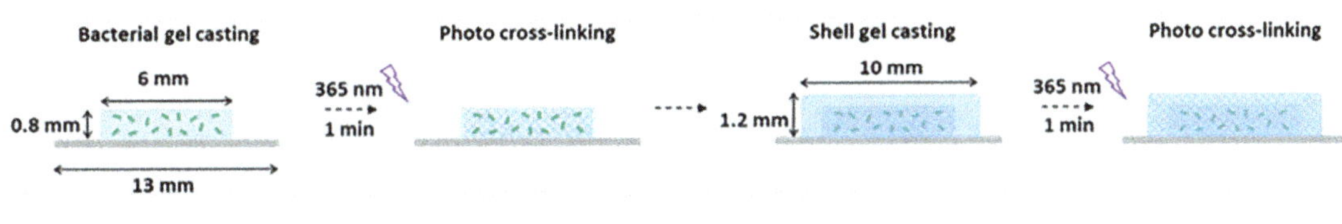

Fig. 2 Graphical schematic portraying the method to fabricate core-shell ELM constructs

Shell fabrication: Take a PDMS mold of 10 mm diameter to cast a shell of polymer which would encapsulate as well as prevent the leakage of bacteria from the core. Place the 10 mm PDMS mold on the coverslip such that core is at the center, pipette 60 μL of Pluronic diacrylate solution covering the entire exposed surface area of the core to eventually make a shell of 10 mm diameter and 1.2 mm thick. Crosslink this setup by UV exposure with the same settings as the core but for 90 s followed by a 10-min incubation at room temperature. Transfer the setup back to the sterile hood, remove the PDMS mold, and place the coverslip with the core-shell ELM in a 24-well-plate well with 350 μL of nutrient medium. Place the 24-well plate in an incubator at 37 °C until analysis.

Analysis and Maintenance of Constructs

The media surrounding the ELM constructs should be checked for pH changes, and the volume of media after evaporation/absorption by polymer and make up these variables by adding the missing volume of media.

Most of the bacterial growth in terms of colonies within the core of the construct occurs during the first 24 h after which bacteria do not grow rapidly due to mechanical constrictions as well as due to its metabolic/growth phases. They remain to be alive and actively secrete the molecule of interest. The constructs have to be visualized by suitable microscopic techniques (preferably bright-field) to check for the outgrowth of bacteria within the core and also to confirm the integrity of the overall construct (*see* **Note 5**) (as seen in Fig. 3).

In the case of ELMs encapsulating optogenetically engineered bacteria, the well plate with ELMs is incubated without any light illumination for the first 12–18 h followed by transfer onto Optowell/any illumination device to irradiate them and induce the protein expression. At certain time points, the medium supernatant is collected to quantify the released protein by Enzyme-linked immunosorbent assay (ELISA) or mass spectrometric techniques.

3.3 Light Responsiveness

3.3.1 Setting up Pulsed Irradiation with Optowell

1. Arrange the ELM constructs in a 24-well plate. Place the well plate on the space provided on Optowell with direct illumination underneath individual wells.
2. Connect the Optowell to a laptop/PC with the Optowell control software and set the desired parameters.
3. Select the wells that are to be illuminated, then select the desired channel based on the desired wavelength followed by setting the percentage of needed intensity. In the experiments conducted with the pDawn vector system to express a therapeutic peptide, Intensities were set at 3% (80 μW/cm^2), 6% (105 μW/cm^2), and 9% (125 μW/cm^2) to study the release profile of peptide in response to light intensity. Intensities in the range of 1–10% seem to be functional w.r.t bacteria and do not oversaturate or exhaust them with excessive light or heat.

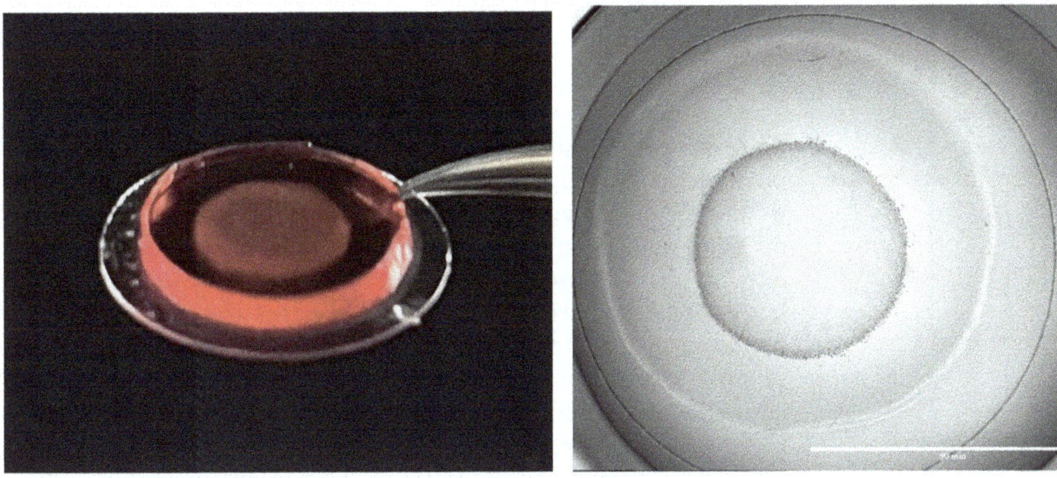

Fig. 3 Snapshot of bilayer engineered living material (ELM) encapsulating bacteria within the core and protected by a shell of pluronic diacrylate (figure to left); Brightfield 10x image of the same ELM on left, clearly marked distinct zones of core and shell fabricated on a 13 mm coverslip can be seen (figure to right)

4. Set the duration of pulses for light illumination in respective channels followed by pulse duration with no light illumination. In our optogenetic ELMs, pulsed irradiation (blue light) was set for 2 s followed by 60 s of darkness for all the wells for a period of 6 days.
5. Set the number of cycles needed based on the extent to which the experiment is planned, and illumination is needed. For example, illumination for 2 s at a specific wavelength followed by a minute of no light straight for 24 h requires approximately 1400 cycles. After setting the number of cycles, export and store the profile.
6. The settings will be saved in the Optowell device after the previous step, disconnect the laptop/PC, place the Optowell and well plate setup in the incubator, and switch on the device. Programmed illumination will start automatically (*see* **Note 6**).

3.3.2 Tuning the Drug Release

1. Tuning the drug release from an engineered optogenetic circuit is done by modulating a few variables such as intensity, duration, and frequency of pulsed irradiation (e.g., 1 s/min, 2 s/min, 5 s/min, 10 s/min).
2. A scan to analyze the dynamic range of intensities (1–10%), which the system responds to is a crucial step. Once the saturation intensity is determined, testing for tunable drug release is tested across a range of varying intensities—0, 80, 105, and 125 μW/cm^2 in the Optowell setup.
3. Duration and frequency (pulsed irradiation/unit time) of light irradiation can also be varied per well to screen for a range of variables to characterize the tunable drug release. For example,

pulsed irradiation at 5 s per minute frequency might induce higher production of target molecule than 1 s per minute frequency.

3.3.3 Switching ON and OFF Drug Release

To test the response of the engineered bacterial system to switching on and off of light, they are irradiated with light at a specific frequency of pulsed irradiation for a definite period followed by no light irradiation for a definite period. This loop can be modulated as per the user to understand the response dynamics. It is known that bacteria produces and secretes the protein or drug when exposed to light, but they also tend to produce low amounts of protein due to the leakiness of the genetic circuit. Optowell allows users to perform high throughput screening of such variables to figure out an ideal range of duration, intensity, and frequency of irradiation for respective samples. The final step after incubating the samples in Optowell is a qualitative and quantitative analysis of supernatants from individual wells containing ELMs. These supernatants can be analyzed by different techniques, such as western blot, mass spectrometry, and ELISA to detect the secreted biomolecule and can even be tested on mammalian cells to see their intended effects.

3.3.4 Characterization by ELISA

ELISA stands for enzyme-linked Immunosorbent assay and is a predominantly used biochemical assay to detect and quantify a molecule of interest using antibodies with specificity toward the molecule to be quantified. The steps involved in ELISA are:

1. Coating step: Dilute a suitable antigen or antibody specific to the molecule of interest in phosphate-buffered saline (PBS) and incubate it in a suitable 96-well plate until it gets adsorbed to the surface of the well (incubate at 4 °C overnight).
2. Blocking step I: Add 100 μL of 2% bovine serum albumin (BSA) and incubate overnight at 4 °C.
3. Washing step: Pipette out the contents of the wells and wash with 100 μL of wash buffer (PBS + 0.1% Tween-20) thrice to remove the unbound BSA.
4. Add 100 μL of a sample containing a molecule of interest to the wells and incubate for 1 h at room temperature. Empty the contents of the well and repeat the washing step.
5. Blocking step II: Perform a second blocking step to minimize nonspecific binding (*see* **Note 7**).
6. Primary antibody treatment: Prepare the primary antibody solution in the ratio of 1:500 (or 1:1000/1:2000, depending on the optimization needed for specific antibody) in 2% bovine serum albumin (BSA) solution (*see* **Note 8**). (In our experiments, an anti-YebF primary antibody is used as the therapeutic peptide is being transported out by YebF carrier protein, and

after the cleavage of the signal peptide of YebF, the final protein containing a domain of YebF is secreted into the bacterial periplasm through sec pathway. It could be a go-to strategy to achieve the secretion of the protein of interest in E. coli by designing a fusion protein wherein the protein of interest is linked to the C-terminal of the YebF carrier protein.) Add 100 μL to wells followed by incubation at room temperature for 1 h and washing step later.

7. Secondary antibody treatment: Prepare secondary antibody conjugated with horseradish peroxidase in the ratio of 1:500 (or 1:1000/1:2000, depending on the optimization for specific antibody) in 1% BSA. Add 100 μL to wells followed by incubation at room temperature for 1 h and washing step later.
8. Substrate treatment: Add 100 μL of Tetramethylbenzidine (TMB) substrate and leave it on shaker for 15 min until the blue coloration reaches its maximum intensity (*see* **Note 9**).
9. Stopping step: Add 100 μL of 1M Hydrochloric acid (HCl) to the wells to stop the reaction and it results in yellow coloration whose absorbance should be measured at 450 nm using a plate reader within 30 min.

4 Notes

1. All the steps of experiments should be performed in a sterile hood except silanisation of coverslips.
2. Be extremely careful while handling SYLGARD™ 184 silicone elastomer base to cast the PDMS molds. Wear thick rubber gloves while handling and be sure to prevent the skin contact as the substance is very sticky and difficult to wash off.
3. It is suggested to use the silanized coverslips within a week after they are prepared. Storage over longer periods can be set under inert conditions like under nitrogen or preferably argon but over time, the reactivity of the layer can decline and affect the integrity of ELMs if used.
4. Perform all the steps involving pluronic diacrylate on ice or at 4 °C. Exposure of pluronic diacrylate solution to 14 °C or higher temperatures results in polymerization.
5. To ensure that bacteria within ELMs are not growing out, they can be examined microscopically. The other way to ensure this is to take around 50–100 μL of supernatant and inoculate it in a 5 mL nutrient broth (LB medium + antibiotic) (or) spot it on an agar plate (LB agar + antibiotic) and incubate it in a shaker overnight. No bacterial growth in the culture/agar plate on the next day would confirm the nonleakiness of constructs.

6. Make sure not to touch or disturb the setup of Optowell while it is running as the device is quite sensitive to perturbations and it can affect the ongoing method.
7. Two steps of blocking are included in the ELISA protocol to make sure of the removal of the binding of non-specific molecules to the wells. In some cases, one blocking step could be enough too, so this depends on the type of sample being quantified.
8. Optimizing the ideal dilution of antibodies (primary and secondary) is crucial for the optimal sensitivity of ELISA. Depending on the antibody, different concentrations need to be tested to identify the right ratio. A standard curve must be established with serial dilutions of the purified protein of interest within the range of concentrations that are expected to be secreted from the ELM. This will be the basis for quantifying levels of the protein that are released from the ELM.
9. Do remember to prepare all the reagents needed to perform ELISA at room temperature. For example, TMB substrate is usually stored at 4 °C, so take it out an hour before and let it warm up to room temperature.

References

1. Gilbert C, Tang T-C, Ott W et al (2021) Living materials with programmable functionalities grown from engineered microbial co-cultures. Nat Mater 20:691–700
2. Sankaran S, del Campo A (2019) Optoregulated protein release from an engineered living material. Adv Biosyst 3:1800312
3. Sankaran S, Becker J, Wittmann C et al (2019) Optoregulated drug release from an engineered living material: self-replenishing drug depots for long-term, light-regulated delivery. Small 15:1804717
4. Bhusari S, Kim J, Polizzi K et al (2023) Encapsulation of bacteria in bilayer Pluronic thin film hydrogels: a safe format for engineered living materials. Biomater Adv 145:213240
5. Liu X, Yuk H, Lin S et al (2018) 3D printing of living responsive materials and devices. Adv Mater 30(4)
6. Bhusari S, Sankaran S, del Campo A (2022) Regulating bacterial behavior within hydrogels of tunable viscoelasticity. Adv Sci Rev 9(17): 2106026
7. Duraj-Thatte AM, Courchesne ND, Praveschotinunt P et al (2019) Genetically programmable self-regenerating bacterial hydrogels. Adv Mater 31:e1901826
8. Dhakane P, Tadimarri VS, Sankaran S (2022) Light-regulated pro-angiogenic engineered living materials. https://www.biorxiv.org/content/10.1101/2022.10.28.514190v1
9. Ohlendorf R, Vidavski RR, Eldar A et al (2012) From dusk till dawn: one-plasmid systems for light-regulated gene expression. J Mol Biol 416:534–542
10. Biase MD, de Leonardis P, Castelletto V et al (2011) Photopolymerization of Pluronic F127 diacrylate: a colloid-templated polymerization. Soft Matter 7:4928–4937

Chapter 9

Light-Controlled Secretion and Injection of Proteins into Eukaryotic Cells by Optogenetic Control of the Bacterial Type III Secretion System

Florian Lindner, Sebastian Grossmann, and Andreas Diepold

Abstract

Gram-negative bacteria can use the type III secretion system (T3SS) to inject effector proteins into eukaryotic target cells. In this chapter, we describe the application of a light-controlled T3SS, based on the targeted sequestration of an essential dynamic T3SS component with the help of optogenetic interaction switches. This method enables to control the secretion or injection into eukaryotic cells for a wide range of protein cargos with high temporal and spatial precision.

Key words Protein translocation, Membrane transport, Cellular control, Bacterial secretion systems, Optogenetic interaction switches, Injectisome, LITESEC-T3SS, iLID, LOVTRAP

1 Introduction

Using bacteria to deliver drugs is a rapidly emerging field in medical science [1]. Some bacterial species have evolved mechanisms for efficient protein delivery, either to the surrounding environment or into eukaryotic target cells. One such system, the bacterial type III secretion system (T3SS), uses a needle-like structure to directly inject proteins into target cells. The hollow conduit of the T3SS spans both bacterial membranes and the eukaryotic membrane, enabling a one-step transport of proteins from the bacterial cytosol into the target cell cytoplasm (Fig. 1a) [2, 3], with high export rates and overall amounts ($>10^6$ translocated proteins per cell) [4, 5]. Many Gram-negative pathogenic bacteria use the T3SS for the manipulation of host cells. However, the T3SS can also be harnessed to translocate heterologous proteins of interest marked with a specific N-terminal T3SS secretion sequence. It can therefore serve as an efficient tool for protein delivery into eukaryotic host

Armin Baumschlager (ed.), *Optogenetics: Methods and Protocols*, Methods in Molecular Biology, vol. 2840,
https://doi.org/10.1007/978-1-0716-4047-0_9,

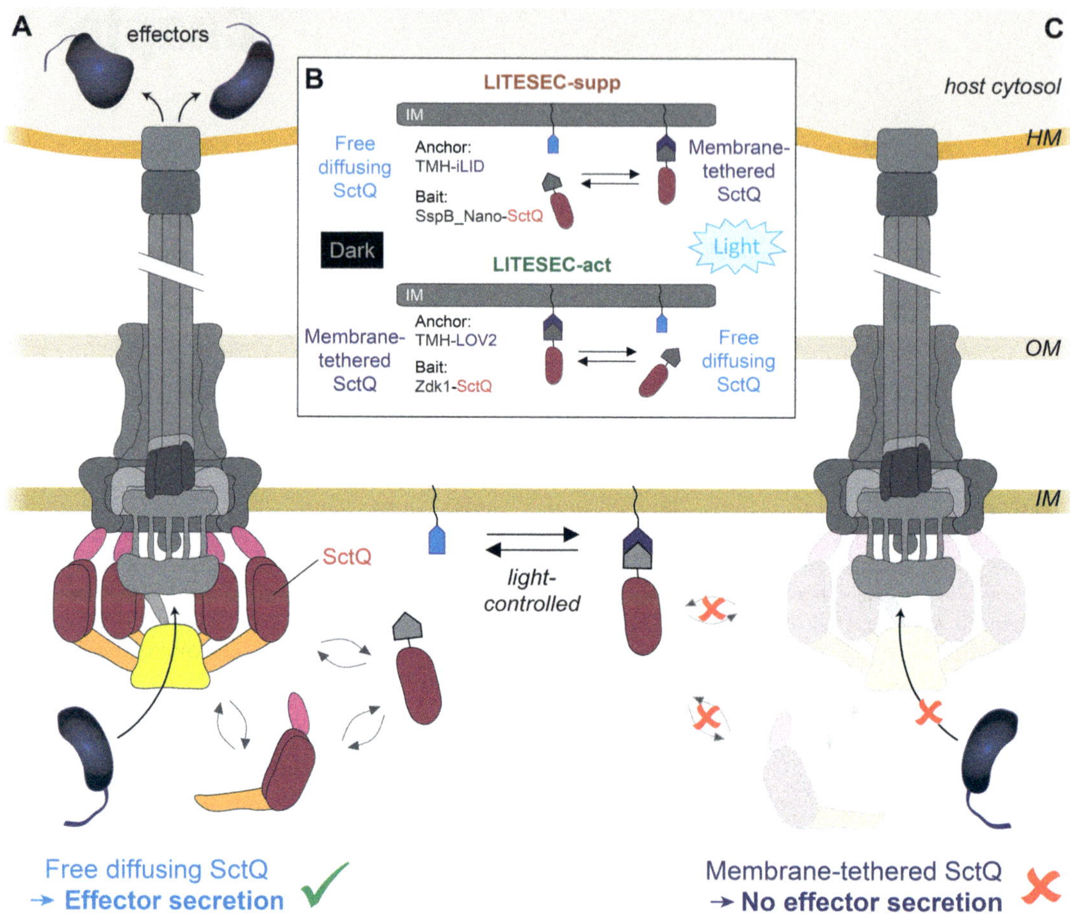

Fig. 1 *Working principle of the LITESEC systems for light-controlled activation and deactivation of protein translocation by the type III secretion system.* (**a**) Schematic representation of the active T3SS injectisome. Effector translocation by the T3SS is licensed by the functional interaction of the unbound bait-SctQ fusion with the T3SS. (**b**) Different states of the bait and anchor proteins in dark and light conditions (also see Table 1). In the light-suppressed LITESEC-supp system (top), the bait protein, a fusion of the interaction switch domain SspB_Nano and the essential T3SS component SctQ, is tethered to the inner membrane (IM) by a membrane anchor, a fusion of a transmembrane helix (TMH) and the corresponding interaction switch domain iLID, in the light, and gets released in the dark. Conversely, in the light-activated LITESEC-act system (bottom), the bait protein, Zdk1-SctQ, is tethered to the membrane anchor, TMH-LOV2, in the dark, and is released by illumination. (**c**) Sequestration of the bait-SctQ fusion protein to the membrane prevents effector secretion. *HM* host membrane, *OM* bacterial outer membrane, *IM* bacterial inner membrane. (Adapted from Ref. [14])

cells for cell culture studies or therapeutic applications, such as cellular reprogramming, induction of apoptosis, or even gene editing [1, 5]. However, the T3SS is nonspecifically activated by contact with eukaryotic target cells, which can lead to off-target effects. To overcome this severe limitation, a method to efficiently and reversibly control the activity of the T3SS is required.

Cellular processes can be controlled by external illumination using optogenetics, a technique based on natural light-sensitive proteins. It allows to regulate biological processes at different levels, ranging from the light-induced control of gene expression to direct protein control at the post-translational level [6]. Its fast, easy, and reversible tunability gives light an advantage over chemical inducers and other environmental triggers, such as temperature or pH, and enables precise spatiotemporal control [7]. Most optogenetic tools are based on modified opsins, rhodopsins, or phototropins, which undergo a conformational change upon irradiation [8]. While optogenetics is well established in eukaryotes, using this technique in prokaryotes is an emerging and rapidly expanding field [8, 9].

Optogenetic interaction switches consist of pairs of interacting proteins whose affinity is strongly altered upon illumination of a specific wavelength. They can be used for the precise, fast, and reversible control of the localization of target proteins by light-induced sequestration to an interaction partner with a specific intracellular localization [10]. In bacteria, the inner membrane can be used as a sequestration site [9, 11]. We have used this approach for SctQ, an essential component of the T3SS that shuttles between the cytosol and a T3SS-bound state [12, 13]. By fusing one optogenetic interaction partner to SctQ (the *bait*) and anchoring the other optogenetic interaction partner to the inner membrane by fusion to a trans-membrane-helix (the *anchor*) (*see* **Note 1**), SctQ can be sequestered to the membrane in a light-dependent manner [14] (*see* **Note 2**). The membrane-bound SctQ can no longer function at its native place of action in the cytosol, which interrupts secretion. Light-induced release of SctQ restores functionality of the T3SS within a short time, allowing the controlled secretion of cargo proteins. The integration of optogenetic interaction switches that either dissociate or associate upon illumination (LOVTRAP [15] and iLID [16] in this case) allows to either activate or inactivate the T3SS with light (Fig. 1 and Table 1).

In this chapter, we describe the application of this system, termed LITESEC-T3SS for **l**ight-**i**nduced **t**ranslocation of **e**ffectors through **s**equestration of **e**ndogenous **c**omponents of the **T3SS**.

For general considerations concerning the design and implementation of light-controlled bacterial systems based on optogenetic switches, *see* **Note 3** and the references listed there. We strongly recommend to check in preliminary experiments that the protein of interest maintains its biological function when fused to the optogenetic interaction switch (*see* **Note 4**), and that the optogenetic interaction switch and membrane sequestration work in the chosen biological system (*see* **Note 5**).

Table 1
Overview of optogenetic systems utilized for LITESEC-T3SS

System	System class and properties	Used *anchor* and bait proteins	Dissociation rate after activation	Reference	Application in LITESEC-T3SS
LOV	Light-released Dark = bound state Light = unbound state	*LOV2V416L* ***(Note 22)*** Zdk1	~ 100 s	[15]	Release of bait protein by blue light → activation of secretion
iLID	Dark-released Dark = unbound state Light = bound state	*iLID* SspB_Nano	~ 90-180 s	[16]	Tethering of bait protein by blue light → suppression of secretion

Optogenetic base systems that were utilized for control of the T3SS function [14]

2 Materials

2.1 Bacterial Strains

Bacterial strains for light-controlled secretion and injection need to express fusions (i) of a dynamic essential component of the secretion system to one partner of an optogenetic interaction switch (*bait*, preferably expressed from the native genetic environment) and (ii) of a localization determinant (*anchor*, e.g., a transmembrane helix (TMH)) to the other partner of the interaction switch (which can be expressed from a plasmid). In such a strain background (described as 1. below), cargo proteins of choice, for example, specific to a readout or application, fused to N-terminal T3SS secretion signals, can be additionally expressed from plasmid (described as 2. below). Plasmid backbones used in our protocol are pBAD/His B (L-arabinose inducible expression, medium copy number) [17] and pACYC184 (constitutive expression, low copy number) [18]. Ensure compatibility of plasmids (e.g., pBAD-based cargo expression plasmids cannot be combined with pBAD-based anchor protein expression plasmids).

1. Strains for optogenetic control of the T3SS, all based on the nonpathogenic reference strain *Yersinia enterocolitica* multi-effector knock-out strain ΔHOPEMTasd [19] (see Table 1):
   (a) Light-activatable secretion:
   LITESEC-act2 (Zdk1-SctQ; mCherry-SctL + pBAD::TMH-FLAG-$LOV2_{V416L}$) or
   LITESEC-act3 (Zdk1-SctQ; mCherry-SctL + pACYC184::TMH-FLAG-$LOV2_{V416L}$).
   (b) Light-suppressible secretion:
   LITESEC-supp1 (SspB_Nano-SctQ; mCherry-SctL + pBAD::TMH-FLAG-iLID) or
   LITESEC-supp2 (SspB_Nano-SctQ; mCherry-SctL + pACYC184::TMH-FLAG-iLID).
   (c) Positive control: Reference strain.
   (d) Negative control: Δ*sctQ*.
2. Cargo expression plasmids for quantification of protein translocations (optional):
   (a) pBAD::$YopE_{1-138}$-Nanoluc-FLAG.
   (b) pBAD::SycE, $YopE_{1-138}$-tBID-FLAG [5] (*see* **Note 6**).
   (c) pBAD::SycE, $YopE_{1-138}$-HiBiT.

2.2 Bacterial Cultivation

In this subheading, we provide the protocol for the application of the *Yersinia enterocolitica*-based LITESEC-T3SS system. As protocols for propagation of bacteria and induction of secretion systems vary, this protocol can be adapted to the specificities of the studied bacterium and secretion system.

All buffers and solutions should be prepared using ultrapure water at room temperature, additives are sterilized by filtration.

1. BHI medium: 37 g/L Brain Heart Infusion Broth, dissolved in ultrapure water and sterilized by autoclaving.
2. Overnight cultivation medium: BHI supplemented with 35 μg/mL nalidixic acid (Nal), 60 μg/mL 2,6-diaminopimelic acid (DAP) (*see* **Note 7**) and corresponding antibiotics (200 μg/mL ampicillin for pBAD-based plasmids or 20 μg/mL chloramphenicol for pACYC184-based plasmids).
3. Secretion-inducing cultivation medium: BHI supplemented with 5 mM ethylene glycol-bis(β-aminoethyl ether)-N,N,N′,N′-tetraacetic acid (EGTA) (*see* **Note 8**), 35 μg/mL Nal, 60 μg/mL DAP, 20 mM $MgCl_2$, 0.4% (w/v) glycerol, and corresponding antibiotics.

4. Non-secretion-inducing cultivation medium: BHI supplemented with 5 mM $CaCl_2$ (*see* **Note 8**), 35 μg/mL Nal, 60 μg/mL DAP, 20 mM $MgCl_2$, 0.4% (w/v) glycerol, and corresponding antibiotics.
5. Aluminum foil.
6. Shaking incubators (28 °C, 37 °C).

2.3 Protein Analysis

1. 100% (w/v) Trichloroacetic acid (TCA), precooled (stored at 4 °C).
2. 100% Acetone, ice-cold (stored at −20 °C).
3. 1x sample buffer: 20 g/L SDS, 100 mM Tris, 10% (v/v) glycerol, 50 mM DTT, adjust pH to 6.8 with HCl.
4. 1x PBS (phosphate-buffered saline): 137 mM NaCl, 2.7 mM KCl, 10 mM Na_2HPO_4, 2 mM KH_2PO_4, pH = 7.4.
5. 12% SDS-PAGE gel.
6. 1x SDS running buffer: 25 mM Tris, 200 mM glycine, 1 g/L SDS.
7. Plate reader able to measure luminescence (for NanoLuc readout assay only).
8. SDS-PAGE running chamber.
9. Nitrocellulose membrane (pore size: 0.2 μm).
10. Blotting buffer: 300 mM Tris, 300 mM glycine, 0.5 g/L SDS.
11. Blocking solution: 5% nonfat dried milk powder in 1x PBS.
12. PBS-T: 1x PBS + 0.2% Tween 20.
13. Primary antibody solution: Rabbit anti-FLAG antibody, 1: 2000 in blocking solution.
14. Secondary antibody solution: HRP-conjugated anti-rabbit antibody, 1:10,000 in blocking solution.
15. HRP detection reagent.
16. Semidry blot transfer system.
17. Luminescent image analyzer.
18. Precooled tabletop centrifuge (4 °C).
19. Preheated thermo block (55 °C).

2.4 Eukaryotic Cell Culture

1. Eukaryotic cell lines are specific for readout assay of choice. Cells should be 80–90% confluent at Day 0; volumes in the protocol refer to propagation in cell culture-compatible 10 cm diameter Petri dishes.
   (a) HeLa cells stably expressing LgBiT [20] for split-luciferase assay.

   (b) Suitable cell lines for alternative readout assay of choice, for example, HEp-2 for beta-lactamase assay or phenotypic apoptosis assay.
2. Cell culture cultivation medium (about 10 mL/strain, prewarm to 37 °C):
   (a) Dulbecco's modified Eagle Medium (DMEM) supplemented with 10% fetal calf serum (FCS) for HeLa cells expressing LgBiT.
   (b) Roswell Park Memorial Institute 1640 medium (RPMI 1640) for HEp-2 cells.
3. Trypsin solution: 0.25% Trypsin-EDTA solution.
4. Cell culture working medium (colorless growth medium): RPMI 1640 medium without phenol red, supplemented with 60 μg/mL DAP and L-arabinose, where required for the expression of anchor or cargo protein (see Subheading 2.1).
5. Infection buffer (about 2 mL/sample): Prewarmed 1x PBS supplemented with 60 μg/mL DAP and L-arabinose, where required for the expression of anchor or cargo protein (see Subheading 2.1).
6. Cell culture cultivation medium containing 100 μg/mL gentamycin.
7. Cell culture-compatible 96-well plate.
8. Plate reader able to measure luminescence (for split-luciferase readout assay only).
9. Tabletop microcentrifuge (37 °C).
10. 10 mL glass or plastic pipette.
11. Hemocytometer for estimation of cell density of resuspension.
12. Stereo microscope, 5x objective.

2.5 Optogenetic Illumination Setup

1. Optogenetic illumination setup allows for the controlled illumination of shaking flasks. An example setup consisting of two blue light sources (light source 1, "Globo lighting 10 W LED 9 V 34118S"; light source 2, "Rolux LED-Leiste DF-7024-12 V 1.5 W") placed around the bacterial cultures is shown in Fig. 2 in a shaking incubator. The light intensity at a wavelength of 488 nm at the culture location should be adjusted to ~1 mW/cm^2.
2. Optogenetic illumination setup during translocation assays: cell culture plates placed ~5–10 cm under light source 2 (Fig. 2), which is attached to the top of the incubator. For dark conditions, plates are wrapped with aluminum foil.

3 Methods

3.1 Secretion and Analysis of Protein Cargos into the Supernatant

1. Start experiment at Day 0 with inoculation of overnight cultures of the bacterial strains to be tested, including control strains, from bacterial glycerol stocks or from single colonies from plate in 5 mL of overnight cultivation medium (or respective growth medium). Cultivate overnight at 28 °C, 200 rpm shaking.
2. The next day (day 1), to compare *light* vs. *dark* conditions, prepare two cultures in parallel for each strain for the following experiment: Inoculate 5 mL secretion-inducing cultivation medium with stationary bacterial overnight culture to an OD_{600} of 0.15 (usually about 80 μL).
3. Incubate the bacterial cultures for 90 min at 28 °C, 200 rpm to restore exponential growth.
4. If required, induce expression of membrane anchor or cargo protein from plasmid by addition of L-arabinose prior to transfer to 37 °C.
5. Transfer bacterial cultures to a shaking incubator preheated to 37 °C (*see* **Note 9**), equipped with controllable blue light sources (see Fig. 2). Cultivate one culture of each strain under blue light and one in the dark (wrap dark culture with aluminum foil) (*see* **Note 10**) for 3 h at 37 °C, 200 rpm.
6. Collect 2 mL of bacterial culture for 10 min at 4 °C, ≥12,000 rcf (for analysis of secreted proteins).
7. In the meantime, measure the OD_{600} of the remaining bacterial culture (important for later normalization).
8. Carefully transfer 1.8 mL of the supernatant (SN) into a fresh microcentrifuge tube, avoiding to transfer parts of the cell pellet. Store total cell (TC) pellet (without remaining SN) at −20 °C for later analysis of expression level and stability of the fusion protein, if required.
9. When working with a cargo fusion to a luciferase, such as NanoLuc, a luciferase assay can be performed at this stage. The following steps describe the alternative readout procedure for native and other substrates by SDS-PAGE analysis of the SN.
10. Add 200 μL of cold 100% TCA (10% final) to SN for protein precipitation.
11. Incubate at 4 °C for 1 h or overnight.
12. Centrifuge TCA solution for 15 min at 4 °C, ≥16,000 rcf.
13. Carefully remove supernatant and wash precipitated protein pellet twice with 900 μL ice-cold acetone (*see* **Note 11**) (centrifuge for 5 min at 4 °C, ≥12,000 rcf after each washing step).

14. After the second washing step, carefully remove acetone as completely as possible without disturbing the protein pellet.
15. Dry protein pellets by incubating microcentrifuge tubes with open lids until pellet has become white (usually between 15 and 30 min) (*see* **Note 12**).
16. For normalization of the samples with respect to the amount of bacteria in the original culture, resuspend pellet in the amount of 1x sample buffer equivalent to 0.6 ODu (1 ODu = 1 mL of a culture at an OD of 1, corresponding to the supernatant of about $5*10^8$ bacteria for *Yersinia*) in 15 μL, based on the OD_{600} value measured in **Step 2** (*see* **Note 13**).
17. Incubate 5 min at 99 °C to denature proteins.
18. Load 15 μL of normalized protein samples on an SDS-polyacrylamide gel and run for 60–90 min at 130 V and 400 mA.
19. Afterward, gels can either be stained directly with Coomassie-based staining solution (**Step 20**) or transferred onto a nitrocellulose membrane for Western blot (outlined in **Steps 21–28** for FLAG-tagged cargo proteins).
20. To visualize the total amount of secreted proteins, stain gel afterward with Coomassie staining solution (2 h—overnight).
21. Transfer proteins from gel to nitrocellulose membrane using a semi-dry blot transfer system (settings for BioRad Trans-blot Turbo: 1.3 A, 25 V, 7 min).
22. Block membrane in 15 mL milk blocking solution on rocking incubator for 1 h at room temperature (RT) or overnight at 4 °C.
23. Wash membrane prior to primary antibody incubation once in 1x PBS and twice in 1x PBS-T, 5 min each.
24. Incubate with primary antibody on rocking incubator overnight at 4 °C.
25. Wash membrane prior to secondary antibody once in 1x PBS and 3–4 times in 1x PBS-T, 5 min each.
26. Incubate with the secondary antibody on shaking incubator at RT for 1 h.
27. Wash membrane prior to HRP detection 4 times with 1x PBS-T and once with 1x PBS.
28. Add 800 μL of HRP detection reagent onto the membrane and image with a luminescent image analyzer.
29. Optimization and troubleshooting: In case of missing or insufficient optogenetic control of secretion, *see* **Note 14**.

3.2 Light-Dependent Translocation of Protein Cargo in Eukaryotic Cells

Depending on the readout method of choice, specific cell lines and bacterial strains carrying the corresponding protein cargo for translocation should be used.

1. One day prior to the experiment (Day 0) prepare eukaryotic cells for infection assay: Carefully aspirate cell culture cultivation medium from the Petri dish used for cell cultivation (see Subheading 2.4).
2. Wash the eukaryotic cells by carefully adding sterile PBS.
3. Aspirate the sterile PBS and add 1 mL of trypsin solution.
4. Incubate the plate at 37 °C and 5% CO_2 for 3 min.
5. After 3 min, check every 30 s with binocular to determine if cells have detached. Carefully tapping the plate can help to visualize detached cells.
6. Once the majority of cells are detached (usually after 4–5 min), immediately add 7–8 mL of fresh cell culture cultivation medium to the dish (*see* **Note 15**).
7. Carefully resuspend cells using a 10 mL pipette. Transfer resuspension into a centrifugation tube and collect cells by centrifugation (5 min, 300 rcf, RT).
8. Aspirate supernatant and carefully resuspend pellet in 1 mL of cultivation medium.
9. Mix thoroughly to disperse any cell aggregates and pipette 20 μL of suspension onto the disposable hemocytometer.
10. Count cells and adjust cell density to 1x 10^5/mL with culture cultivation medium.
11. Seed 200 μL of cells per well in a 96-well plate to obtain 20,000 cells per well (*see* **Note 16**).
12. Place in the 37 °C incubator with 5% CO_2 overnight.
13. On Day 0, inoculate overnight cultures of the bacterial strains to be tested, including control strains, from bacterial glycerol stocks or from single colonies from plate in 5 mL of overnight cultivation medium (or respective growth medium). Cultivate overnight at 28 °C, 200 rpm.
14. The next day (Day 1), inoculate 5 mL nonsecretion-inducing cultivation medium with stationary bacterial overnight culture to an OD_{600} of 0.12 (usually about 60 μL).
15. Incubate the bacterial cultures for 90 min at 28 °C, 200 rpm to restore exponential growth.
16. If required, induce expression of membrane anchor or cargo protein from plasmid by addition of L-arabinose prior to transfer to 37 °C (see **Note 14**).
17. Transfer bacterial cultures to a shaking incubator preheated to 37 °C (*see* **Note 9**), equipped with controllable blue light sources (see Fig. 2). Cultivate cultures under illumination

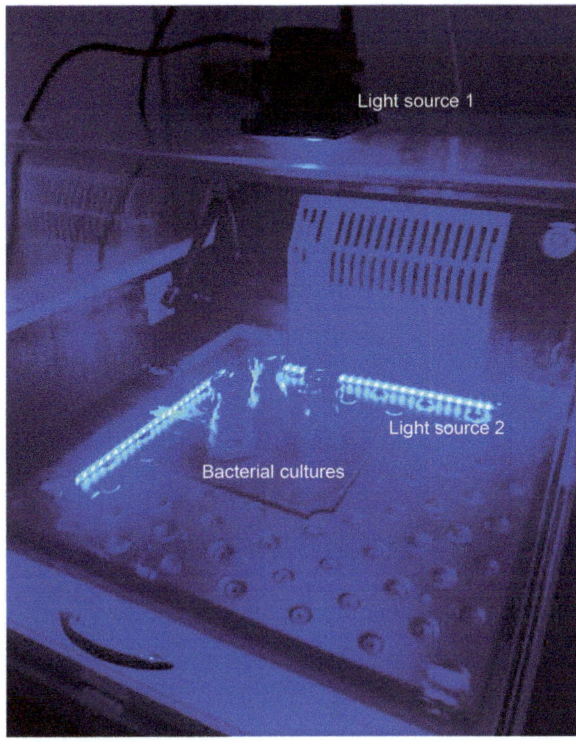

Fig. 2 *Optogenetic experimental setup*. The optogenetic experimental setup consists of two blue light sources that are placed around the bacterial cultures. The light intensity at 488 nm at the site of the cultures is around 1 mW/cm^2. Bacteria are cultivated at 37 °C, to induce T3SS assembly

conditions activating the T3SS (e.g., *light* for LITESEC-act; *dark*, for example, wrap culture with aluminum foil, for LITESEC-supp) (*see* **Note 10**) for 2 h at 37 °C and 200 rpm (*see* **Note 17**).

18. In the meantime, prewarm infection buffer and plate reader to 37 °C.
19. Aspirate medium from eukaryotic cells in 96-well plate and add 120 μL cell culture working medium per well. To minimize differences in evaporation between wells, add 120 μL PBS to the wells surrounding the measured wells.
20. After 2 h incubation at 37 °C, measure OD_{600} and then harvest 0.5 ODu (corresponding to about 2.5*10^8 bacteria for *Yersinia*) of bacterial culture by centrifugation for 4 min at 2400 rcf, 37 °C.
21. Remove supernatant and carefully resuspend bacterial pellet in 750 μL of preheated infection buffer (washing step to remove antibiotics).
22. Collect bacteria again by centrifugation (4 min at 2400 rcf, 37 °C) and resuspend in 1 mL of preheated infection buffer.

23. Incubate bacterial suspension at 37 °C under illumination conditions that suppress secretion (e.g., *light* for LITESEC-supp, *dark* for LITESEC-act) for 15 min (*see* **Note 18**).
24. Add bacteria to eukaryotic cells in 96-well plate (from **Step 1**) at the desired multiplicity of infection (MOI). The addition of 8 μL of the bacterial resuspension corresponds to an MOI of 100 (referring to the time point of seeding the eukaryotic cells, actual MOI will be lower), which may serve as a good starting point for many assays. Carefully resuspend twice to mix. The adherent cells should not be detached.
25. Incubate under blue light for 60 min at 37 °C in 5% CO_2 (*see* **Note 19**).
26. After 1 h of infection, remove the cell culture medium and add 300 μL of RPMI medium containing gentamycin (*see* **Note 20**).
27. At this stage, the assay can be read out (*see* **Note 21**).
28. Optimization and troubleshooting: In case of missing or insufficient optogenetic control of protein translocation, *see* **Note 14** for optimization and troubleshooting.

4 Notes

In the following notes section, we provide additional information from personal experience in the application of optogenetic interaction switches to control protein secretion in bacteria.

1. As a transmembrane helix (TMH) domain, we have utilized the N-terminus of *Escherichia coli* TatA, extended by Val-Leu for stable integration in the membrane and followed by a FLAG tag for detection and a short flexible linker [14]. The complete TMH-FLAG-linker protein sequence is `MGGISIWQLLIIA-VIVVLLVLFGTKKLGSDYKDDDDKGGAGG`.
2. For optogenetic interaction switches that consist of two interacting domains with different sizes, the choice of which interaction domain is fused to either the target protein or the membrane anchor may influence the success of the experiment. In most previous studies, the smaller interacting domain (e.g., Zdk1, SspB_Nano, PIF) was fused to the target protein (which can be sterically affected by larger fused domains) and acts as a bait, while the larger interacting domain (AsLOV2, iLID, PhyB) was used as the anchor domain [9, 11, 14, 21].
3. Important prerequisites for establishing optogenetic control of intracellular dynamic processes in bacteria are (a) a dynamic and at least temporary presence of the target protein in the cytosol to allow for membrane sequestration, (b) a loss of function of

the target protein when tethered to the anchor protein, and (c) maintenance of functionality and stability of a fusion of the target protein to one of the optogenetic interaction domains.

As optogenetic interaction switches widely differ in binding affinity, kinetics, and other properties, the choice of a suitable base system is important, see [9] for further information. We recommend integrating the optogenetic bait fusion protein onto the genome of the biological vector by modifying the wildtype target gene, rather than expressing the fusion protein from a plasmid vector with the wildtype protein still present, to ensure a tight control of the function.

Short and flexible glycine-rich peptide linkers of 5–20 amino acids can be used to link the optogenetic bait protein to the target protein while retaining the function of both [9]. The same applies to the anchor protein, where linker length can influence the binding affinity and resulting membrane sequestration. In our setup, we have used a GGAGG linker between TMH domain and optogenetic anchor protein (LOV2/iLID) and a GGSGGSGG linker for the fusion of the target protein to the optogenetic bait domain (Zdk1/SspB_Nano).

4. We recommend to check the expression level and stability of the created fusion protein by Western blot prior to functional assays, as flexible linker domains can be subject to proteolysis in some bacteria. If possible, also check that the fusion protein maintains its biological function.
5. This can be tested by coexpressing the used *anchor* protein and a fusion of the *bait* protein with a fluorescence protein (e.g., mCherry) and then testing the relocalization of the fluorescence to/from the membrane upon activation of the optogenetic interaction switch by fluorescence microscopy [14].
6. Co-expression of the corresponding chaperone (e.g., SycE for effector signal YopE) was shown to enhance the efficiency of protein translocation into eukaryotic cells [5].
7. The used *Yersinia enterocolitica* multi-effector knockout strains are auxotrophic for DAP due to a deletion in the aspartate-β-semialdehyde dehydrogenase (*asd*) gene [19] and resistant to nalidixic acid.
8. Secretion of effector proteins in *Yersinia* is natively induced by host-cell contact, which can be mimicked by low Ca^{2+} levels, corresponding to the intracellular environment of the target cells. Secretion can therefore be induced by chelation of Ca^{2+} with 5 mM EGTA; nonsecreting conditions are maintained by addition of 5 mM $CaCl_2$. In case addition of $CaCl_2$ leads to precipitation, filter medium before adding the other components.

9. A faster temperature increase of the culture, which may enhance secretion, can be achieved by incubation in a shaking 37 °C water bath; flasks can then be transferred to the illuminated 37 °C shaker after 5 min.
10. Depending on the optogenetic interaction switch, light pulses can be sufficient to control secretion; however, constant illumination achieves the tightest control in our experience. Ambient light seems to have an effect on the optogenetic control of the system (in our experience, more strongly for the LOVTRAP-based LITESEC-act system than for the iLID-based LITESEC-supp [14]; therefore, try to avoid unspecific illumination as much as possible to have tighter regulation).
11. Work on ice to keep acetone as cold as possible to avoid dissolution of protein pellet. Add acetone by pipetting carefully at the edge of the microcentrifuge tubes and not directly onto the protein pellet to prevent disruption.
12. To speed up the procedure, protein pellets can also be dried in a heat block at up to 50 °C.
13. The required resuspension volume V_{res} can be calculated as follows: $V_{res} = (OD_{600} * V_{sample} * V_{loaded})/ODu_{loaded}$, e.g.: Supernatant from 1.8 mL to get the equivalent of 0.6 ODu in 15 μL sample: $(OD_{600} * 1.8\ mL * 15\ \mu L)/0.6\ mL$. (1 ODu = 1 mL of a culture at an OD_{600} of 1, corresponding to about $5*10^8$ bacteria for *Yersinia*.) When resuspending protein pellets in 1x sample buffer, try to avoid bubbles. If buffer turns yellow during resuspension of protein pellet, this is an indication for remaining TCA in the sample. The pH can be adjusted by addition of 1M NaOH in 1 μL steps, until the sample is blue. Note that solubilization may be impaired in these samples and take into account the volume change in these samples.
14. *Anchor*-to-*bait* protein expression ratios were shown to be crucial for a functional cellular control using an optogenetic interaction switch [14]. Insufficient anchor/bait ratios may prevent the complete sequestration of the bait, leading to secretion even under nonactivating conditions. Contrary, high anchor/bait ratios may lead to sequestration of the bait even under activating conditions, preventing full secretion. Anchor expression levels therefore have to be adapted to the bait protein levels. We recommend expressing the anchor protein from an inducible expression vector. This allows to vary the induction level, and therefore to screen for the optimal anchor/bait ratio.

    In addition, mutant variants of optogenetic interaction switch proteins with different affinities have been developed for many optogenetic base systems [11, 15, 22], which allows to change the dynamic range and kinetics of the system and the eventual outcome of the experiment. Utilizing such a mutant (LOV_{V416L}) [15] in the anchor protein and optimizing its expression was essential for the creation of the light-activated LITESEC-act system [14].

15. The addition of the serum inhibits further activity of the trypsin at this point.
16. We recommend using triplicates of each biological sample, including the control strains, as well as a triplicate of wells to which no bacteria are added and a triplicate of wells to which no assay substrate is added. To compare *light* and *dark* conditions, two such plates should be prepared. As an example, for a comparison of 6 bacterial strains, $(6 + 2) * 3 = 24$ wells should be seeded in two plates. Preferably use the center of the plate to minimize differences in evaporation between wells.
17. Incubation at illumination that allows for T3SS activation ensures that the T3SS injectisome is assembled and needles are formed prior to the actual experiment [23, 24].
18. This step ensures tight control by illumination during infection assays by preventing preactivation prior to the infection assay.
19. Incubation duration can be changed according to the speed and affinity of bacterial attachment, the efficiency of protein injection, and the sensitivity of the readout assay.
20. Replacement of cell culture working medium with fresh RPMI medium containing gentamycin minimizes ongoing injection by inhibiting bacterial protein biosynthesis.
21. Suitable readout assays depend on the nature of the injected protein. Examples are reporter assays such as the split-NanoLuc assay [20] or a FRET-based beta-lactamase assay [25], or phenotypic analysis, e.g., an apoptosis assay upon translocation of a proapoptotic protein [5].
22. The application of the V416L mutant of the LOV2 domain, which has a lower affinity to Zdk1 [15], was essential to reach full activation of the T3SS in our experiments.

Acknowledgments

The authors thank Katja Langenfeld for expert support on bacterial cell culture and infection. Work in the Diepold lab is supported by the Max Planck Society.

References

1. Bai F, Li Z, Umezawa A, Terada N, Jin S (2018) Bacterial type III secretion system as a protein delivery tool for a broad range of biomedical applications. Biotechnol Adv 36(2): 482–493. https://doi.org/10.1016/j.biotechadv.2018.01.016
2. Cornelis GR (2006) The type III secretion injectisome. Nat Rev Microbiol 4(11): 811–825. https://doi.org/10.1038/nrmicro1526
3. Deng W, Marshall NC, Rowland JL, McCoy JM, Worrall LJ, Santos AS, Strynadka NCJ, Finlay BB (2017) Assembly, structure, function and regulation of type III secretion systems. Nat Rev Microbiol 15(6):323–337. https://doi.org/10.1038/nrmicro.2017.20
4. Schlumberger MC, Müller AJ, Ehrbar K, Winnen B, Duss I, Stecher B, Hardt W-D (2005) Real-time imaging of type III secretion: Salmonella SipA injection into host cells. Proc Natl Acad Sci USA 102(35):12548–12553. https://doi.org/10.1073/pnas.0503407102
5. Ittig SJ, Schmutz C, Kasper CA, Amstutz M, Schmidt A, Sauteur L, Vigano MA, Low SH, Affolter M, Cornelis GR, Nigg EA, Arrieumerlou C (2015) A bacterial type III secretion-based protein delivery tool for broad applications in cell biology. J Cell Biol 211(4): 913–931. https://doi.org/10.1083/jcb.201502074
6. Deisseroth K (2011) Optogenetics. Nat Methods 8(1):26–29. https://doi.org/10.1038/nmeth.f.324
7. Toettcher JE, Voigt CA, Weiner OD, Lim WA (2011) The promise of optogenetics in cell biology: interrogating molecular circuits in space and time. Nat Methods 8(1):35–38. https://doi.org/10.1038/nmeth.f.326
8. Baumschlager A, Khammash M (2021) Synthetic biological approaches for optogenetics and tools for transcriptional light-control in bacteria. Adv Biol 5(5). https://doi.org/10.1002/adbi.202000256
9. Lindner F, Diepold A (2022) Optogenetics in bacteria—applications and opportunities. FEMS Microbiol Rev 46(2). https://doi.org/10.1093/femsre/fuab055
10. Endo M, Ozawa T (2017) Strategies for development of optogenetic systems and their applications. J Photochem Photobiol C: Photochem Rev 30:10–23. https://doi.org/10.1016/j.jphotochemrev.2016.10.003
11. Tang R, Yang S, Nagel G, Gao S (2021) mem-iLID, a fast and economic protein purification method. Biosci Rep 41(7). https://doi.org/10.1042/BSR20210800
12. Diepold A, Kudryashev M, Delalez NJ, Berry RM, Armitage JP (2015) Composition, formation, and regulation of the cytosolic C-ring, a dynamic component of the type III secretion injectisome. PLoS Biol 13(1):e1002039. https://doi.org/10.1371/journal.pbio.1002039
13. Diepold A, Sezgin E, Huseyin M, Mortimer T, Eggeling C, Armitage JP (2017) A dynamic and adaptive network of cytosolic interactions governs protein export by the T3SS injectisome. Nat Commun 8(1). https://doi.org/10.1038/ncomms15940
14. Lindner F, Milne-Davies B, Langenfeld K, Stiewe T, Diepold A (2020) LITESEC-T3SS—light-controlled protein delivery into eukaryotic cells with high spatial and temporal resolution. Nat Commun 11(1). https://doi.org/10.1038/s41467-020-16169-w
15. Wang H, Vilela M, Winkler A, Tarnawski M, Schlichting I, Yumerefendi H, Kuhlman B, Liu R, Danuser G, Hahn KM (2016) LOVTRAP: an optogenetic system for photoinduced protein dissociation. Nat Methods 13(9):755–758. https://doi.org/10.1038/nmeth.3926
16. Guntas G, Hallett RA, Zimmerman SP, Williams T, Yumerefendi H, Bear JE, Kuhlman B (2015) Engineering an improved light-induced dimer (iLID) for controlling the localization and activity of signaling proteins. Proc Natl Acad Sci USA 112(1):112–117. https://doi.org/10.1073/pnas.1417910112
17. Guzman LM, Belin D, Carson MJ, Beckwith J (1995) Tight regulation, modulation, and high-level expression by vectors containing the arabinose PBAD promoter. J Bacteriol 177(14):4121–4130. https://doi.org/10.1128/jb.177.14.4121-4130.1995
18. Chakravartty V, Cronan JE (2015) A series of medium and high copy number arabinose-inducible *Escherichia coli* expression vectors compatible with pBR322 and pACYC184. Plasmid 81:21–26. https://doi.org/10.1016/j.plasmid.2015.03.001
19. Kudryashev M, Stenta M, Schmelz S, Amstutz M, Wiesand U, Castaño-Díez D, Degiacomi MT, Münnich S, Bleck CKE, Kowal J, Diepold A, Heinz DW, Dal Peraro M, Cornelis GR, Stahlberg H (2013) In situ structural analysis of the Yersinia enterocolitica injectisome. elife 2. https://doi.org/10.7554/eLife.00792

20. Westerhausen S, Nowak M, Torres-Vargas CE, Bilitewski U, Bohn E, Grin I, Wagner S (2020) A NanoLuc luciferase-based assay enabling the real-time analysis of protein secretion and injection by bacterial type III secretion systems. Mol Microbiol 113(6):1240–1254. https://doi.org/10.1111/mmi.14490
21. Toettcher JE, Gong D, Lim WA, Weiner OD (2011) Light control of plasma membrane recruitment using the Phy-PIF system. Methods Enzymol 497:409–423. https://doi.org/10.1016/B978-0-12-385075-1.00017-2
22. Zimmerman SP, Hallett RA, Bourke AM, Bear JE, Kennedy MJ, Kuhlman B (2016) Tuning the binding affinities and reversion kinetics of a light inducible dimer allows control of transmembrane protein localization. Biochemistry 55(37):5264–5271. https://doi.org/10.1021/acs.biochem.6b00529
23. Diepold A (2020) Assembly and post-assembly turnover and dynamics in the type III secretion system. Curr Top Microbiol Immunol 427:35–66. https://doi.org/10.1007/82_2019_164
24. Milne-Davies B, Helbig C, Wimmi S, Cheng DWC, Paczia N, Diepold A (2019) Life after secretion—Yersinia enterocolitica rapidly toggles effector secretion and can resume cell division in response to changing external conditions. Front Microbiol 10. https://doi.org/10.3389/fmicb.2019.02128
25. Wolters M, Zobiak B, Nauth T, Aepfelbacher M (2015) Analysis of Yersinia enterocolitica effector translocation into host cells using Beta-lactamase effector fusions. J Vis Exp 104. https://doi.org/10.3791/53115

Chapter 10

Synthesis and Application of Photocaged Isopropyl β-D-1-Thiogalactopyranoside for Light-Mediated Control of Bacterial Gene Expression

Fabienne Knapp, Fabian Hogenkamp, Seung-Hyun Paik, Karl-Erich Jaeger, Jörg Pietruszka, and Thomas Drepper

Abstract

Photocaged compounds are chemical conjugates that are designed to release an active molecule upon exposure to light of a specific wavelength. In recent years, photocaged inducer molecules such as caged isopropyl β-D-1-thiogalactopyranoside (cIPTG) have been increasingly used as a powerful tool for light-driven gene expression in bacteria, allowing researchers to precisely and noninvasively tune the expression of specific target genes. In this chapter, we present a guideline for the synthesis of 6-nitropiperonyl photocaged IPTG (NP-cIPTG) as well as its in vivo application as an optochemical on-switch of gene transcription in *Escherichia coli* and other bacteria.

Key words Caged compounds, Gene expression, Optogenetics, Photochemistry, Optochemical control, Synthetic biology, Light regulation

1 Introduction

Reprogramming and optimizing bio(techno)logical processes often requires precise control of gene expression and the underlying regulatory networks. In the last decades, many different methods have been developed to engineer such processes in bacteria. However, the implementation of optogenetic tools for these purposes has only emerged in recent years. In contrast to conventional regulatory systems, light-responsive switches enable noninvasive control of cellular functions with high spatiotemporal resolution [1].

In principle, light-driven gene expression in bacteria can be achieved either by light-sensitive proteins or by light-activatable (bio)molecules [2, 3]. In contrast to genetically encoded light-sensitive photoreceptor proteins, light-activatable molecules

Armin Baumschlager (ed.), *Optogenetics: Methods and Protocols*, Methods in Molecular Biology, vol. 2840, https://doi.org/10.1007/978-1-0716-4047-0_10,

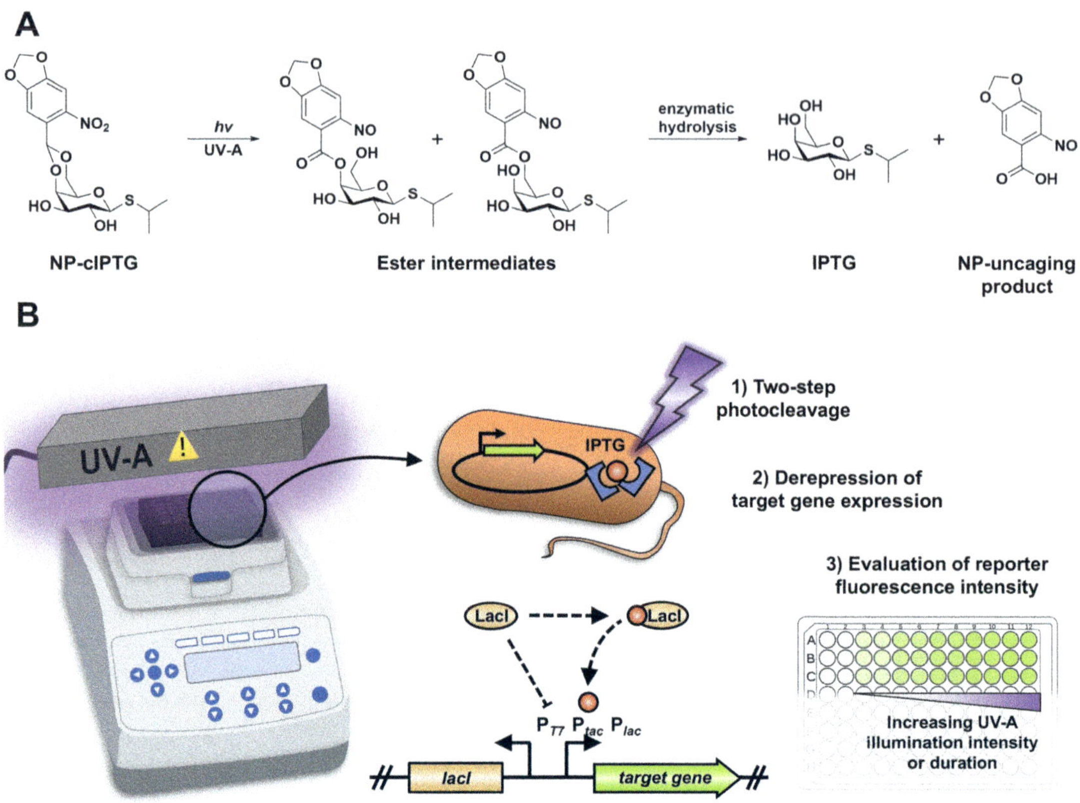

Fig. 1 In vivo application of NP-cIPTG for light-mediated control of gene expression. (**A**) The light-sensitivity of NP-cIPTG is mediated by the photolabile nitropiperonyl (NP) protecting group that inhibits the biological function of the attached inducer molecule IPTG until the group is cleaved off in a two-step process that is initiated by UV-A light irradiation ($\lambda_{max} = 365$ nm) followed by intracellular hydrolysis. (**B**) For the application of NP-cIPTG, the caged inducer has to be added to an expression culture containing a bacterial host that harbors a respective target/reporter gene under control of an IPTG-inducible promoter prior to cultivation. The photocaged inducer needs to be uncaged via UV-A light irradiation to induce target gene expression. The light intensity or illumination time can be varied to optimize the performance of the light-responsive expression system

consist of a bioactive moiety and a photo-removable protecting group that renders the molecule inactive. Light irradiation at a specific wavelength restores the bioactivity of the molecule by photochemically initiated covalent bond cleavage and subsequent release of the photocage (Fig. 1) [4]. This optochemical approach has previously been used to control various processes, including cell signaling [5–7], drug delivery [8], and gene expression [9–11]. In this context, especially 2-nitrobenzyl-(NB)-based photocaging groups and their derivatives such as 6-nitropiperonyl (NP) have been frequently employed for an appropriate and well-characterized UV-A light-mediated release of bioactive molecules [12–14].

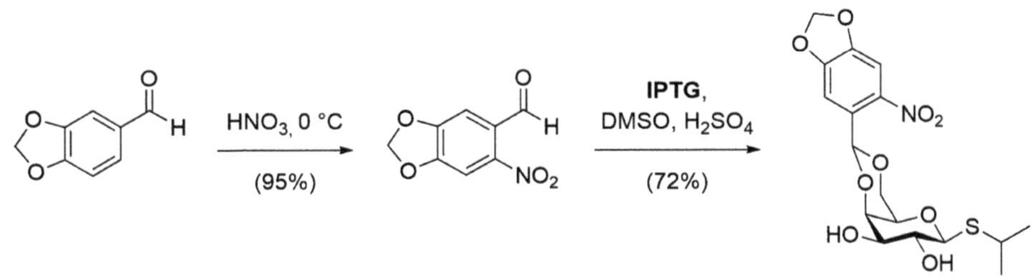

Fig. 2 Two-step synthesis of NP-cIPTG from commercially available piperonal

To broaden the applicability of caged compounds as versatile optochemical switches, a variety of photolabile protecting groups has been developed focusing on redshifted absorption spectra [5, 15, 16], improved solubility [16–18] as well as some general aspects including a pronounced stability toward enzymatic hydrolysis, good biocompatibility, and low overall toxicity of caged compounds (also including the photolysis products) [19]. Furthermore, several photocaged inducer molecules have been established as suitable for different well-characterized bacterial promoter systems [10, 20]. These caged inducers have been successfully used for optochemical regulation of target gene expression in various bacterial hosts including *Escherichia coli*, *Pseudomonas putida*, *Bacillus subtilis*, *Corynebacterium glutamicum*, and *Rhodobacter capsulatus* [16, 21, 22]. In addition, various biotechnological applications could benefit from optochemical regulation, for example, optimization of production processes in a high-throughput screening approach [23] or transcriptional regulation on a single cell level in a microfluidic setup [24].

In the following chapter, we will present an overview of methods for (i) synthesis (Fig. 2) and in vitro analysis of the NP photocaging group and its respective photocaged compound NP-cIPTG (see Subheading 3.1) and (ii) for its in vivo application as light-responsive on-switch for T7-RNA polymerase-dependent gene expression in *E. coli* Tuner(DE3) (see Subheading 3.2 and Fig. 1B). Additionally, we will state several limitations in the synthesis of functional NP-cIPTG and its biological applications in other bacterial hosts.

2 Materials

2.1 Synthesis of Photocaged IPTG

2.1.1 General Remarks

All chemicals for synthesis were obtained from commercial suppliers and used without further purification unless stated otherwise. Solvents were reagent grade and were dried as well as purified by common methods.

2.1.2 Synthesis of NP-cIPTG

1. Reactions: Round-bottom and Schlenk flasks with a magnetic stirring bar.
2. Product separation: Filter funnel to separate solids.
3. Product extraction: Separation funnel and Erlenmeyer flask.
4. Purification: Silica gel (Merck silica gel 60; 0.063–0.200 μm) and distilled solvents for flash chromatography: petroleum ether, ethyl acetate, dichloromethane, *n*-pentane.
5. Rotatory evaporator.

2.1.3 Analysis of Synthesis Products

1. Thin-layer chromatography (TLC): Precoated silica gel plates (e.g., Polygram® SIL G/UV, Macherey-Nagel), UV-light lamp ($\lambda_{max} = 365$ nm).
2. NMR spectrometer (e.g., Advance/DRX 600 spectrometer, Bruker, USA).
3. Deuterated solvents ($CDCl_3$).
4. UV/VIS Spectrophotometer (e.g., Shimadzu UV-1800 (CPS-240A) UV/VIS Spectrophotometer).
5. In vitro uncaging experiments: Quartz cuvette; uncaging light source (e.g., LUMOS 43® from Atlas Photonics at 375 nm); Thermal Power Sensor (S302C, Thorlabs Inc., USA, *see* **Note 1**) for quantification of light intensity; HPLC system (e.g., Jasco, Germany; column: Hyperclone 5 μ ODS (C18) 120 (Phenomenex) combined with an UV/Vis-detector).

2.2 *In Vivo* Application of Photocaged IPTG in *E. coli*

2.2.1 General Remarks

All culture media were sterilized in an autoclave after preparation (121 °C, 20 min, 200 kPa). Heat-labile components such as antibiotics or IPTG were sterile-filtered (sterile syringe filter, 0.2 μM consisting of polyethersulfone (PES); VWR International GmbH, Germany) and added to the sterile and cooled medium.

2.2.2 Cultivation and Illumination of E. coli *Cells*

1. Chemically competent cells of *E. coli* Tuner(DE3) (Novagen, Germany).
2. Expression plasmid with suitable expression system and appropriate reporter gene (*see* **Note 2**) (*lacI*/P_{T7}/P_{tac}/P_{lac} for use of NP-cIPTG in *E. coli*, for example, pRhotHi-2-lacI-eYFP [16]).
3. Flower plates (m2plabs, now Beckman Coulter Life Sciences, Germany).
4. Gas permeable sealing foils (m2plabs, now Beckman Coulter Life Sciences, Germany).
5. LB medium (e.g., LB Broth (Luria/Miller), Roth, Germany).
6. Appropriate antibiotics (e.g., 10 mg mL^{-1} kanamycin stock solution).

7. IPTG stock solution: 50 mM isopropyl-β-D-thiogalactopyranosid (IPTG) dissolved in 100% ethanol or DMSO.
8. NP-cIPTG, presolved in 100% ethanol or DMSO for improved solubility.
9. Incubator for cell growth (e.g., Eppendorf ThermoMixer C, Eppendorf, Germany or BioLector system, m2plabs, now Beckman Coulter Life Sciences, Germany).
10. UV-A lamp (e.g., VL-315.BL hand lamp 45 W, $\lambda_{max} = 365$ nm, Vilber Lourmat, France; distance to Flowerplate®: 1.5 cm, approx. 0.9 mW cm^{-2}); *see* **Note 1**.
11. Device for quantifying the light intensity of the used light source (e.g., Thermal Power Sensor S302C, Thorlabs Inc., USA) (*see* **Note 1**).

2.2.3 Evaluation of Light-Controlled Gene Expression

1. 100 mM Tris-HCl buffer; pH 8.0.
2. 96-well clear flat bottom microtiter plates (e.g., Greiner Bio-One GmbH, Germany).
3. Microplate reader (e.g., Infinite® M1000 pro, Tecan Group LTC., Switzerland) for cell density measurements and reporter fluorescence measurements.

3 Methods

3.1 Synthesis and In Vitro Analysis of Photocaged Compounds

3.1.1 Synthesis of 6-Nitropiperonal [25]

1. Place 70% nitric acid (56.0 mL, 0.87 mol, 13.0 eq.; *see* **Note 3**) into a 250 mL round bottom flask equipped with a magnetic stirring bar and cool it to 0 °C.
2. Add piperonal (10.0 g, 66.6 mmol, 1.00 eq.) in portions keeping the temperature always at 0 °C. Subsequently, stir the reaction mixture for 4 h.
3. After complete conversion (as judged by TLC considering the individual R_f-value of 6-nitropiperonal—see Subheading 3.1.3), pour the reaction mixture onto ice water. Filtrate the resulting precipitate off and wash it with ice water until the pH is neutral.
4. Dissolve the solid in 250 mL of cold dichloromethane, and wash it with saturated sodium bicarbonate solution.
5. Dry the organic phase over magnesium sulfate.
6. Filter off the magnesium sulfate through a filter funnel into a preweighed round bottom flask.
7. Remove the solvent under reduced pressure by using a rotatory evaporator.

8. Determine the yield by weighing the round bottom flask and comparison of the final product amount with the theoretical amount of a full conversion ($M = 195.12$ g mol^{-1}): 12.4 g (63.3 mmol, 95%) as yellow solid.

3.1.2 Synthesis of NP-cIPTG [14]

1. Dissolve IPTG (100 mg, 420 μmol, 1.00 eq.) and 6-nitropiperonal (245 mg, 1.26 mmol, 3.00 eq.) in 1 mL dry dimethyl sulfoxide at 0 °C in a 50 mL Schlenk flask protected from light.
2. Place the flask in an ice bath and slowly add the sulfuric acid (150 μL, 2.97 mmol, 7.00 eq., *see* **Notes 3** and **4**) in a dropwise manner.
3. After the addition is complete, remove the ice bath and continuously stir the solution for 24 h at room temperature.
4. After complete conversion (as judged by TLC considering the individual R_f-value of NP-cIPTG—see Subheading 3.1.3), pour the reaction mixture onto ice water.
5. Extract the product from the reaction mixture with ethyl acetate (3 × 30 mL).
6. Dry the combined organic phase over magnesium sulfate.
7. Filter off the magnesium sulfate through a filter funnel into a preweighed round bottom flask.
8. Remove the solvent under reduced pressure to obtain the crude product.
9. Purify the crude product by column chromatography (mobile phase: EtOAc:*n*-pentane, 70:30) by fractionation of the elution volume and combination of the desired fractions (as judged by TLC considering the individual R_f-value of NP-cIPTG—see Subheading 3.1.3).
10. Remove the solvent of the combined product fractions in a preweighed round bottom flask under reduced pressure to obtain the final product.
11. Determine the yield by weighing the round bottom flask and comparison of the final product amount with the theoretical amount of a full conversion ($M = 415.41$ g mol^{-1}): 123 mg (296 μmol, 72%) as colorless solid.
12. Store the final product at −20 °C in the absence of light.

3.1.3 In Vitro Analysis of Synthesis Products via TLC

1. Apply 1 μL of the reaction solution near to the edge of a precoated silica gel plate using a capillary.
2. Apply 1 μL of the reference compound (reactant) to the same silica gel plate using a capillary.

3. Place the silica gel plate in a saturated thin layer chromatography chamber filled with the corresponding solvent mixture (filling height approx. 1 cm) and wait until the solvent front reaches the upper end of the TLC plate.
4. Take the silica gel plate out of the chamber and mark the solvent front with a pencil.
5. Wait until the silica gel plate has dried.
6. Visualize the spots either via oxidative staining or via UV-light ($\lambda_{max} = 365$ nm), mark them with a pencil, and calculate the R_f-values (*see* **Note 5**).
7. Compare the calculated R_f-values with theoretical values of the final product and reference compound (*see* **Note 5**).

3.1.4 Structural Analysis of Synthesis Products—NMR Spectroscopy

1. Dissolve ~20 mg of 6-nitropiperonal or NP-cIPTG in 0.5 mL deuterated solvents ($CDCl_3$).
2. Record ^1H- and ^{13}C-NMR spectra on the 600 Ultra ShieldTM NMR instrument (Bruker, USA) at 20 °C (*see* **Note 6** for exemplary data). The chemical shifts are given in ppm relative to the solvent (^1H: $CDCl_3$ = 7.26 ppm/^{13}C: $CDCl_3$ = 77.16 ppm). Signals were assigned by means of H-COSY-, HSQC-, and HMBC experiments, and splitting patterns are reported as singlet (s), doublet (d), triplet (t), multiplet (m), and broad singlet (brs).

3.1.5 Measurement of Absorption Spectra

1. Fill a quartz cuvette with 1 mL of methanol and perform a background measurement.
2. Dissolve NP-cIPTG in MeOH (4.2 mg in 10 mL) to receive a 1 mM stock solution (*see* **Note 7**).
3. Pipette 200 μL of the stock solution into a quartz cuvette and dilute with 800 μL of methanol.
4. Record an absorbance spectrum in the wavelength range of 200–800 nm on a Shimadzu UV-1800 (CPS-240A) UV/VIS Spectrophotometer (Shimadzu, Germany) (*see* **Note 8**).
5. Determine the absorption maxima and compare them to literature data (*see* **Note 9**).

3.1.6 Quantification of Uncaging Half-Life Times

1. Dissolve NP-cIPTG in MeOH (4.2 mg in 10 mL) to receive a 1 mM stock solution.
2. Transfer 1 mL of the stock solution into an HPLC vial and analyze the sample by reverse phase HPLC in a Jasco HPLC system [column: Hyperclone 5 μ ODS (C18) 120 (Phenomenex); detector: UV; mobile phase:methanol:water 30:70, 1 mL min^{-1}, 25 °C; $t_R = 22.0$ min].

3. Determine the peak area (in mAU min^{-1}) of the NP-cIPTG at the absorption maximum (336 nm).
4. Irradiate 1 mL of the stock solution in a cuvette at room temperature using the LUMOS 43® (375 nm, 6.2 mW cm^{-2}) for 1 min.
5. Repeat **Step 2** for samples with different irradiation times (2, 3, 4, 5, 10, 20, and 30 min).
6. Transfer the irradiated solutions into HPLC vials and analyze each sample by reverse phase HPLC in a Jasco HPLC system [column: Hyperclone 5 μ ODS (C18) 120 (Phenomenex); detector: UV; mobile phase:methanol:water 30:70, 1 mL min^{-1}, 25 °C; $t_R = 22.0$ min]. Determine the peak area of the NP-cIPTG (*see* **Note 10**) at the absorption maximum (336 nm).
7. Measure the reduction of the NP-cIPTG concentration by a UV detector (Jasco MD-2010-Plus) at the absorption maximum (*see* **Notes 11** and **12**).
8. For the data evaluation plot the concentration decrease against the irradiation time.
9. Plot the data with a first-order exponential fit (*see* **Note 13**).
10. Determine the half-life time $t_{0.5}$ by using the parameters determined by the exponential fit (*see* **Note 14**).

3.2 *In Vivo* Application of Photocaged IPTG in E. coli

To evaluate and apply NP-cIPTG for light-mediated induction of bacterial gene transcription, bacterial cultivations need to be performed. In the following, all cultivation procedures will be exemplarily described for T7-RNA-polymerase-dependent target gene expression using *E. coli* Tuner(DE3) as expression host (see Fig. 1B). This *E. coli* strain is derived from *E. coli* BL21(DE3) and lacks the LacY permease encoding gene and thus, is able to take up IPTG exclusively via diffusion, but not via active import resulting in a gradual and homogeneous induction response [14]. NP-cIPTG is a hydrophobic NB-based photocaged IPTG derivative that is well-established for the light-mediated control of bacterial gene expression [14, 16]. *See* **Note 15** to transfer the NP-cIPTG-based light control of target gene expression to other bacterial hosts (Table 1).

3.2.1 Cultivation and Illumination of E. coli Cells

1. Transform chemically competent *E. coli* cells with approx. 100 ng of a respective expression plasmid via a heat shock at 42 °C for 90 s (e.g., pRhotHi-2-lacI-eYFP [14] for the use of NP-cIPTG). For comments on plasmid selection, please *see* **Note 16**. After a sufficient resting period for development of the antibiotic resistance at 37 °C (e.g., 90 min for kanamycin), cultivate the resulting expression strain overnight under

Table 1
Bacterial hosts and conditions for NP-cIPTG-mediated control of gene expression

Host organism	Expression plasmid and promoter system	Cultivation conditions	NP-cIPTG and DMSO final conc[a]	Induction time	Induction factor[b]	Refs.
E. coli Tuner (DE3)	pRhotHi-2-lacI-eYFP; *lacI*/P_{T7}	30 °C, Flowerplate, LB medium	50 μM 0.25% (v/v)	2–2.5 h	115	[16]
P. putida KT2440	pVLT33-GFPmut3; *lacI*/P_{tac}	30 °C, Flowerplate, LB medium	1 mM 0.5% (v/v)	3 h	3	
B. subtilis DB430	pHT01-sfGFP; *lacI*/P_{grac}	30 °C, Flowerplate, LB medium	1 mM 0.5% (v/v)	5 h	20	
C. glutamicum ATCC 13032	pEKEx2-EYFP *lacI*/P_{tac}	30 °C, Flowerplate, BHI or CGXII medium	100 μM 0.25% (v/v)	6–8 h	60–220	[21]
R. capsulatus SB1003	pRholtHi-eYFP *lacI*/P_{tac}	30 °C, RCV medium, (IR illumination for phototrophic growth at λ_{max} = 850 nm)	1 mM 0.5% (v/v)	6 h	5	[22]

[a]Final concentration of NP-cIPTG and DMSO in the cultivation medium
[b]Compared to basal expression level of uninduced cultures

appropriate selection pressure and temperature (e.g., on LB agar plates supplemented with 50 μg mL^{-1} kanamycin at 37 °C).

2. Store transformation plates at 4 °C until use.
3. Inoculate the precultures with a single bacterial colony from a transformation plate in different wells of a 48-well Flowerplate® containing 800 μL LB medium supplemented with appropriate antibiotics for plasmid maintenance (e.g., 50 μg mL^{-1} kanamycin) and incubate them at 30 °C and 1200 rpm in a ThermoMixer C in the dark for 18–20 h (*see* **Note 15**).
4. Calculate the required photocaged compound amount in consideration of the total culture volume as well as the targeted end concentration and weigh it under low light conditions (*see* **Notes 17** and **20**).
5. Presolve NP-cIPTG under low light conditions in a reasonable amount of ethanol or DMSO to enhance solubility in the cultivation medium and store it wrapped in aluminum foil at 4 °C until use. For *E. coli* Tuner(DE3), a final concentration of 0.25% DMSO (v/v) in the cultivation medium was adjusted (*see* **Notes 18–20**).
6. Inoculate the *E. coli* main cultures to a cell density corresponding to an OD_{580nm} of 0.05 in a Flowerplate® containing 800 μL LB medium supplemented with appropriate

antibiotics (e.g., 50 μg μmL^{-1} kanamycin). As the last step, quickly add the required amount of presolved photocaged compound (i.e., 50 μM NP-cIPTG for *E. coli* Tuner(DE3) and P_{T7} promoter) and grow the cultures at 30 °C and 1200 rpm in a ThermoMixer C in the dark (*see* **Notes 15**, **20**, and **21**).

7. After approx. 2 h (or during early logarithmic growth), induce the (reporter) gene expression by illumination of respective *E. coli* cultures using a UV-A lamp (320–400 nm with emission maximum at 365 nm and ~0.9 mW cm^{-2}; VL- 315.BL 45-W lamp) for 30 min or addition of 50 μM conventional IPTG immediately after illumination (*see* **Note 22**).
8. Cultivate the cells for additional 18 h at 30 °C and 1200 rpm in a ThermoMixer C in the dark.

3.2.2 Evaluation of Light-Controlled Gene Expression

1. Transfer 20 μL of each expression culture (see Subheading 3.2.1) into a microtiter plate and add 80 μL of 1x Tris-HCl buffer pH 8.0 to dilute the cell suspension five-fold (*see* **Note 23**).
2. Analyze the cell suspension in a microplate reader (e.g., Infinite M1000 Pro, Tecan Group LTG., Switzerland) by measuring (i) the cell density (absorption at 580 nm) and the corresponding eYFP fluorescence intensities (λ_{ex} = 488 nm, λ_{em} = 527 nm).
3. For an evaluation or optimization of the optochemical system, you may compare the fluorescence values normalized to the respective cell density of expression cultures treated at varying exposure times or intensities (Fig. 1B, *see* **Note 24**).

4 Notes

1. Use a Light meter (e.g., Thermal Power Sensor S302C, Thorlabs) to determine the light intensity of the applied light source. For this purpose, determine the light intensity at a realistic distance (from the surface of the used cultivation vessel to the light source). All light-absorbing parts of the experimental setup, for example, microtiterplate foils, must be considered for this measurement.
2. Use reporter genes encoding fluorescent proteins to characterize light-mediated induction of gene expression during cell growth under individual cultivation and illumination conditions (e.g., media, temperature, light intensity, and exposure time).
3. Handle with care, nitric acid, and sulphuric acid are strong oxidative acids.

4. DMSO has a melting point of 18 °C, therefore start slow dropwise addition immediately. Otherwise, the solution will solidify as it cools down. If this is the case, set the ice bath aside briefly until the solution is liquid again. Place the ice bath underneath again and continue the dropwise addition.
5. R_f-values are calculated using the following equation:

$$R_f - \text{value} = \frac{\text{migration distance of compound spot [cm]}}{\text{migration distance of solvent [cm]}}$$

Reference values:

6-Nitropiperonal—R_f (petroleum ether:EtOAc, 60:40) = 0.51.

NP-cIPTG—R_f(petroleum ether:EtOAc, 85:15) = 0.40.

6. ^{1}H- and ^{13}C-NMR data of 6-nitropiperonal and NP-cIPTG.

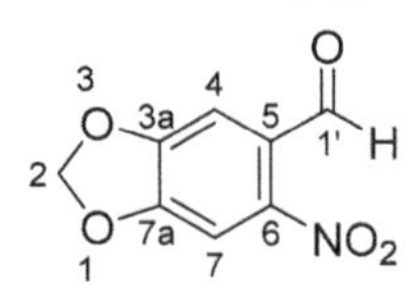

6-Nitropiperonal

6-Nitropiperonal—^{1}H-NMR (600 MHz, $CDCl_3$)—δ_H [ppm] = 10.30 (s, 1 H, 1′-H), 7.53 (s, 1 H, 4-H), 7.34 (s, 1 H, 7-H), 6.24 (s, 2 H, 2-H_2); **^{13}C-NMR** (151 MHz, $CDCl_3$)—δ_C [ppm] = 186.9 (C-1′), 152.3 (C-7a/C-3a), 151.5 (C-3a/C-7a), 146.1 (C-6), 128.2 (C-5), 107.58 (C-4), 105.2 (C-7), 103.9 (C-2).

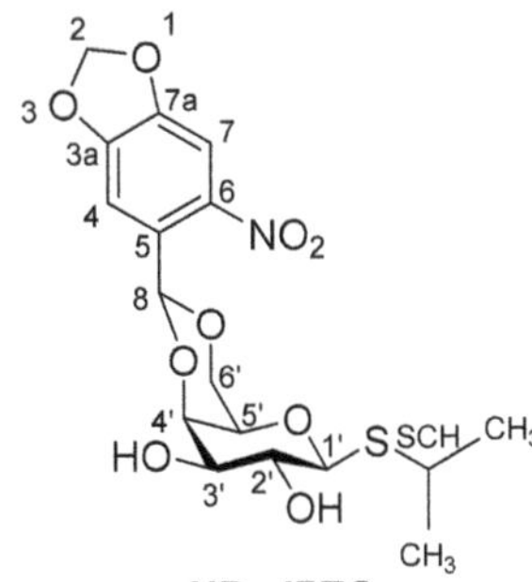

NP-cIPTG

NP-cIPTG—^{1}H-NMR (600 MHz, $CDCl_3$): δ [ppm] = 1.31 (d, $^3J_{CH3,SCH}$ = 6.7 Hz, 3 H, CH_3), 1.33 (d, $^3J_{CH3,SCH}$ = 6.7 Hz, 3 H, CH_3), 2.90 (brs, 2 H, 2-OH, 3-OH), 3.23 (septet, $^3J_{SCH,CH3}$ = 6.7 Hz, 1 H, SCH), 3.49 (s, 1 H, 5′-H), 3.64–3.74 (m, 2 H, 2′-H, 3′-H), 4.04 (dd, $^2J_{6'a,6'b}$ = 12.6 Hz, $^3J_{6'a,5'}$ = 1.9 Hz, 1 H, 6′-H_a), 4.25 (dd, $^2J_{6'b,6'a}$ = 12.6 Hz, $^3J_{6'a,5'}$ = 1.5 Hz, 1 H, 6′-H_b), 4.27 (d, $^3J_{4',3'}$ = 3.1 Hz, 1 H, 4′-H), 4.40 (d, $^3J_{1',2'}$ = 8.6 Hz, 1′-H),

6.07–6.11 (m, 2 H, 2-H), 6.13 (s, 1 H, 8-H), 7.33 (s, 1 H, 4-H), 7.38 (s, 1 H, 7-H); ^{13}**C-NMR** (151 MHz, $CDCl_3$): δ [ppm] = 24.0 (CH_3), 24.2 (CH_3), 35.3 (SCH), 69.7 (C-6′), 70.0 (C-5′), 70.1 (C-2′), 73.8 (C-3′), 76.2 (C-4′), 85.5 (C-1′), 96.7 (C-8), 103.2 (C-2), 105.3 (C-7), 107.7 (C-4), 128.9 (C-5), 142.3 (C-6), 148.2 (C-7a), 151.7 (C-3a).

See Fig. 3 for exemplary ^{1}H- (**A**) and ^{13}C- (**B**) NMR spectra of NP-cIPTG in $CDCl_3$.

7. Use this stock solution if you want to quantify the uncaging half-life time.
8. The maximum absorbance should be between a value of 0.5 and 1.
9. UV/Vis (MeOH)—λ_{max} (ε) = 241 nm (11,078 ε/dm^3 mol^{-1} cm^{-1}), 336 (4027).
10. Due to the exponential decay of NP-cIPTG occurring as a result of irradiation, the integral of the peak area decreases as the irradiation increases.
11. To determine the relative decrease in concentration, it is correlated to the peak area.
12. Perform in triplicates for an increase in measurement accuracy.
13. Perform the analysis of the data with a suitable software tool (e.g., Origin version 2018b) using the first-order exponential fit:

$$c(t) = A \cdot e^{\left(-\frac{t}{\tau}\right)} + y_0$$

First-order exponential decay function (ExpDec1). [A = amplitude; τ = time constant; y_0 = offset]

14. The half-life time $t_{0.5}$ can be derived directly from the time constant τ.

$$t_{0.5} = \tau \cdot \ln(2)$$

Equation for calculating the half-life time. [τ = time constant].

15. The here-described cultivation conditions are applicable for *E. coli* Tuner(DE3) in combination with *lacI*/P_{T7}-based expression systems according to [16]. Adapt the media compositions, cultivation conditions, and inducer concentrations to the individual requirements of the applied host strain, photocaged compound, and promoter system (for an overview of NP-cIPTG applications, see Table 1). Note that each variation of the host strain, expression plasmid, and target gene requires the evaluation of suitable UV-A-light exposure conditions (i.e., light intensity and/or exposure time) for

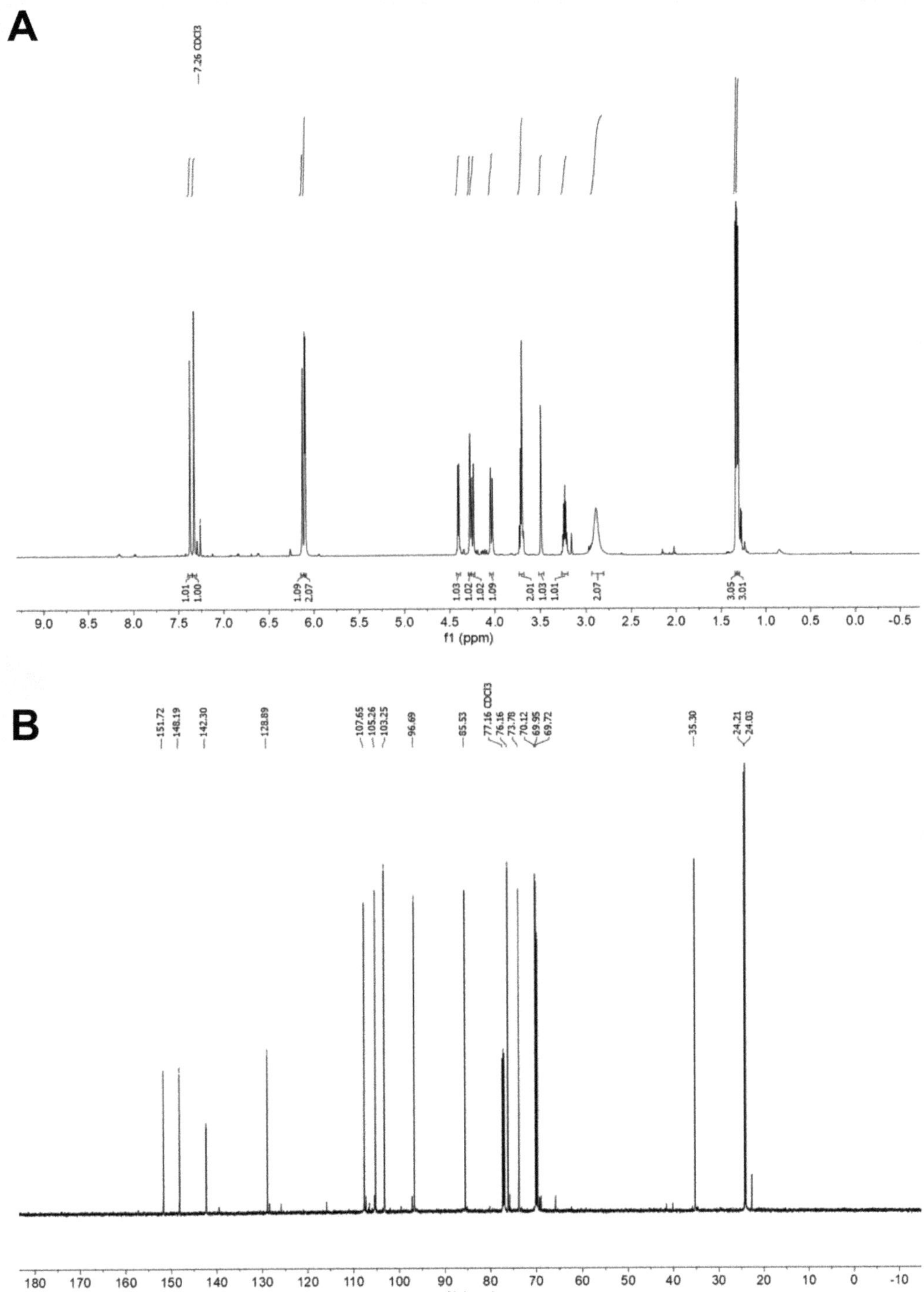

Fig. 3 Exemplary ^{1}H- (**A**) and ^{13}C- (**B**) NMR-spectra of NP-cIPTG in $CDCl_3$ (600 MHz/151 MHz). Detailed information about the spectra is given in *see* **Note 6**

efficient light induction of gene transcription. Furthermore, the toxic effects of UV-A light irradiation as well as caged inducers and respective photoproducts to new host strains must initially be tested. For this purpose, the growth of cells at increasing UV-A exposure times/intensities and NP-cIPTG concentrations (before and after complete photocleavage) should be compared with the growth of cultures without light exposure and cIPTG addition.

16. You may use other P_{T7}-based expression plasmids or plasmids carrying alternative promoters with appropriate LacI binding sites (*lac* operator) such as pET22b (Novagen).

17. Prepare fresh cIPTG stock solutions immediately before use, as it is most stable in their solid form.

18. Because of the low water-solubility of the NP caging group, the solubility of NP-cIPTG is limited as well (highest concentration in water: 0.7 mM [16]). Pre-solution of this caged compound in an organic solvent such as ethanol or DMSO is desirable before adding it to the cultivation medium at higher concentrations. Please note that NP-cIPTG will form emulsion-like structures at elevated concentrations after addition of presolved solution to LB medium, which requires both a thorough blending of the emulsion before distribution into the Flowerplate® wells and continuous mixing during cultivation.

19. Make sure that the concentration of the organic solvent for preparation of the fresh NP-cIPTG stock solution does not exceed a certain level to avoid toxic side effects and growth impairment of the used expression strain. For *E. coli* Tuner (DE3), DMSO was used in an amount that resulted in a final concentration of 0.25% (v/v) in the expression culture.

20. To avoid undesirable uncaging of the photocages during sample preparation and cultivation process, keep the photocaged compound in the dark (especially as soon as it is (pre-) solved in organic solvent or medium), for example, by wrapping the stock solution and cultivation vessel in aluminum foil and reduce the lab light to a minimum during main culture preparation.

21. Use a culture of *E. coli* cells supplemented with NP-cIPTG or equimolar conventional inducer, which is nonirradiated, as well as a culture without photocaged compound or conventional inducer supplementation but with irradiation for appropriate control experiments.

22. Do not apply exposure times of over 30 min for the uncaging process to avoid unwanted side effects of UV-A light on the cell population due to radiation damage and/or overheating.

Ensure a homogeneous light intensity over the entire plate during irradiation.

23. Dilution may vary depending on the host strain and the applied promoter system.
24. In our experimental setup, time intervals of 5 min were suitable to determine the conditions for optimal target gene expression. Create light gradients, for example, via dimming with varying layers of diffusion foils (e.g., White Diffusion LEE216, LEE Filters, USA).

Acknowledgments

The work was supported by grants from (i) the German Federal Ministry of Education and Research (BMBF, FKZ: 031A167), (ii) the Deutsche Forschungsgemeinschaft (DFG, German Research Foundation)—Project ID 458090666/CRC1535/1, and (iii) the Bioeconomy Science Center. The scientific activities of the Bioeconomy Science Center were financially supported by the Ministry of Culture and Science of the German federal state of North Rhine-Westphalia within the framework of the NRW Strategieprojekt BioSC (No. 313/323-400-00213).

References

1. Hartmann D, Smith JM, Mazzotti G et al (2020) Controlling gene expression with light: a multidisciplinary endeavour. Biochem Soc Trans 48:1645–1659. https://doi.org/10.1042/BST20200014
2. Hughes RM (2018) A compendium of chemical and genetic approaches to light-regulated gene transcription. Crit Rev Biochem Mol Biol 53:453–474. https://doi.org/10.1080/10409238.2018.1487382
3. Baumschlager A, Khammash M (2021) Synthetic biological approaches for optogenetics and tools for transcriptional light-control in bacteria. Adv Biol 5:2000256. https://doi.org/10.1002/adbi.202000256
4. Adams SR, Tsien RY (1993) Controlling cell chemistry with caged compounds. Annu Rev Physiol 55:755–784. https://doi.org/10.1146/annurev.ph.55.030193.003543
5. Bardhan A, Deiters A (2019) Development of photolabile protecting groups and their application to the optochemical control of cell signaling. Curr Opin Struct Biol 57:164–175. https://doi.org/10.1016/j.sbi.2019.03.028
6. Liu J, Hemphill J, Samanta S et al (2017) Genetic code expansion in zebrafish embryos and its application to optical control of cell signaling. J Am Chem Soc 139:9100–9103. https://doi.org/10.1021/jacs.7b02145
7. Kolarski D, Sugiyama A, Breton G et al (2019) Controlling the circadian clock with high temporal resolution through photodosing. J Am Chem Soc 141:15784–15791. https://doi.org/10.1021/jacs.9b05445
8. Silva JM, Silva E, Reis RL (2019) Light-triggered release of photocaged therapeutics—where are we now? J Control Release 298:154–176. https://doi.org/10.1016/j.jconrel.2019.02.006
9. Young DD, Deiters A (2007) Photochemical activation of protein expression in bacterial cells. Angew Chem Int Ed Engl 46:4290–4292. https://doi.org/10.1002/anie.200700057
10. Binder D, Bier C, Grünberger A et al (2016) Photocaged arabinose: a novel optogenetic switch for rapid and gradual control of microbial gene expression. Chembiochem 17:296–299. https://doi.org/10.1002/cbic.201500609
11. Bier C, Binder D, Drobietz D et al (2016) Photocaged carbohydrates: versatile tools for

controlling gene expression by light. Synthesis 49:42–52. https://doi.org/10.1055/s-0035-1562617

12. Ankenbruck N, Courtney T, Naro Y et al (2018) Optochemical control of biological processes in cells and animals. Angew Chem Int Ed Engl 57:2768–2798. https://doi.org/10.1002/anie.201700171
13. Corrie JET, Furuta T, Givens R et al (2005) Photoremovable protecting groups used for the caging of biomolecules. In: Goeldner M, Givens R (eds) Dynamic studies in biology. Wiley, Weinheim, pp 1–94. https://doi.org/10.1002/3527605592.ch1
14. Binder D, Grünberger A, Loeschcke A et al (2014) Light-responsive control of bacterial gene expression: precise triggering of the *lac* promoter activity using photocaged IPTG. Integr Biol 6:755–765. https://doi.org/10.1039/C4IB00027G
15. Vorobev AY, Moskalensky AE (2020) Long-wavelength photoremovable protecting groups: on the way to in vivo application. Comput Struct Biotechnol J 18:27–34. https://doi.org/10.1016/j.csbj.2019.11.007
16. Hogenkamp F, Hilgers F, Knapp A et al (2021) Effect of photocaged isopropyl β-D-1-thiogalactopyranoside solubility on the light responsiveness of LacI-controlled expression systems in different bacteria. Chembiochem 22:539–547. https://doi.org/10.1002/cbic.202000377
17. Russell AG, Sadler MJ, Laidlaw HJ et al (2012) Photorelease of tyrosine from α-carboxy-6-nitroveratryl (αCNV) derivatives. Photochem Photobiol Sci 11:556–563. https://doi.org/10.1039/c2pp05320a
18. Noguchi M, Skwarczynski M, Prakash H et al (2008) Development of novel water-soluble photocleavable protective group and its application for design of photoresponsive paclitaxel prodrugs. Bioorg Med Chem 16:5389–5397. https://doi.org/10.1016/j.bmc.2008.04.022
19. Pelliccioli AP, Wirz J (2002) Photoremovable protecting groups: reaction mechanisms and applications. Photochem Photobiol Sci 1: 441–458. https://doi.org/10.1039/b200777k
20. Hogenkamp F, Hilgers F, Bitzenhofer NL et al (2022) Optochemical control of bacterial gene expression: novel photocaged compounds for different promoter systems. Chembiochem 23. https://doi.org/10.1002/cbic.202100467
21. Binder D, Frohwitter J, Mahr R et al (2016) Light-controlled cell factories: employing photocaged isopropyl- β-D-1-thiogalactopyranoside for light-mediated optimization of lac-based gene expression and (+)-valencene biosynthesis in *Corynebacterium glutamicum*. Appl Environ Microbiol 82:6141–6149. https://doi.org/10.1128/AEM.01457-16
22. Hilgers F, Hogenkamp F, Klaus O et al (2022) Light-mediated control of gene expression in the anoxygenic phototrophic bacterium *Rhodobacter capsulatus* using photocaged inducers. Front Bioeng Biotechnol 10. https://doi.org/10.3389/fbioe.2022.902059
23. Wandrey G, Bier C, Binder D et al (2016) Light-induced gene expression with photocaged IPTG for induction profiling in a high-throughput screening system. Microb Cell Factories 15:63. https://doi.org/10.1186/s12934-016-0461-3
24. Burmeister A, Akhtar Q, Hollmann L et al (2021) (Optochemical) control of synthetic microbial co-culture interactions on a microcolony level. ACS Synth Biol 10:1308–1319. https://doi.org/10.1021/acssynbio.0c00382
25. Duarte CD, Tributino JLM, Lacerda DI et al (2007) Synthesis, pharmacological evaluation and electrochemical studies of novel 6-nitro-3,4-methylenedioxyphenyl-*N*-acylhydrazone derivatives: discovery of LASSBio-881, a new ligand of cannabinoid receptors. Bioorg Med Chem 15:2421–2433. https://doi.org/10.1016/j.bmc.2007.01.013

Chapter 11

Optogenetic Clustering and Stimulation of the T Cell Receptor in Nongenetically Modified Human T Cells

Anja Armbruster, Maximilian Hörner, and Wilfried Weber

Abstract

Methods for the precise temporal control of cell surface receptor activation are indispensable for the investigation of signaling processes in mammalian cells. Optogenetics enables such precise control, but its application in primary cells is limited by the imperative for genetic manipulation of target cells. We here describe a method that overcomes this obstacle and enables the precise activation of the T cell receptor in nongenetically engineered human T cells by light. Our optogenetic receptor activation system OptoREACT employs a TCR-specific scFv fused to PIF6 that interacts with tetramerized PhyB in a light-dependent manner and thereby clusters and activates the T cell receptor in response to red light. OptoREACT not only omits genetic manipulation of the target cell but, because of its modular nature, is likely applicable to a broad range of oligomerization-activated cell surface receptors.

Key words Clustering, Extracellular optogenetics, Primary cells, Receptor activation, T cells

1 Introduction

The perception and integration of dynamic biological signals are fundamental processes in cells and organisms. As vital signal perceivers, cell surface receptors that rely on dimerization or oligomerization, such as receptor tyrosine kinases or immune receptors such as the T cell receptor (TCR), have been studied extensively in order to gain insight into signaling processes [1, 2]. The investigation of signaling has spawned a large repertoire of synthetic biological tools for the meticulous control of receptor activation, of which optogenetics has distinguished itself in particular [3–6]. While the optogenetic tools developed are most appropriate for various signaling studies, their application is limited by the need for genetic manipulation of target cells. This impedes their application in primary cells, which are notoriously difficult to engineer. We therefore strived to

Armin Baumschlager (ed.), *Optogenetics: Methods and Protocols*, Methods in Molecular Biology, vol. 2840,
https://doi.org/10.1007/978-1-0716-4047-0_11,

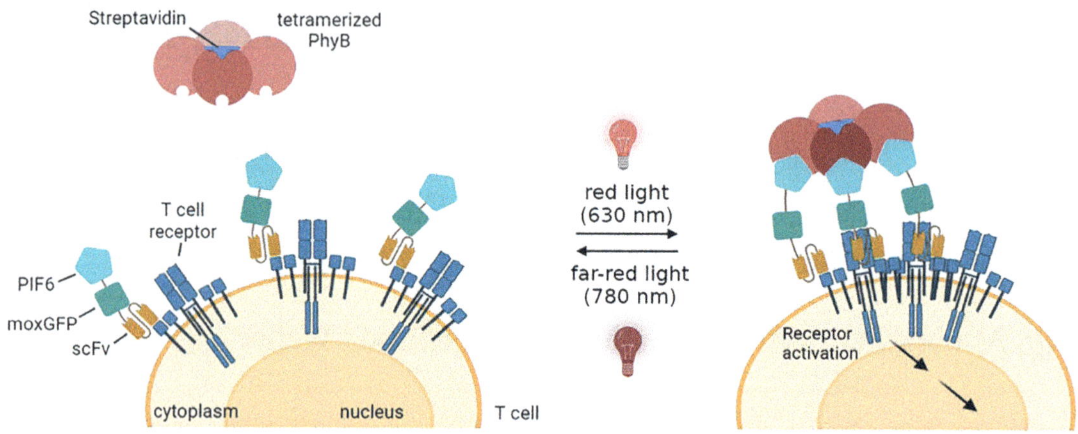

Fig. 1 *Light-dependent stimulation of the T cell receptor by clustering in nongenetically modified T cells.* A fusion protein of an scFv of the anti-CD3ε antibody UCHT1 and phytochrome interacting factor 6 (PIF6) binds the T cell receptor of nongenetically engineered T cells. Stimulation of the TCR by clustering is induced upon binding of the TCR-bound scFv-PIF6 fusion protein by tetramerized phytochrome B in response to red light (~630 nm) illumination. Adapted with permission from Armbruster et al., 2024 [7]. Copyright 2024 American Chemical Society

overcome this limitation by employing extracellular optogenetic techniques in order to create a system for the reversible stimulation of cell surface receptors in nongenetically engineered cells by light. The molecular switch of this technique is the *Arabidopsis thaliana* protein phytochrome B (PhyB), which interacts with its phytochrome interacting factor 6 (PIF6) under red light illumination. Importantly, this interaction can be reversed within seconds by illumination with far-red light. Our system, termed OptoREACT (optogenetic REceptor ACTivation), is composed of (i) monomeric PIF6 fused to a single-chain variable fragment (scFv) of the anti-CD3ε antibody UCHT1 (termed $scFv_{PIF6}$) that binds the T cell receptor and (ii) tetramerized PhyB (termed PhyBt) (see Fig. 1) [7]. When applied to nongenetically engineered T cells, illumination with red light (~630 nm) induces heterodimerization between PhyB and PIF6, thereby leading to clustering and activation of receptors, while far-red (~780 nm) illumination inactivates the system. In this manuscript, we describe the production and purification of the scFv-PIF6 fusion protein, its validation in cell culture, and the final assembly and application of the OptoREACT system for the light-dependent stimulation of the T cell receptor of nongenetically manipulated T cells.

2 Materials

2.1 Production, Purification, and Characterization of $scFv_{PIF6}$

2.1.1 Transformation and Expression of $scFv_{PIF6}$ Expression Plasmid in E. coli

1. *Escherichia coli* expression strain T7 Shuffle Express (New England Biolabs, cat. No. C3029J).
2. Expression plasmid pMH71 for $scFv_{PIF6}$.
3. LB medium: 10 gL^{-1} Tryptone, 5 gL^{-1} yeast extract, 10 gL^{-1} NaCl, pH 7.0, autoclave.
4. Ampicillin, 1 mg mL^{-1} in ddH_2O, sterile filter, aliquot and store at −20 °C.
5. Isopropyl-β-D-thiogalactopyranoside (IPTG) stock solution, dissolved to 1 M in ddH_2O. Aliquot and store at −20 °C.

2.1.2 Immobilized Metal-Ion Affinity Chromatography (IMAC) Purification of $scFv_{PIF6}$

1. Lysis buffer: 50 mM NaH_2PO_4, 300 mM NaCl, 10 mM imidazole, pH 8.0.
2. Wash buffer: 50 mM $NaH_2PO_4 \times 2H_2O$, 300 mM NaCl, 20 mM imidazole, pH 8.0.
3. Elution buffer: 50 mM $NaH_2PO_4 \times 2H_2O$, 300 mM NaCl, 250 mM imidazole, pH 8.0.
4. PBS: 2.68 mM KCl, 1.47 mM KH_2PO_4, 8.03 mM NaH_2PO_4, 137 mM NaCl, pH 7.2.
5. Ni-NTA Superflow agarose resin (Qiagen, cat. No. 30410).
6. 5x SDS-loading buffer: 50% (v/v) glycerol, 0.3125 M Tris-HCl pH 6.8, 0.05% (w/v) bromophenol blue, 10% (w/v) SDS, 12.5% (v/v) 2-mercaptoethanol, in ddH_2O. Prepare under fume hood. Slowly solve SDS pellets in Tris-HCl buffer until clear, then add glycerol, 2-mercaptoethanol, and bromophenol blue. Aliquot and store at −20 °C.
7. Pierce Disposable Columns 10 mL, for gravity-flow procedures (Thermo Fisher Scientific, cat. No. 29924).
8. Pierce Dextran Desalting Column (5K MWCO) 10 mL (Thermo Fisher Scientific, cat. No. 43233).
9. Sonicator, for example, Bandelin Sonopuls HD3100.
10. Equipment and reagents for SDS-PAGE and Coomassie staining.

2.1.3 Purification of $scFv_{PIF6}$ by Size Exclusion Chromatography (SEC)

1. Fast protein liquid chromatography system (e.g., Äkta Explorer, Cytiva) with size exclusion chromatography column (e.g., Superdex 200 Increase 10/300 GL, Cytiva) and fraction collector.
2. Spin concentrator with MWCO of 30 kDa or lower (e.g., Corning, cat. No. 431489).
3. 1 mL syringe with low dead volume and semi-blunt cannula.

4. 96-well plate for fraction collection.
5. PBS: 2.68 mM KCl, 1.47 mM KH_2PO_4, 8.03 mM NaH_2PO_4, 137 mM NaCl, pH 7.2. Filter and de-gas.
6. Liquid nitrogen.

2.2 Validation of the Binding of $scFv_{PIF6}$ to the TCR

1. Jurkat E6.1 cells (ATCC, cat. No. TIB-152).
2. PBS: 2.68 mM KCL, 1.47 mM KH_2PO_4, 8.03 mM NaH_2PO_4, 137 mM NaCl, pH 7.2. Autoclave and keep sterile.
3. Fetal calf serum (FCS), sterile. Aliquot and store at −20 °C.
4. Staining buffer: PBS supplemented with 1% (v/v) FCS.
5. Uncoated 96-well V-bottom plate.
6. Centrifuge with swing-out rotor for multiwell plates.
7. Flow cytometer with a 488 nm laser for GFP detection (e.g., Attune NxT, Thermo Fisher Scientific).

2.3 Light-Dependent Activation of T Cells by TCR Clustering

1. Purified $scFv_{PIF6}$ protein from Subheading 2.1.
2. Purified PhyBt conjugates (see [8, 9] and Subheading 3.3, **Step 1**).
3. Jurkat E6.1 cells.
4. RPMI-1640 medium without biotin (US Biological, cat. No. R9002-01).
5. Penicillin and streptomycin solution (10,000 U mL^{-1} penicillin and 10 mg mL^{-1} streptomycin).
6. Fetal calf serum (FCS), sterile. Aliquot and store at −20 °C (*see* **Note 1**).
7. PBS: 2.68 mM KCl, 1.47 mM KH_2PO_4, 8.03 mM NaH_2PO_4, 137 mM NaCl, pH 7.2. Autoclave and keep sterile.
8. Staining buffer: PBS supplemented with 1% (v/v) FCS.
9. Tris-HCl(2-carboxyethyl)phosphine (TCEP) hydrochloride.
10. Fresh 0.5 M $NaHCO_3$ solution, pH 9.0.
11. 100 mM TCEP stock solution. Weight 2.87 g of Tris-HCl (2-carboxyethyl)phosphine (TCEP) hydrochloride and fill up to 100 mL with 0.5 M $NaHCO_3$ (pH 9.0) solution. Adjust the pH to 8.0 with NaOH or HCL and filter through a 0.22 μm PES filter. Aliquot and store at −20 °C, do not refreeze thawed aliquots.
12. Anti-CD3ε (UCHT1) antibody (BioLegend, cat. No. 300438).
13. PE-conjugated anti-CD69 antibody for flow cytometry (BD Pharmingen, cat. No. 555531) (other conjugates possible, but should not interfere with PhyB and GFP fluorescence).

14. Uncoated 96-well V-bottom plate.
15. OptoPlate-96 illumination device [10] and optoConfig-96 software [11].
16. Cell culture incubator at 37 °C, 5% CO_2 and a humidified atmosphere.
17. Centrifuge with swing-out rotor for multiwell plates.
18. Flow cytometer with a 488 nm laser for GFP, 637 nm laser for PhyB, and 561 nm laser for PE (or other laser and conjugate) detection (e.g., Attune NxT, Thermo Fisher Scientific).
19. Cell culture lab that can be darkened completely and illuminated with green light.

3 Methods

3.1 Production, Purification, and Characterization of $scFv_{PIF6}$

This subheading describes the production and purification of the $scFv_{PIF6}$ fusion protein from the bacterial expression plasmid pMH71. The vector constitutes the coding sequence for an scFv of the anti-CD3ε antibody UCHT1 (in VH-VL orientation), moxGFP [12], PIF6(1-100, C9S, C10S) [5], and a C-terminal His6-Tag under the control of a T7 promoter in the pRSET backbone with ampicillin resistance.

3.1.1 Transformation and Expression of $scFv_{PIF6}$ Expression Plasmid in *E. coli*

1. Thaw a 50 μL aliquot of chemically competent *E. coli* T7 Shuffle Express for 10 min on ice.
2. Add 1 μL of pMH71 plasmid DNA (approx. 1–100 ng) and mix carefully, incubate for 20 min on ice.
3. Heat shock bacteria at 42 °C for 45 s, transfer to ice immediately, and incubate for 2–5 min.
4. Add 400 μL LB medium and incubate for 1 h at 37 °C while shaking at 700 rpm.
5. Inoculate an overnight culture of 100 mL LB medium supplemented with 100 μg mL^{-1} ampicillin with the transformed cells and incubate at 37 °C and 150 rpm overnight.
6. Inoculate 1 L of LB medium supplemented with 100 μg mL^{-1} ampicillin with the overnight culture to a final OD_{600} of 0.1.
7. Grow the culture at 30 °C and 150 rpm to an OD_{600} of 0.7.
8. Induce expression by adding 1 mM isopropyl-β-D-thiogalactopyranoside (IPTG) and incubate at 18 °C overnight (*see* **Note 2**).
9. Harvest cells by centrifugation at 6,000 *g* and 4 °C for 10 min and resuspend the cell pellet in 25 mL lysis buffer (*see* **Note 3**).

3.1.2 Immobilized Metal-Ion Affinity Chromatography (IMAC) Purification of $scFv_{PIF6}$

1. Lyse the cells by sonication for 10 min with 60% amplitude and 0.5 s pulse and 1 s pause intervals (*see* **Note 4**).
2. Clarify the lysate by centrifugation at 30,000 *g* and 4 °C for 30–60 min.
3. Prepare the gravity-flow purification column. Fill the column with 20% ethanol and place the lower frit at the bottom of the column. Pipet 2 mL of Ni-NTA agarose slurry onto the frit and let settle for 30 min (*see* **Note 5**). When the resin is settled, place the second frit on top of the resin and leave a gap of around 1 mm. Let the ethanol drain out of the column (*see* **Note 6**).
4. Equilibrate the column with 10 mL (10 column volumes, CV) lysis buffer.
5. Take a small sample (40 μL) of the clarified lysate for analysis by SDS-PAGE and keep on ice. Then add the clarified lysate onto the column and collect the flow-through.
6. Take a sample of the flow-through for SDS-PAGE and keep on ice.
7. Wash the column twice with 20 mL wash buffer, collect the wash fractions, and take a sample of each fraction for SDS-PAGE.
8. For elution, place a 15 mL falcon tube underneath the column, then add 5 mL of elution buffer. Take a sample of the eluate for SDS-PAGE. The eluate can be stored at 4 °C overnight.
9. Denature samples for SDS-PAGE by adding 5x SDS-loading buffer (add 10 μL of 5x loading buffer to 40 μL of sample) and boil for 10 min at 95 °C.
10. Perform SDS-PAGE and Coomassie staining of the samples collected during protein purification and analyze protein purity of the elution fraction.

3.1.3 Purification of $scFv_{PIF6}$ by Size Exclusion Chromatography (SEC)

Due to the ability of the scFv to form dimers with other scFvs, it is necessary to separate dimers and aggregates from the desired monomeric protein by SEC. Dimeric $scFv_{PIF6}$ proteins will activate the TCR in a light-independent manner. This protocol is for the Cytiva Äkta Explorer system equipped with a Superdex 200 Increase 10/300 GL column but can be adapted to other fast protein liquid chromatography systems with minor changes.

1. Prepare 1 L of PBS. Sterile filter using a bottle top filter and de-gas.
2. Concentrate the eluate by centrifugation at 8,000 *g* (fixed-angle rotor) using the spin concentrator to a volume appropriate for the sample loop and chromatography system used, for example, 550 μL for a 500 μL sample loop (*see* **Note 7**).

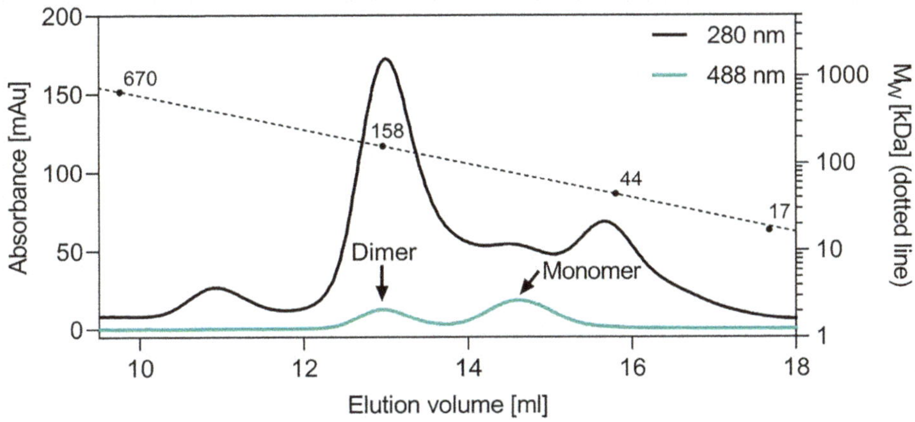

Fig. 2 *Purification of monomeric scFv$_{PIF6}$ by size exclusion chromatography.* scFv$_{PIF6}$ was purified by size exclusion chromatography to separate monomers from dimers which are formed by interlinking heavy and light chains of two scFv fragments. Total protein and GFP were monitored at 280 nm and 488 nm, respectively. Peaks of monomeric and dimeric scFv$_{PIF6}$ were identified by comparison with an SEC standard and are marked. Adapted with permission from Armbruster et al., 2024 [7]. Copyright 2024 American Chemical Society

3. Wash the tubing of the chromatography system with 50 mL PBS and equilibrate the column with 35 mL PBS at a flow rate of 0.5 mL min^{-1} (*see* **Note 8**).
4. Load the concentrated protein into a low dead-volume syringe using a semi-blunt-ended cannula (*see* **Note 9**).
5. Start a size exclusion run with 25 mL run volume and a flow rate of 0.5 mL min^{-1}. During the run, monitor protein absorbance at 280 nm and GFP absorbance at 488 nm. Collect fractions of 200 μL starting from 5 mL elution volume.
6. Determine the fractions that constitute the monomer peak of the scFv$_{PIF6}$ according to the 280 nm and 488 nm curves and pool the corresponding fractions (see Fig. 2).
7. Determine the concentration of the purified protein by Bradford assay or by comparison to a GFP standard.
8. Aliquot protein, shock-freeze in liquid nitrogen, and store at −80 °C (*see* **Note 10**).

3.2 Validation of the Binding of scFv$_{PIF6}$ to the TCR

This subheading describes the validation of the binding of scFv$_{PIF6}$ to the TCR of Jurkat T cells. This protocol is optimized for the use of Jurkat E6.1 cells but can be adjusted to other cells with minor changes.

1. Seed Jurkat E6.1 cells into an uncoated V-bottom 96-well plate at a density of 1 × 10^5 cells/well in 60 μL staining buffer.
2. Add 20 μL of scFv$_{PIF6}$ in staining buffer to a final concentration of 100 nM per well and mix carefully (*see* **Note 11**).

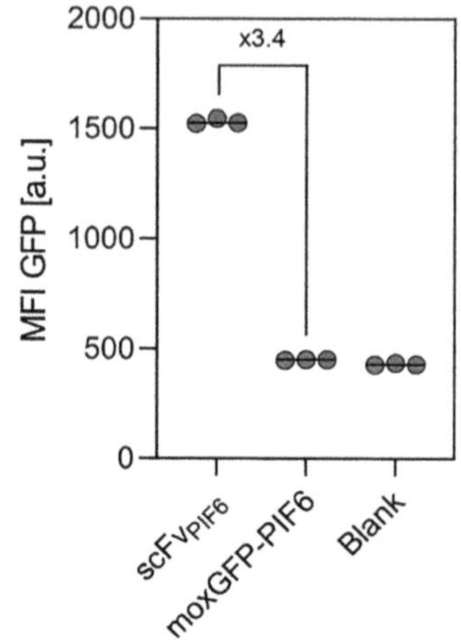

Fig. 3 *Validation of the binding of scFv$_{PIF6}$ to the TCR.* Binding of scFv$_{PIF6}$ to the T cell receptor. Jurkat T cells were incubated with 100 nM of scFv$_{PIF6}$ for 60 min. Afterward, cells were washed and binding of scFv$_{PIF6}$ to the TCR was analyzed by measuring cellular GFP fluorescence by flow cytometry. As control, cells were incubated with a moxGFP-PIF6 fusion protein. Increase in fluorescence of samples with scFv$_{PIF6}$ is compared to the control with only moxGFP-PIF6. Adapted with permission from Armbruster et al., 2024 [7]. Copyright 2024 American Chemical Society

3. Incubate the cells for 60 min on ice.
4. Pellet the cells by centrifugation at 300 *g* for 5 min using a swing-out rotor and discard the supernatant by quickly flipping the plate over a waste container (*see* **Note 12**). Resuspend the cells in 200 μL staining buffer and repeat the washing step.
5. Resuspend the cells in 200 μL staining buffer and transfer them to an appropriate plate or tubes for subsequent analysis by flow cytometry to detect GFP fluorescence. Compare GFP fluorescence of cells incubated with or without scFv$_{PIF6}$ (*see* **Note 13**) (see Fig. 3).

3.3 Light-Dependent Activation of T Cells by TCR Clustering

This subheading describes the generation of tetramerized phytochrome B as well as the light-dependent stimulation of Jurkat E6.1 cells and the subsequent analysis of the early T cell activation marker CD69 by flow cytometry. This protocol can be adjusted to other cells with minor changes.

1. Generate tetramerized phytochrome B (PhyBt). Two comprehensive protocols describing the production and purification of phytochrome B [8] and the generation of PhyB tetramers [9] were published elsewhere. The reader is therefore kindly referred to those.

2. Split Jurkat E6.1 cells the evening before the experiment in a way that the culture reaches 3–6 × 10^5 cells mL^{-1} at the time of the experiment (*see* **Note 14**).
3. Seed Jurkat E6.1 cells into an uncoated black 96-well plate with a clear flat bottom at a density of 2 × 10^5 cells/well in 60 μL RPMI-1640 medium without biotin, supplemented with 5% (v/v) FCS, 1% (v/v) penicillin/streptomycin and 10 mM HEPES.
4. Rest cells for 2 h in a cell culture incubator at 37 °C and 5% CO_2 in a humidified atmosphere.
5. Program an optoPlate-96 with optoConfig-96 for pulsed (2 min ON and 28 min OFF) 630-nm illumination at 10 μmol m^{-2} s^{-1} and constant 780-nm illumination at 450 μmol m^{-2} s^{-1} for the first 3 min and 20 μmol m^{-2} s^{-1} afterward (*see* **Note 15**).
6. Thaw one aliquot of PhyBt and add TCEP to a final concentration of 0.5 mM from a freshly thawed 100 mM TCEP aliquot. Incubate at RT for 10 min in the dark (*see* **Note 16**).
7. Inactivate PhyB by illuminating the sample prepared in **Step 6** with 780-nm light for 2–5 min on an optoPlate-96 (*see* **Note 17**).
8. From now on, work only under green light to prevent activation of PhyB (*see* **Note 18**).
9. Prepare a master mix for the stimulation condition and for each control containing the different proteins in PBS supplemented with 10% (v/v) FCS (*see* **Note 19**). Each master mix should be sufficient for at least 3 replicates at 630-nm and 780-nm illumination. From the mastermix, 20 μL is added to each well containing 60 μL of cells (prepared in **Step 3**). Therefore, the proteins in the mastermix must be 4x concentrated compared to the final concentration (*see* **Note 20**).
   (a) Stimulation mastermix: 320 nM $scFV_{PIF6}$ and 80 nM PhyBt (final concentrations: 80 nM and 20 nM, respectively).
   (b) $scFv_{PIF6}$ control mastermix: 320 nM $scFv_{PIF6}$ (final concentration: 80 nM).
   (c) PhyBt control mastermix: 80 nM of PhyBt (final concentration: 20 nM).
   (d) UCHT1 control mastermix: 40 nM of UCHT1 antibody (final concentration 10 nM).
   (e) Blank: only PBS supplemented with 10% (v/v) FCS.
10. Add 20 μL of the mastermixes to the respective wells and mix by carefully pipetting up and down.

11. Place the 96-well plate on the optoPlate-96, switch on the illumination, and illuminate in a cell culture incubator at 37 °C and 5% CO_2 for 6 h.
12. Resuspend the cells and transfer into a V-bottom 96-well plate. Pellet the cells by centrifugation at 300 *g* for 5 min using a swing-out rotor and discard the supernatant by quickly flipping the plate over a waste container.
13. Resuspend the cells in 30 μL of CD69 staining solution (CD69-PE antibody diluted 1:100 in staining buffer), and incubate 30 min on ice.
14. Add 150 μL of staining buffer and pellet cells by centrifugation (300 g for 5 min).
15. Discard the supernatant and wash cells twice with 200 μL of staining buffer.
16. Resuspend the cells in 200 μL of staining buffer and transfer to an appropriate plate or tubes for subsequent analysis by flow cytometry to detect PE, PhyB, and GFP fluorescence (*see* **Note 21**) (see Fig. 4).

4 Notes

1. Free biotin in FCS can interfere with the binding of biotinylated PhyB to streptavidin. This can be prevented by incubating FCS with streptavidin beads to bind free biotin and subsequent removal of the streptavidin beads by centrifugation and sterile filtration of the FCS.
2. Protein expression was found to yield higher protein amounts at 18 °C. In case a refrigerated incubator is not available, expression at 25 °C also yields reasonable results.
3. At this step, the resuspended pellet can be shock-frozen in liquid nitrogen and stored at −80 °C for multiple years. For purification, thaw the pellet in a water bath at 37 °C before sonication.
4. During sonication, the cell suspension should be cooled to avoid protein degradation from generated heat. Placing the Falcon tube in a beaker filled with ice water provides optimal cooling.
5. The amount of Ni-NTA agarose should be adjusted to the respective amount of expression culture and cell pellet. 2 mL of slurry yielded optimal results for 1 L of expression culture in our experience.
6. It is crucial to avoid any air bubbles in the resin bed of the column. To remove air from frits before column assembly, place them in a Falcon tube filled with 20% ethanol, shake or

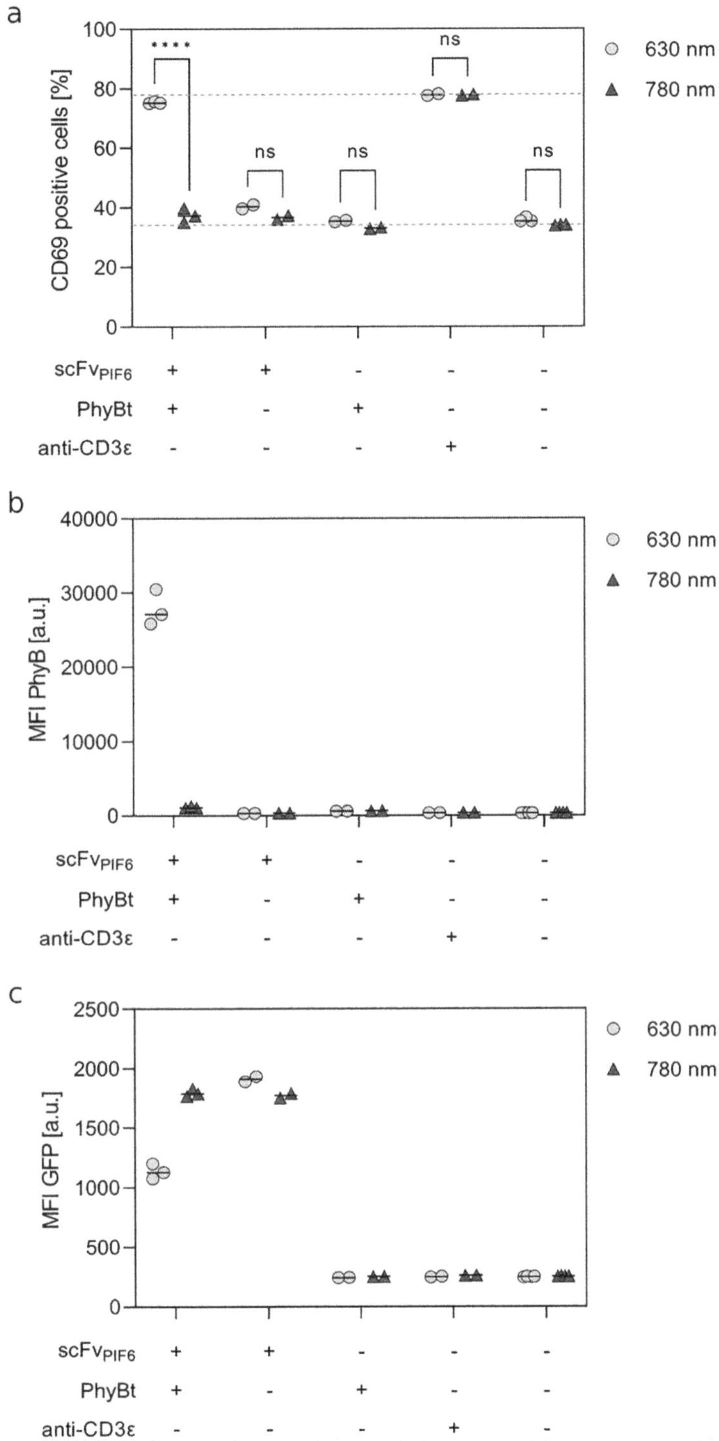

Fig. 4 *Light-dependent stimulation of the TCR of nongenetically manipulated T cells.* Red-light-dependent stimulation of the T cell receptor on Jurkat cells. The extracellular optogenetic OptoREACT system was used to stimulate the TCR on nongenetically manipulated T cells in a light-dependent manner. Jurkat T cells were incubated as indicated with 80 nM $scFv_{PIF6}$ and/or 20 nM PhyB tetramers (PhyBt) and illuminated for 6 h with 630 or 780-nm light. As reference for T cell stimulation, 10 nM soluble anti-CD3ε antibody was used. Afterward, cells were analyzed by flow cytometry for expression of the early T cell activation marker CD69 (**a**) and PhyB (**b**) and GFP fluorescence (**c**) (*ns* no significant difference, ****$P \leq 0.0001$, multiple unpaired t tests). Adapted with permission from Armbruster et al., 2024 [7]. Copyright 2024 American Chemical Society

flick until air is released, and then add them to the column. If air bubbles form in the slurry after settling, gently stir the slurry with a pipet tip and let the slurry settle again.

7. Concentrating the protein extensively can cause precipitation. This can be avoided by centrifugation at room temperature instead at 4 °C and interrupting centrifugation every few minutes to invert the concentrator tube to mix the protein solution.
8. The flow rate can be increased to match the maximum column pressure if desired.
9. Avoid taking up air into the syringe and remove any air bubbles formed by pulling back the plunger of the syringe moderately fast and slowly pushing it forward again while holding the syringe vertically up. Do not load air onto the chromatography column.
10. It is crucial to store proteins at −80 °C to conserve their functionality. Storage at −20 °C is not sufficient.
11. Proteins should be thawed quickly at 37 °C (e.g., in a water bath) and transferred to ice immediately after.
12. Alternatively, the supernatant can be removed using a pipette, but this increases the risk of detaching the cell pellet.
13. Ideally, a control with only moxGFP is included to assess unspecific binding of moxGFP to cells.
14. Jurkat cells are easily activated, resulting in increased CD69 expression. Incubating them overnight at a lower density and resting them after seeding allows them to recover from stress and return to a low basal CD69 expression level.
15. A high intensity of 780-nm light is used to inactivate all PhyB before switching to a lower intensity suitable for illumination for a longer duration of time.
16. As a light-sensing protein, PhyB should be protected from light while handling, for example, by wrapping the tube with aluminum foil. TCEP is added to maintain PhyB in the reduced state, in which it exhibits its maximum functionality.
17. Place the tube on the optoPlate-96 and cover it with aluminum foil. This way, the sample is more evenly illuminated.
18. PhyB is activated within seconds by room light. Therefore, handle PhyB upon inactivation with 780-nm light only under dim green light to prevent its unintended activation. Green light is suited best for handling of PhyB as this wavelength range is in the minimum of the absorbance spectrum of PhyB.

19. Using PBS containing FCS for the mastermixes ensures that the reducing agent TCEP from the PhyB solution is not negatively impacting scFv folding.
20. Mastermixes should be prepared at a larger volume than the minimum volume needed for the respective number of replicates to account for pipetting errors. For 6 replicates with 20 μL per replicate, we suggest preparing at least 130 μL of mastermix.
21. PhyB fluorescence can be measured by flow cytometry using a 637-nm excitation laser and an 670/14-nm emission filter.

Acknowledgments

We thank the technical workshop of the Faculty of Biology for the design and construction of illumination devices and acknowledge the scientific assistance of the Core Facility Signalling Factory for their assistance with size exclusion experiments. This work was supported by the German Research Foundation (Deutsche Forschungsgemeinschaft, DFG) under Germany's Excellence Strategy—CIBSS, EXC-2189, Project ID: 390939984 and under the Excellence Initiative of the German Federal and State Governments—BIOSS, EXC-294, and in part by the Ministry for Science, Research and Arts of the State of Baden-Württemberg.

References

1. Heldin C-H (1995) Dimerization of cell surface receptors in signal transduction. Cell 80: 213–223. https://doi.org/10.1016/0092-8674(95)90404-2
2. Cochran JR, Aivazian D, Cameron TO, Stern LJ (2001) Receptor clustering and transmembrane signaling in T cells. Trends Biochem Sci 26:304–310. https://doi.org/10.1016/S0968-0004(01)01815-1
3. Allen ME, Zhou W, Thangaraj J et al (2019) An AND-gated drug and photoactivatable Cre-loxP system for spatiotemporal control in cell-based therapeutics. ACS Synth Biol 8: 2359–2371. https://doi.org/10.1021/acssynbio.9b00175
4. Tischer DK, Weiner OD (2019) Light-based tuning of ligand half-life supports kinetic proofreading model of T cell signaling. elife 8: e42498. https://doi.org/10.7554/eLife.42498
5. Yousefi S, Günther M, Hörner M et al (2019) Optogenetic control shows that kinetic proofreading regulates the activity of the T cell receptor. https://doi.org/10.7554/eLife.42475.001
6. O'Donoghue GP, Bugaj LJ, Anderson W et al (2021) T cells selectively filter oscillatory signals on the minutes timescale. Proc Natl Acad Sci 118:e2019285118. https://doi.org/10.1073/pnas.2019285118
7. Armbruster A, Ehret AK, Russ M, et al (2024) OptoREACT: Optogenetic Receptor Activation on Nonengineered Human T Cells. ACS Synth Biol. https://doi.org/10.1021/acssynbio.3c00518
8. Hörner M, Yousefi OS, Schamel W, Weber W (2020) Production, purification and characterization of recombinant biotinylated phytochrome B for extracellular optogenetics. Bio Protoc 10. https://doi.org/10.21769/bioprotoc.3541
9. Yousefi OS, Hörner M, Wess M et al (2020) Optogenetic tuning of ligand binding to the human T cell receptor using the opto-ligand-TCR system. Bio Protoc 10. https://doi.org/10.21769/bioprotoc.3540
10. Bugaj LJ, Lim WA (2019) High-throughput multicolor optogenetics in microwell plates. Nat Protoc 14:2205–2228. https://doi.org/10.1038/s41596-019-0178-y

11. Thomas OS, Hörner M, Weber W (2020) A graphical user interface to design high-throughput optogenetic experiments with the optoPlate-96. Nat Protoc 15:2785–2787. https://doi.org/10.1038/s41596-020-0349-x
12. Costantini LM, Baloban M, Markwardt ML et al (2015) A palette of fluorescent proteins optimized for diverse cellular environments. Nat Commun 6. https://doi.org/10.1038/ncomms8670

Chapter 12

Extracellular Optogenetics to Interrogate Unmodified Primary Cell Functions

Morgane Jaeger, Renaud Vincentelli, and Rémi Lasserre

Abstract

Optogenetics often requires genetic modification of the target cells to enable the expression of specific optogenetic tools, making it difficult to study primary cells in their native state. We have recently generated a fully extracellular optogenetic system for reversible light control of T cell receptor (TCR) activation on murine naïve T cells, a cell model that is very difficult to manipulate genetically. This molecular system is very versatile and can be easily modified to study different cell systems in different species. Here we describe how to produce and use this extracellular optogenetic system to manipulate primary T cell activation.

Key words Extracellular optogenetics, Red/far-red optogenetics, Phytochrome, Nonengineered primary cell, Receptor activation, T cell receptor, T cells

1 Introduction

Optogenetics is a powerful tool that has been largely developed in the field of neuroscience to control the activity of specific cells or circuits in the nervous system with light [1]. Today, these technologies are in use in a wide range of biological systems for basic and translational research [2]. Traditionally, optogenetics has been achieved by genetically modifying cells to express light-sensitive proteins, but more recently, extracellular optogenetic systems have been developed to apply an accurate spatiotemporal control of cell stimulations [3–6]. The use of purified recombinant photoreceptors and proteins allows the precise control of biological systems with light without the need for invasive genetic modification.

One of the key advantages of extracellular optogenetic systems is that they can be applied to a wide range of cells and tissues, regardless of their genetic background. This means that researchers can study biological systems of interest in different model organisms, including cell types that cannot be easily genetically manipulated. Another advantage of extracellular optogenetic systems is

Armin Baumschlager (ed.), *Optogenetics: Methods and Protocols*, Methods in Molecular Biology, vol. 2840,
https://doi.org/10.1007/978-1-0716-4047-0_12,

their low invasiveness. Indeed, in conventional optogenetic systems, the overexpression of light-sensitive compounds linked to bioactive protein domains in target cells can alter cell homeostasis and fine regulation. Extracellular optogenetic approaches are based on the addition of compounds to cell media and don't require cell manipulation.

Overall, extracellular optogenetic systems provide a powerful and versatile tool for manipulating cellular signaling and have the potential to be used in a range of applications in basic and applied biomedical research.

LiTE (Light-inducible T cell engager) technology is a recent development in the field of immunology that allows for precise control over the activation of T cells. This technology was described by Jaeger et al. and uses synthetic proteins that activate T cells only when they are exposed to red light [6]. Indeed, by combining a light-inducible protein with a T cell engager, they have been able to reversibly control T cell receptor stimulation with light. In addition, they showed that this technology can be exploited to create a powerful tool to direct tumor cell lysis by the T cells in vitro. Beyond its interest in basic immunology research, such a technology could set the basis for future light-targeted cancer immunotherapies with minimized off-target effects and reduced toxicity to healthy tissues. The following sections of this chapter describe the materials and procedures necessary for the production of the LiTe system, and its use in primary T cell manipulation.

2 Materials

2.1 Production of HoloPhyB

2.1.1 Reagents

1. Bacteria BL21(DE3) *Escherichia coli* (Sigma Aldrich).
2. PhyB plasmids: In order to produce HoloPhyB, three plasmids that encode the heme oxygenase 1 (Ho1), phytochromobilin (PΦB) synthase lacking the transit peptide (ΔHY2), and Phytochrome B ($PhyB_{1\text{-}651}$) are needed. We used the plasmids described by the laboratory of Dr. K Rosen [7]. The latter was modified to add a Strep-tag at the C terminus of $PhyB_{1\text{-}651}$. Alternatively, we used the plasmid pMH1105 [8], a gift from Dr. Wilfried Weber (Addgene, plasmid # 131864; http://n2t.net/addgene:131864; RRID:Addgene_131864) to produce a biotinylated form of PhyB; this plasmid encodes three required genes to produce HoloPhyB: $PhyB_{1\text{-}651}$, HO1 and PCB:ferredoxin oxidoreductase (PcyA). It enables higher production yields for HolophyB.
3. Streptomycin sulfate salt.
4. LB medium.
5. Isopropyl-β-D-1-thiogalactopyranoside (IPTG).

6. Biotin solution: Dissolve 150 mg biotin (Sigma Aldrich) in 125 mL warm PBS (microwaved), filter sterile into an autoclaved bottle. 12.5 mL IPTG stock (1 M) and 1 mL streptomycin (50 mg/mL) can be added directly to this biotin solution to induce PhyB production and biotinylation when added to the BL-21 culture.

2.1.2 Equipment and Consumables

1. Thermostated orbital shaking incubator at a shaking speed of 200.
2. 50 mL and 2 L glass flasks.
3. Spectrophotometer (OD 600).
4. Centrifuge.
5. Liquid nitrogen.

2.2 Purification of HoloPhyB

2.2.1 Reagents

1. Buffers for Nickel affinity purification: (buffer A) 50 mM HEPES, 500 mM NaCl, 5% glycerol, 0.5 mM tris (2-carboxyethyl) phosphine (TCEP), 20 mM imidazole, pH 8. Elution buffer (buffer B): buffer A + 500 mM Imidazole.
2. Bacterial lysis buffer: buffer A + 0.5 mg/mL of lysozyme.
3. DNAse (10 μg/mL solution).

2.2.2 Equipment and Consumables

1. Water bath.
2. Sonicator.
3. Centrifuge.
4. 1 mL Histrap ff crude (Cytiva).
5. HiPrep 26/10 desalting column (Cytiva).
6. Äkta Xpress chromatography system (Cytiva).

2.3 SDS-PAGE and Western Blot Analysis of HoloPhyB Production

2.3.1 Reagents

1. 2X Laemmli buffer: 4% SDS, 20% glycerol, 10% 2-mercaptoethanol, 0.004% bromophenol blue, and 0.125 M Tris HCl.
2. Imperial protein stain.
3. Bovine serum albumin (BSA).
4. TBS/Tween buffer: 137 mM NaCl, 20 mM Tris pH 7.6, 0.1% Tween 20.
5. Anti-6xHis antibody (clone F24-796, BD Biosciences).
6. Goat anti-mouse IgG antibody HRP (Thermo Fischer).
7. PierceTM ECL Western kit (Thermo ScientificTM).

2.3.2 Equipment and Consumables

1. Dry heating bath at 95 °C.
2. 10% SDS–PAGE gel such as Bio-Rad 10% Mini-PROTEAN® TGX™ Precast Protein Gels.
3. Gel Electrophoresis unit (Bio-Rad).

4. Gel Transfer unit (Bio-Rad).
5. Electrophoresis power supply.
6. Nitrocellulose membrane.
7. Azure biosystems 300 imaging system.

2.4 Spectral Analysis of PhyB Coupling to PCB and of Its Photoswitching

1. Optoplate [9].
2. NanoDrop™ One (Thermo Scientific™).
3. Green safe Light for Phytochrome manipulation.

2.5 Production of the LiTe Protein

2.5.1 Reagents

1. Human embryonic kidney 293 EBNA cell line (HEK293T).
2. HEK culture medium: Dulbecco's modified Eagle medium supplemented with 10% Fetal calf serum (FCS), 1 mM sodium pyruvate (NaPy), 2 mM L-glutamine (L-GLu) and Geneticin.
3. DNA: two plasmids for protein expression in mammalian cells are required to produce the LiTE protein: A plasmid encoding the Heavy Chain of the H57-597 Fab fragment linked to the Phytochrome Interacting Factor 6 (PIF6), and a plasmid encoding the Light Chain of the H57-597 Fab fragment.
4. Polyethylenimine (1 mg/mL) (PEI, Polysciences).
5. Culture medium: DMEM supplemented with 10% FBS.

2.5.2 Equipment and Consumables

1. Cell culture incubator set at 37 °C with 5% CO_2.
2. T175 flask.
3. Vortex.
4. FACS tubes.

2.6 Purification of the LiTe Protein

2.6.1 Reagents

1. Protein expression medium: DMEM supplemented with 2% Fetal bovine serum (FBS), 0.5% Tryptone TN1, and 1.25 mM Valproic acid.
2. Penicillin-streptomycin (Pen-Strep).
3. Phosphate buffer saline (PBS).
4. Loading buffer for liquid phase chromatography (Buffer A): 50 mM Tris-HCl, 300 mM NaCl, 10 mM Imidazole in H20 MQ at PH = 8.
5. Elution buffer for liquid phase chromatography (Buffer B): 50 mM Tris-HCl, 300 mM NaCl, 500 mM Imidazole in H20 MQ at PH = 8.

2.6.2 Equipment and Consumables

1. Centrifuge.
2. 5 mL Histrap ff crude (Cytiva).
3. Äkta Xpress chromatography system (Cytiva).
4. Slide-A-Lyzer™ Dialysis Cassette (Thermo Fisher Scientific).

2.7 Analysis of the Binding of the LiTE Protein to the TCR by Flow Cytometry

2.7.1 Reagents

1. RPMI medium: Roswell Park Memorial Institute medium (RPMI) supplemented with 5% Fetal bovine serum, 1 mM sodium pyruvate, and 10 mM HEPES buffer solution with 10% CO_2 at 37 °C.
2. Murine T cell line model such as the Do-11-10 hybridoma.
3. Phosphate buffer saline (PBS).
4. FACS buffer: PBS 1X supplemented with 0.5% FCS.
5. AF647 anti-6xhistidine (clone AD1.1.10, Bio-Rad).
6. AF488-H57 Ab (clone H57-597, BD Pharmingen).

2.7.2 Equipment and Consumables

1. Cell culture incubator set at 37 °C with 7.5% CO_2.
2. BD FACSCantoTM.
3. FlowJo software (Treestar Inc., CA).

2.8 Photo-stimulations and T-Cell Activation Analyses by Flow Cytometry and ELISA Test

2.8.1 Reagents

1. Dynabeads™ Untouched™ mouse T cell kit (Invitrogen™) or EasySep™ mouse T cell isolation kit (STEMCELL Technologies).
2. Complete DMEM/F-12 medium: Dulbecco's Modified Eagle Medium/Nutrient Mixture F-12 (DMEM/F-12) supplemented with 10% Fetal bovine serum, 1 mM Sodium pyruvate, 10 mM HEPES buffer solution, 50 U/mL Penicillin-streptomycin, 0.05 mM β2-mercapto-ethanol.
3. Phosphate buffer saline (PBS).
4. Streptavidin-coated polystyrene beads (5.0–5.9 μm, Spherotech).
5. HoloPhyB protein.
6. LiTE protein.
7. Monoclonal antibodies anti-CD28 (clone H37.51, Thermo Fisher).
8. FACS buffer: PBS supplemented with 0.5% FBS.
9. APC anti-CD69 (clone REA937, BioLegend).
10. FITC anti-CD62L (clone MEL-14, BioLegend).
11. FITC anti-CD25 (clone 7D4, BD Biosciences).

2.8.2 Equipment and Consumables

1. Cell culture incubator set at 37 °C with 7.5% CO_2.
2. Black clear-bottom 96-well plates (Greiner).
3. Optoplate.
4. Ice bucket and ice.
5. Centrifuge.
6. BD FACSCanto™.
7. Tecan INFINITE M1000 PRO.

2.9 Photo-stimulations of T Cell and Calcium Imaging

The following reagents, consumables, and equipment are required in addition to those listed above.

2.9.1 Reagents

1. Dynabeads™ Untouched™ mouse T cell kit (Invitrogen™) or EasySep™ mouse T cell isolation kit (STEMCELL Technologies).
2. Complete DMEM/F-12 medium: Dulbecco's Modified Eagle Medium/Nutrient Mixture F-12 (DMEM/F-12) supplemented with 10% Fetal bovine serum (FBS), 1 mM Sodium pyruvate, 10 mM HEPES buffer solution, 50 U/mL Penicillin-streptomycin, 0.05 mM β2-mercapto-ethanol.
3. PBX calcium kit (BD Bioscience).
4. LiTe protein.
5. Streptavidin-coated polystyrene beads (5.0–5.9 μm, Spherotech).
6. PBS 1X: Dilute 50 mL of 10X PBS in 450 mL of H20 MQ.
7. HoloPhyB protein.
8. Tris (2-carboxyethyl) phosphine (TCEP).

2.9.2 Equipment and Consumables

1. Cell culture incubator set at 37 °C with 10% CO_2.
2. Green safe light.
3. Lab-TekTM II chamber coverglass (Thermo FisherTM).
4. 656 nm and 740 nm LED system (MIGHTEX).
5. Video microscope.

3 Methods

3.1 Production of HoloPhyB

WARNING: All PhyB manipulations should be performed under a green-safe light to limit its photodamage.

1. Transform BL21(DE3) *E. coli* bacteria with the plasmid pMH1105 following the manufacturer's protocol and plate the colonies on agar plates supplemented with streptomycin (100 μg/mL).
2. Pick one colony (with a pipet tip) and transfer it to 10 mL of LB (with streptomycin (100 μg/mL)) in a 50 mL flask. Grow cells overnight at 37 °C at 200 rpm in the dark. This is the preculture for one flask. Increase the preculture volume for multiple flasks.
3. The next morning, inoculate 400 mL LB (streptomycin (100 μg/mL)) in 2 L flask with one preculture/flask.
4. Grow the transfected bacteria at 30 °C at 200 rpm in the dark.

5. Induce protein production with 1 mM Isopropyl-β-D-1-thio-galactopyranoside (IPTG) when the culture reaches an OD at 600 nm between 0.6 and 0.8 (*see* **Note 1**). Drop the temperature to 18 °C at that point. For the production of biotinylated PhyB, biotin (5 μM) is added to the culture medium at the same time as IPTG.
6. Grow the bacteria for another 20 h.
7. Harvest the bacteria (it should be green) by centrifugation for 10 min at 6500 *g* and at 4 °C (*see* **Note 2**).
8. Resuspend the pellets in lysis buffer. To calculate the volume of lysis buffer, we apply the formula: $V = (\text{OD at 600 nm}) \times (\text{volume of bacteria culture})/100$.
9. Perform shock freezing of the bacteria in liquid nitrogen. Pellets can be stored at −80 °C or − 20 °C.

3.2 Purification of HoloPhyB

1. Thaw the bacteria (10 min at 37 °C), add DNAse (10 μg/mL) and $MgSO_4$ (20 mM) in the solution, and put on a wheel at 4 ° C for 20 min.
2. Sonicate sample: 5 cycles of 30 s. on/30 s. Off (30% power) on ice to finish to break the bacteria. Collect a sample for SDS-PAGE (total fraction).
3. Clear lysate from debris by centrifugation at 30,000 *g* and 4 °C for 30 min.
4. Load the supernatant onto a 1 mL FF crude Nickel column on AKTA Xpress. Wash outbound E. coli proteins with 10% buffer B (50 mM Imidazole) and elute the protein in 50% buffer B (250 mM Imidazole).
5. Exchange the purified protein buffer to PBS (+0.5 mM TCEP) using a HiPrep 26/10 desalting column.
6. Aliquot, shock freeze, and store the protein at −80 °C. (avoid freeze-thaw cycles). This can be kept for months.

   See **Note 3**.

3.3 SDS-PAGE and Western Blot Analysis of HoloPhyB Production

1. Denature 5 min at 95 °C purified proteins in 1X Laemmli loading buffer in reducing conditions.
2. Load denatured proteins on a 10% SDS–PAGE gel for electrophoresis. After migration, the gel can be directly stained with Imperial Protein Stain according to manufacturer protocol. If further characterization of the purified protein by western blot is required, duplicate the electrophoresis gel and proceed as follows after migration:
3. For western blot analysis, transfer the proteins from the gel to a nitrocellulose membrane.
4. Block membranes with 4% bovine serum albumin (BSA) in TBS/T buffer.

5. Detect PhyB with a primary anti-6xHis antibody (0.5 μg/mL in TBS 1% BSA), then with a secondary goat anti-mouse IgG antibody coupled to HRP (0.1 μg/mL in TBS 1% BSA).
6. Reveal by chemiluminescence with the PierceTM ECL Western kit and Azure biosystems 300 imaging system.

3.4 Spectral Analysis of PhyB Coupling to PCB and of Its Photoswitching

In order to evaluate the proportion of holoPhyB (PhyB coupled to PCB) in the purified protein and verify its capacity to be photo-switched in response to specific wavelength of light:

1. An absorption spectrum of the purified PhyB is measured with a NanoDrop from 260 to 800 nm. The ratio of the absorption at 280 nm, which detects the total amount of PhyB proteins, and at 680 nm, which detects PΦB, reflects the yield of coupling of PhyB to PΦB.
2. The absorption spectrum of HoloPhyB can be measured directly, or after 3 min under light exposure at 680 nm (photo-conversion in the Pfr form). Photoconversion of HoloPhyB in its Pr form can be evaluated after exposure to 740 nm light for 15 min, before recording a new absorption spectrum.

3.5 Production of the LiTe Protein

3.5.1 Cell Culture

1. Culture human embryonic kidney 293-EBNA cell line (HEK293-EBNA) at 37 °C in HEK culture medium with 7.5% CO_2.
2. Seed 9.10^6 HEK293 adherent cells in T175 flask in DMEM 10% FBS culture medium (without antibiotics) the day before transfection so that they are 60–70% confluent on the day of transfection.

3.5.2 Transfection and Production

1. For each T175 flask, dilute 40 μg of DNA (pYD7-H57Fab-HC-PIF plasmid and pTT22-H57Fab-LC plasmid (ratio 1:3)) in 2 mL of serum-free medium (DMEM).
2. Then, dilute 80 μL of PEI in 2 mL of serum-free medium (DMEM) and mix it with the diluted DNA solution.
3. Vortex the solution for 2–3 s, then incubate 15 min at room temperature.
4. Add the 4 mL of DNA/PEI mixture drop-wise onto the cells plated in 40 mL of serum-containing culture medium and gently rock the flask to ensure an even distribution of the complexes (do not swirl the flask) (*see* **Note 4**).
5. Incubate overnight at 37 °C in the cell culture incubator, and then replace culture medium with expression medium. Maintain cells at 37 °C with 5% CO_2 for 7 days.
6. Harvest supernatant and centrifuge it at 3000 RCF for 15 min at 4 °C to remove cell debris.

3.6 Purification of the LiTe Protein

1. Purify the protein by Nickel affinity chromatography (6%—94% buffer B-A for washing and 50%–50% buffer B-A for elution).
2. Exchange the purified protein buffer to PBS using a Slide-A-Lyzer™ Dialysis Cassette.
3. Verify the purity of the LiTE protein on SDS-PAGE, and its TCR binding capacity by flow cytometry (see below).
4. Aliquot, freeze, and store the protein at −80 °C (avoid freeze-thaw cycles).

See **Note 5**.

3.7 Analysis of the Binding of LiTe Protein to T Cell Lines by Flow Cytometry

1. Wash the T cells with FACS buffer (5.10^4 cells per condition).
2. Incubate the T cells with 10 μg/mL of LiTE protein 45 min at 10 °C.
3. Wash the cells 3 times in FACS buffer.
4. Incubate the cells for 45 min at 10 °C with AF647 anti-histidine in FACS buffer.
5. Wash three times with FACS buffer.
6. Resuspend cells in FACS buffer.
7. Analyze by flow cytometry on a BD FACSCantoTM.
8. Analyze data with FlowJo software.

3.8 Photo-stimulations and T-Cell Activation Analyses by Flow Cytometry and ELISA Test

3.8.1 Cell Culture

1. Isolate primary T lymphocytes from lymph nodes of C57BL/6 mice using the Dynabeads™Untouched™ mouse T cell kit or the EasySep™ mouse T cell isolation kit by negative selection.
2. Resuspend the isolated T cells in complete DMEM/F-12 medium and place them in a cell culture incubator at 37 °C with 7.5% CO_2.

3.8.2 PhyB-Coated Beads Preparation

To perform under green-safe light.

1. Wash streptavidin-coated polystyrene beads three times with PBS (15 μL of beads per stimulation condition) (*see* **Note 6**).
2. Add 100 μL of a 20 μg/mL solution of PhyB protein in PBS supplemented with 0.5 mM TCEP to the washed streptavidin-coated polystyrene beads.
3. Incubate for 2 h on ice in the dark.
4. Then wash the beads with PBS supplemented with 0.5 mM TCEP and resuspend them in complete DMEM/F-12 medium.

3.8.3 Photostimulations

To perform under green-safe light and see **Note 7**.

1. Distribute the PhyB-coated beads into black clear-bottom 96-well plates.
2. Illuminate the 96-well plates on an optoPlate for 15 min under 780 nm light to switch all PhyB proteins in Pr state before adding the T cells (*see* **Note 8**).
3. Add 5 10^5 T cells/well in the presence of 0.3 μg/mL of LiTe protein and optionally 2 μg/mL of anti-CD28 antibody.
4. Incubate the plates on the optoPlate placed in a cell culture incubator, and apply the photostimulation programs you designed (*see* **Notes 9** and **10**).
5. At the end of the stimulation phase, harvest the cells and wash them twice with FACS buffer.
6. Then incubate them for 45 min at 10 °C with dedicated fluorescent-labeled antibodies to analyze activation markers such as APC anti-CD69, and FITC anti-CD25 antibodies in FACS buffer.
7. After three washes, resuspend cells in FACS buffer and analyze them by flow cytometry on a BD FACSCanto™ the cell surface expression of these activation markers.
8. Analyze the data with FlowJo software (Treestar Inc., CA).
9. In parallel, cell supernatants can be analyzed by ELISA or multiplex assay (Biolegend) to measure cytokines secretion.

3.9 Photo-stimulations of T Cell and Calcium Imaging

3.9.1 Cell Culture

1. Isolate primary T lymphocytes from lymph nodes of C57BL/6 mice using the Dynabeads™ Untouched™ mouse T cell kit by negative selection or using the EasySep™ mouse T cell isolation kit by negative selection.
2. Place the isolated T cells in complete DMEM/F-12 medium with 10% CO_2 at 37 °C.
3. Load primary T lymphocytes with PBX calcium reporter according to manufacturer's instructions.
4. Wash the primary T lymphocytes three times with DMEM/F12 medium without phenol red.
5. Incubate the primary T lymphocytes with 20 μg/mL of LiTe protein per conditions during 30 min at room temperature.
6. Wash the cells three times with complete DMEM/F12 medium without phenol red.

3.9.2 Beads Preparation

To perform under green-safe light.

1. Wash streptavidin-coated polystyrene beads three times with PBS (30 μL per condition).

2. Add a 20 μg/mL solution of PhyB protein to the washed streptavidin-coated polystyrene beads.
3. Incubate them for 1 h on ice in the dark.
4. Then wash the beads with PBS supplemented with 0.5 mM TCEP and resuspend in complete DMEM/F-12 medium without phenol-red.

3.9.3 Photostimulations

1. Resuspend the beads and distribute them in Lab-TekTM II chamber coverglass under a green safe light.
2. Illuminate the beads under the microscope for 5 min under 740 nm light to switch all PhyB proteins in Pr state (*see* **Note 11**).
3. Drop the primary T cells in the well and let them settle 15 min under 740 nm light exposure.
4. Illuminate the samples with a 656 nm or 740 nm LED system.
5. Image the calcium flux by videomicroscopy, at 37 °C with a frame rate ranging from 60 to 2 frames/minute.

NB: light power as low as 0.14 mW.cm^{-2} for red light, and 2 mW.cm^{-2} for far-red light are sufficient to manipulate PhyB conformation. However, in the case of alternate 680 nm/740 nm exposure, 10 s of high-power exposure ensures a rapid and global PhyB photoswitching (up to 2 and 10 mW.cm^{-2}, respectively). For all experiments, conditions without the LiTE molecule and without PhyB are required to control the specificity of the calcium responses observed. Also, during acquisition, cover the microscope screen to limit any external light that might interfere with your experiments.

4 Notes

1. During PhyB production, once bacteria are induced with IPTG, bacteria have to be maintained in the dark and experiments have to be performed under green safe light.
2. The pellet should be green/blue if HoloPhyB production has been correctly induced.
3. The average yield of purified HoloPhyB is 0.3 mg/L. To scale up PhyB production, we strongly recommend following the procedure described by Hörner et al. [10].
4. When transfecting cells to produce the LiTe protein, it is necessary to transfer the tube containing the PEI into the one containing the DNA and not the other way around and then to vortex quickly.
5. The average yield of purified LiTE is 0.8 mg/L.

6. Use of siliconized tubes for washing streptavidin-coated polystyrene beads to minimize bead loss during the washing procedure.
7. For all PhyB manipulations, it is essential to add antioxidant reagents such as TCEP or beta-mercapto-ethanol to the buffers and cell media to preserve it from oxidation, which impairs its functionality.
8. This step is essential to ensure that all PhyB proteins are in the Pr conformation at the start of the experiment. It avoids unwanted stimulation of the T cells in unactivated conditions.
9. The optoplate programming interface developed by the Dr. Weber team is very convenient for this purpose [11].
10. For pulsed photostimulations, it is necessary to deliver the 780 nm or the 630 nm light at maximum power for 10 s to ensure a rapid and massive photoswitch of the PhyB molecules.
11. For photostimulation of cells under videomicroscope, it is very important to hide the screens to avoid any stray light that could alter the results of the experiments.

References

1. Fenno L, Yizhar O, Deisseroth K (2011) The development and application of optogenetics. Annu Rev Neurosci 34:389–412
2. Bansal A, Shikha S, Zhang Y (2022) Towards translational optogenetics. Nat Biomed Eng 7 (4):349–369. https://doi.org/10.1038/s41551-021-00829-3
3. Yousefi OS et al (2020) Optogenetic tuning of ligand binding to the human T cell receptor using the opto-ligand-TCR system. Bio Protoc 10(5):e3540
4. Yousefi OS et al (2021) Cross-TCR antagonism revealed by optogenetically tuning the half-life of the TCR ligand binding. Int J Mol Sci 22(9):4920
5. Tischer DK, Weiner OD (2019) Light-based tuning of ligand half-life supports kinetic proofreading model of T cell signaling. elife 8: 42498
6. Jaeger M et al (2023) Light-inducible T cell engagers trigger, tune, and shape the activation of primary T cells. Proc Natl Acad Sci USA 120(39):e2302500120
7. Leung DW et al (2008) Genetically encoded photoswitching of actin assembly through the Cdc42-WASP-Arp2/3 complex pathway. Proc Natl Acad Sci USA 105(35):12797–12802
8. Horner M et al (2019) Light-controlled affinity purification of protein complexes exemplified by the resting ZAP70 interactome. Front Immunol 10:226
9. Bugaj LJ, Lim WA (2019) High-throughput multicolor optogenetics in microwell plates. Nat Protoc 14(7):2205–2228
10. Horner M et al (2020) Production, purification and characterization of recombinant biotinylated phytochrome B for extracellular optogenetics. Bio Protoc 10(5):e3541
11. Thomas OS, Horner M, Weber W (2020) A graphical user interface to design high-throughput optogenetic experiments with the optoPlate-96. Nat Protoc 15(9):2785–2787

Chapter 13

Alternating Cellular Functions by Optogenetic Control of Organelles

Kangqiang Qiu, Xiuqiong Xu, Kai Zhang, and Jiajie Diao

Abstract

Organelles play essential roles in cellular homeostasis and various cellular functions in eukaryotic cells. The current experimental strategy to modulate organelle functions is limited due to the dynamic nature and subcellular distribution of organelles in live cells. Optogenetics utilizes photoactivatable proteins to enable dynamic control of molecular activities through visible light. This modality has been rapidly expanded for the dynamic regulation of organelle functions. This chapter describes a method by optical modulation of the mitochondria-lysosome contacts (MLCs). Detailed procedures of transfection, optogenetic MLCs, mitochondrial morphology, and functional analysis are described. Optogenetic control of organelles in live cells offers an innovative paradigm for cell engineering and synthetic biology.

Key words Optogenetics, Mitochondria-lysosome contacts, Mitochondrial fission, Regulation of cellular functions, Subcellular manipulation

1 Introduction

Organelles are integral components of eukaryotic cells, each serving unique roles that collectively contribute to cellular homeostasis and function [1]. For example, the nucleus, containing genomic DNA, governs genetic information [2, 3]. The endoplasmic reticulum (ER) synthesizes proteins and lipids, facilitating their transport to the Golgi apparatus, where modification, sorting, and packaging occur [4, 5]. Mitochondria, renowned as the cell's powerhouse, produce ATP through oxidative phosphorylation, regulate calcium signaling, and participate in apoptosis [6, 7]. Lysosomes, harboring hydrolytic enzymes, facilitate the degradation of cellular waste, macromolecules, and pathogens [8, 9]. Cytoskeletal proteins, such as microtubules, microfilaments, and intermediate filaments, maintain the cellular structure, facilitate intracellular transport, and mediate cell motility [4, 10]. Interactions among organelles, such as ER-Golgi interaction and mitochondria-lysosome contact,

Armin Baumschlager (ed.), *Optogenetics: Methods and Protocols*, Methods in Molecular Biology, vol. 2840,
https://doi.org/10.1007/978-1-0716-4047-0_13,

enable efficient communication and coordination of cellular processes [5, 11]. These interdependent relationships are essential for fundamental cellular functions, including metabolism, protein synthesis, signaling, and maintenance of cellular integrity. Controlling organelles can manipulate cellular functions, and methods to regulate the dynamics of organelles are highly sought-after. However, the current strategy to control organelle functions is limited due to the dynamic nature and subcellular distribution of organelles in live cells.

The emerging optogenetic technology combines light and genetics to control protein interactions in live cells. It offers superior spatiotemporal resolution and reversibility. This method has been applied to various functions in cells and organisms, including neuronal activity, signal transduction, gene expression, posttranslational modification, programmed cell death, and organelle transport. Optogenetics has revolutionized biological research by providing precise control over molecular activity and cellular processes through light [12–14]. This chapter describes the method of optogenetic control for mitochondria-lysosome contacts (MLCs) [15]. The system utilizes blue-light-sensitive dimerzer, cryptochrome (CRY2), and CRY2-interacting bHLH 1 (CIB), to control the association of mitochondria and lysosomes. The proteins LAMP-mCherry-CRY2 and TOM20-CIB-GFP anchor CRY2 and CIB on lysosomes and mitochondria, respectively. Upon blue light illumination, the photoactivated CRY2 interacts with CIB, leading to the formation of MLCs. To study the change of mitochondria, a method for mitochondrial morphology analysis is introduced. The mitochondrial morphology is associated with their functions, and the seahorse experiments are employed to measure the functions of mitochondria after the optogenetic treatment. This protocol outlines the methods for optogenetically manipulating MLCs to induce mitochondrial fission and analyzing the morphology and functions of mitochondria following the optogenetic treatments.

2 Materials

2.1 Transfection

1. Glass-bottomed culture dishes (P35G-1.5-14-C).
2. Serum-free medium: Dulbecco's modified Eagle's medium (DMEM) contains 4.5 g/L glucose and L-glutamine without sodium pyruvate. Store at 4 °C.
3. DMEM-10: DMEM supplemented with 10% fetal bovine serum (FBS), 100 U/mL penicillin, and 100 μg/mL streptomycin. Store at 4 °C.
4. Sterile phosphate buffer saline (PBS): 137 mM NaCl, 2.7 mM KCl, 10 mM Na_2HPO_4, and 2 mM KH_2PO_4. Store at 4 °C.

5. LAMP–mCherry–CRY2 (Expression plasmid provided by Professor Dr. Bianxiao Cui [12]). Store at −80 °C.
6. TOM20–CIB–GFP (Expression plasmid provided by Professor Dr. Robert Hughes [16]). Store at −80 °C.
7. TurboFect™ Transfection Reagent: a lipid-based transfection can help to deliver DNA plasmids and expression vectors into eukaryotic cells (Thermo Fisher Scientific, USA). Store at 4 °C.
8. HeLa cell line (ATCC, USA).
9. 1.5 mL Tubes.
10. Vortex mixer.
11. Cell incubator with 5% CO_2 and 100% humidity at 37 °C.

2.2 Optogenetic Control of Organelles

1. Geneticin™ selective antibiotic: 50 mg/mL active Geneticin in water. Store at 4 °C.
2. MitoTracker™ Deep Red FM (MTDR): 1 mM MTDR in DMSO. Store at −20 °C.
3. A homemade blue light emitting diode (LED) array (300 μW/cm^2) [14].
4. Dry battery (AAA).
5. A box (the box for 1 mL tips).
6. Aluminum foil.
7. Structural illumination microscope (N-SIM, version AR5.11.00 64bit).
8. Sterile phosphate buffer saline (PBS): 137 mM NaCl, 2.7 mM KCl, 10 mM Na_2HPO_4, and 2 mM KH_2PO_4. Store at 4 °C.
9. DMEM-10: DMEM supplemented with 10% fetal bovine serum (FBS), 100 U/mL penicillin, and 100 μg/mL streptomycin. Store at 4 °C.

2.3 Mitochondrial Morphology Analysis

1. ImageJ (NIH).
2. Excel (Microsoft office).

2.4 Mitochondrial Functional Analysis

1. Geneticin™ selective antibiotic: 50 mg/mL active Geneticin in water. Store at 4 °C.
2. XFe-96 cell culture plate (Agilent Technologies, USA).
3. XF DMEM: DMEM containing 1 mM sodium pyruvate, 10 mM glucose, and 2 mM L-glutamine. Store at 4 °C.
4. Aluminum foil.
5. XF Cell Mito Stress Test Kit (Agilent Technologies, USA).
6. Cell incubator with 5% CO_2 and 100% humidity at 37 °C.

7. DMEM-10: DMEM supplemented with 10% fetal bovine serum (FBS), 100 U/mL penicillin, and 100 μg/mL streptomycin. Store at 4 °C.
8. A homemade blue light emitting diode (LED) array (300 μW/ cm^2).
9. Cell incubator with 5% CO_2 and 100% humidity at 37 °C.

3 Methods

3.1 Transfection

Perform the transfection using TurboFect™ transfection reagent following the manufacturer's protocol [17].

1. Seed the HeLa cells on glass-bottomed culture dishes with DMEM-10 in a cell incubator (*see* **Note 1**).
2. After 24 h to adhere, wash the cells with PBS for 3 times, and culture the cells in serum-free DMEM (1 mL) for 30–60 min before transfection.
3. Add plasmids (1.25 μg LAMP–mCherry–CRY2 and 1.25 μg TOM20–CIB–GFP) in a tube with 120 μL serum-free DMEM. Pipette up and down 3–4 times to mix (*see* **Note 2**).
4. Add the TurboFect™ transfection reagent (7.2 μL) in another tube with 120 μL serum-free DMEM. Pipette up and down 3–4 times to mix (*see* **Note 3**).
5. Add the diluted transfection reagent to the diluted plasmids solution all at once (*see* **Note 4**).
6. Mix immediately by pipetting (up and down 3–4 times) or vortexing (10 s).
7. Set the mixtures at room temperature for 15–20 min to allow plasmids/TurboFect complexes to form (*see* **Note 5**).
8. Add the plasmids/Turbofect complexes to the culture medium in the dish drop-wise and swirl gently to distribute the complexes.
9. Treat the cells with the complexes for 6 h in the incubator.
10. After the treatment, replace the transfection medium with fresh DMEM-10. Set the cells in the incubator.
11. Check transfection efficiency 24–48 h post-transfection.

3.2 Optogenetic Control of Mitochondria-Lysosome Contacts (MLCs)

1. For optogenetic control of MLCs, during the transfection **Step 10**, treat the cells with Geneticin™ selective antibiotic at 500 ng/mL to select the expressed cells.
2. Wash the cells with PBS for 3 times after 48 h of the addition for geneticin, followed by MTDR (100 nM) staining for 30 min (*see* **Notes 6** and **7**).

3. Light up the homemade blue (LED) array (300 μW/cm^2) in a box by dry battery (*see* **Note 8**).
4. Wash the cells with PBS for 3 times and treat the cells with fresh DMEM-10.
5. Set the cells under the blue light for 20 min (*see* **Note 9**).
6. Start up the Structured Illumination Microscope (SIM). Select the Apochromat 100×/1.49 numerical aperture oil-immersion objective lens. Select the green, red, and deep red channels for 3D-SIM.
7. After 20 min for blue light exposure, image the cells under SIM. Record the MLCs and the mitochondrial morphology (Fig. 1, *see* **Notes 10** and **11**).
8. For real-time tracking the MLCs, image the cells under the blue LED array after **Step 4** (*see* **Note 12**).

3.3 Analyzing the Mitochondrial Morphology

1. After obtaining SIM fluorescent images of mitochondria, import the images into ImageJ (*see* **Notes 13** and **14**).
2. Convert the image to an 8-bit format, and adjust the threshold.
3. Utilizing the "set measurements" function to select area and shape descriptors and restricting them to the threshold. Obtain the results by using the "analyze particles" function.
4. The "AR" column in the data represents the ratio of the length-to-width (L/W) for mitochondria (*see* **Note 15**).
5. Based on the "L/W" ratio, classify mitochondria into four categories with Excel: round, when $1.0 \leq L/W < 1.5$; intermediate, when $1.5 \leq L/W < 2.0$; tubular, when $2.0 \leq L/W < 5.0$; hyperfused, when $5.0 \leq L/W$ (Fig. 2).

3.4 Seahorse Experiment

The seahorse experiment is performed according to the manufacturer's instruction [18].

1. For seahorse experiment, during the Transfection **Step 10**, treat the cells with Geneticin™ selective antibiotic at 500 ng/mL to select the transfected cells.
2. Two days after the addition of antibiotic, seed the expressed cells in a XFe-96 cell culture plate at 0.8×10^4 cells/well in DMEM-10 for 48 h in a cell incubator.
3. After 48 h, replace the medium with the warm XF DMEM.
4. Expose the cells to blue LED assay (300 μW/cm^2) for 20 min in the incubator (*see* **Note 16**).
5. Employ the XF Cell Mito Stress Test Kit to assess the oxygen consumption rate (OCR) of the cells, following the manufacturer's instructions [18]. In certain cases (as shown in Fig. 3), subject the cells with specific treatments (*see* **Note 17**).

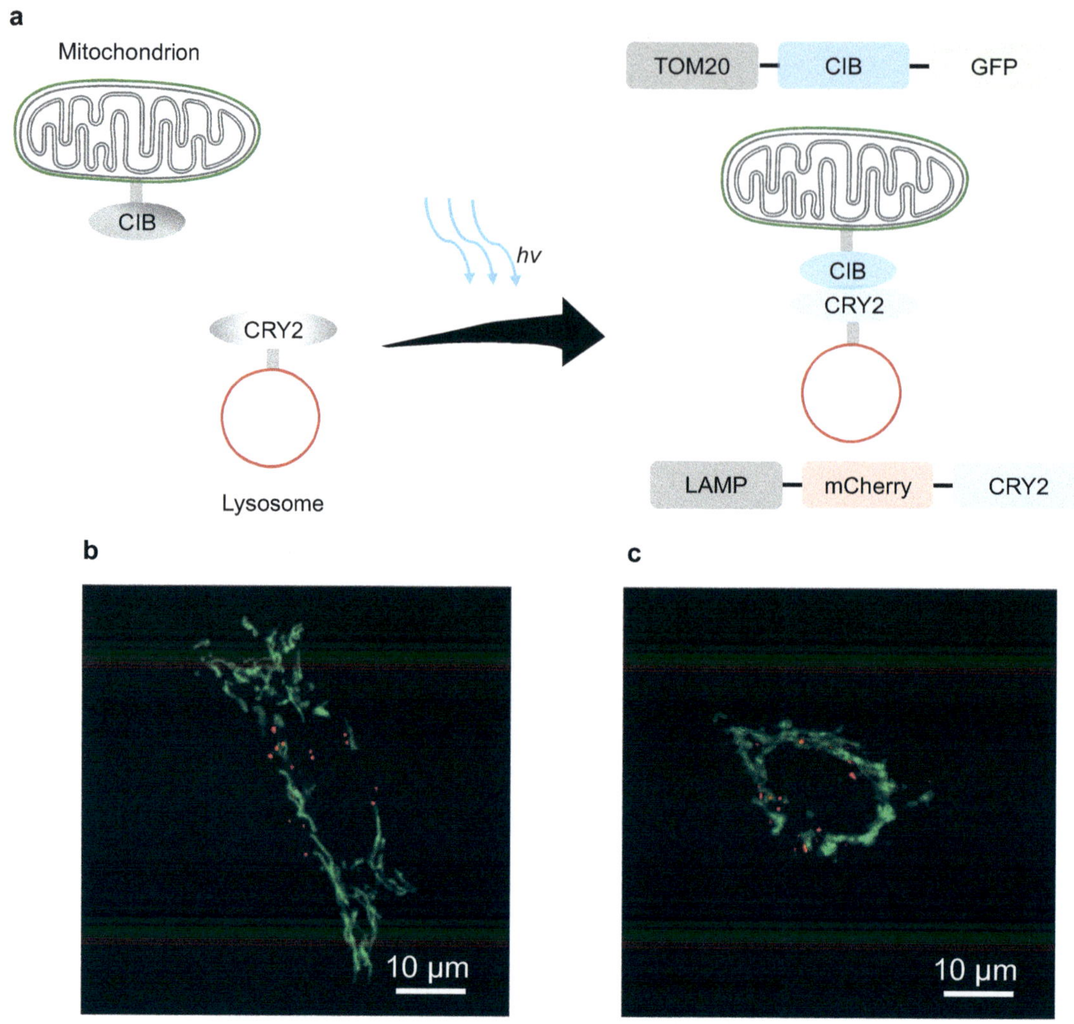

Fig. 1 *Optogenetic induction of MLCs.* (**a**) Schematic of blue light-triggered MLCs. In this system, two light-sensitive proteins, CRY2 and CIB, were strategically tethered to lysosomes and mitochondria, respectively. This anchoring was achieved using specific organelle-targeting transmembrane domains known as LAMP (lysosome-associated membrane protein) and TOM20 (translocase of the outer mitochondrial membrane 20), respectively. Upon illumination with blue light, the light-sensitive proteins, CRY2 and CIB, interacted with each other to form CRY2–CIB complexes. To visualize the expression and localization of these complexes, GFP (Green Fluorescent Protein) and mCherry were utilized as markers, providing fluorescence signals corresponding to the system's expression in the cells. The SIM images of HeLa cells expressing LAMP–mCherry–CRY2 and TOM20–CIB–GFP without (**b**) or with blue light exposure for 20 min (**c**)

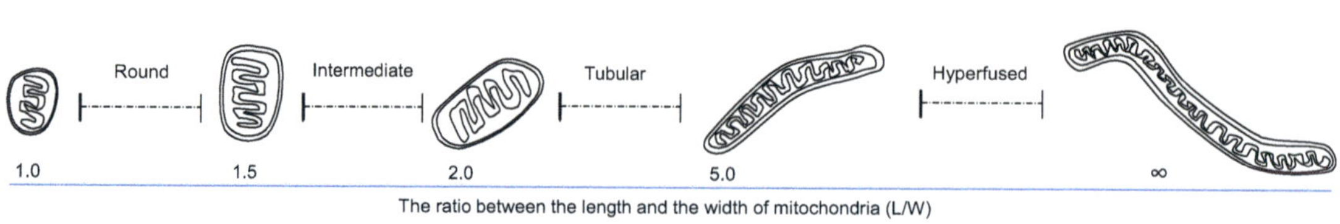

Fig. 2 *Schematic representation of mitochondria divided into four categories*: round, when $1.0 \leq L/W < 1.5$; intermediate, when $1.5 \leq L/W < 2.0$; tubular, when $2.0 \leq L/W < 5.0$; hyperfused, when $5.0 \leq L/W$. *L* stands for length, *W* stands for width, *L/W* stands for dividing the length by the width

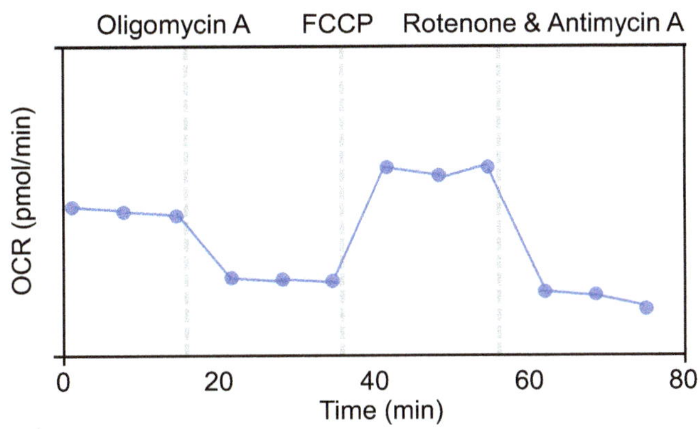

Fig. 3 *The real-time OCR (oxygen consumption rate) curves of cells were measured after various treatments.* Different points in the OCR curve represent specific stages of the experiment. Here is a breakdown of the OCR curve interpretation for each treatment: basal respiration, OCR before the addition of oligomycin A; proton leak, OCR after the addition of oligomycin A; ATP-linked respiration, difference of OCR before and after the addition of oligomycin A; maximal mitochondrial respiration capacity, OCR after the injection of FCCP; nonmitochondrial respiration, OCR after the injection of rotenone and antimycin A

4 Notes

1. Cells should be plated as such so that the monolayer cell density reaches to the optimal 70 ~ 90% confluency at time of transfection.
2. Before using the plasmids, they should be vortexed briefly.
3. Before using the transfection reagent, it should be vortexed briefly.
4. Do not add multiple times, which would result in a bigger size of complex particles.
5. Never keep the mixtures longer than 30 min.
6. PBS was used to wash out the dead cells.
7. The expression level of proteins varies between cells. The dye MTDR can help to visualize the mitochondria in all cells.
8. Pack the LED array in a box wrapped in aluminum foil to prevent blue light from affecting other cells in the same incubator.
9. The optoMLC also works under blue light at room temperature.

10. To record the MLCs, the imaging must be performed within 10 min. The photoactivated proteins cannot work without blue light illumination, and the contacts of the mitochondria and lysosomes could disconnect.
11. To record the mitochondrial morphology, the imaging must be done within 30 min. Without the optogenetic MLCs, the split mitochondria can fuse by nature.
12. The 488 nm laser for the green channel also can activate the protein. To search the focus of the cells, use the deep red channel.
13. The fluorescent intensity for mitochondria must be bright.
14. Localize TOM20–CIB–GFP on the mitochondrial membrane, and the GFP fluorescence it emits provides information about the mitochondrial profile rather than a complete view. To capture a comprehensive representation of mitochondria, utilize images obtained from a commercial dye called MTDR for the analysis.
15. "L" refers to the length measurement, "W" represents the width measurement, and "L/W" indicates the aspect ratio.
16. To avoid the blue light illumination, the other wells in the plate can be covered with aluminum foil.
17. Include these treatments: 1 μM oligomycin A, an inhibitor of electron transport; 2 μM carbonyl cyanide 4-(trifluoromethoxy)phenylhydrazone (FCCP), an accelerator of the electron chain transport (ETC); 500 nM rotenone, an inhibitor of complex I; and 1 μM antimycin A, an inhibitor of complex III.

Acknowledgments

This work was supported by the National Institutes of Health (NIH R35GM128837 to J.D.; R01GM132438 and R01MH124827 to K.Z.).

References

1. Satori CP, Henderson MM, Krautkramer EA et al (2013) Bioanalysis of eukaryotic organelles. Chem Rev 113:2733–2811
2. Alber F, Dokudovskaya S, Veenhoff LM et al (2007) The molecular architecture of the nuclear pore complex. Nature 450:695–701
3. Schumacher B, Pothof J, Vijg J et al (2021) The central role of DNA damage in the ageing process. Nature 592:695–703
4. Fang H, Hu L, Chen Q et al (2023) An ER-targeted "reserve-release" fluorogen for topological quantification of reticulophagy. Biomaterials 292:121929
5. Lin S, Meng T, Huang H et al (2021) Molecular machineries and physiological relevance of ER-mediated membrane contacts. Theranostics 11:974–995

6. Chen Q, Fang H, Shao X et al (2020) A dual-labeling molecular probe to track functional mitochondria-lysosome interactions in live cells. Nat Commun 11:6290
7. Qiu K, Chen Y, Rees TW et al (2019) Organelle-targeting metal complexes: from molecular design to bio-applications. Coord Chem Rev 378:66–86
8. Qiu K, Seino R, Han G et al (2022) De novo design of a membrane-anchored probe for multi-dimensional quantification of endocytic dynamics. Adv Healthc Mater 11:2102185
9. Qiu K, Yadav A, Tian Z et al (2022) Ultralong-term super-resolution tracking of lysosomes in brain organoids by near-infrared noble metal nanoclusters. ACS Mater Lett 4:1565–1573
10. Fletcher DA, Mullins RD (2010) Cell mechanics and the cytoskeleton. Nature 463:485–492
11. Wong YC, Ysselstein D, Krainc D (2018) Mitochondria-lysosome contacts regulate mitochondrial fission via RAB7 GTP hydrolysis. Nature 554:382–386
12. Duan L, Che D, Zhang K et al (2015) Optogenetic control of molecular motors and organelle distributions in cells. Chem Biol 22: 671–682
13. Zhang K, Cui B (2015) Optogenetic control of intracellular signaling pathways. Trends Biotechnol 33:92–100
14. Khamo JS, Krishnamurthy VV, Chen Q et al (2019) Optogenetic delineation of receptor tyrosine kinase subcircuits in PC12 cell differentiation. Cell Chem Biol 26:400–410.e3
15. Qiu K, Zou W, Fang H et al (2022) Light-activated mitochondrial fission through optogenetic control of mitochondria-lysosome contacts. Nat Commun 13:4303
16. Godwin WC, Hoffmann GF, Gray TJ et al (2019) Imaging of morphological and biochemical hallmarks of apoptosis with optimized optogenetic tools. J Biol Chem 294:16918–16929
17. Thermo Scientific (2018) User guide: TurboFect transfection reagent. https://assets.thermofisher.com/TFS-Assets%2FLSG%2Fmanuals%2FMAN0013147_TurboFect_Transfection_Reagent_UG.pdf. Accessed 8 Oct 2023
18. Agilent Technologies (2019) Seahorse XF Cell Mito Stress Test Kit user guide. https://www.agilent.com/cs/library/usermanuals/public/XF_Cell_Mito_Stress_Test_Kit_User_Guide.pdf. Accessed 8 Oct 2023

Chapter 14

Light-Inducible Deformation of Mitochondria in Live Cells

Yutong Song, Peiyuan Huang, and Liting Duan

Abstract

Mitochondria are dynamic organelles with constantly changing morphologies. Despite recent reports indicating that mechanical cues modulate mitochondrial morphologies and functions, there is a lack of methods that can exclusively and precisely exert mechanical forces to and deform mitochondria in live cells. Therefore, how mitochondria sense and respond to mechanical forces remains largely elusive. Optogenetic methods open up new venues for remote and precise manipulation of intracellular activities using light, providing an unprecedented opportunity to establish targeted mechano-stimulation toward mitochondria. This chapter describes the development of a novel optogenetic approach to optically mechanostimulate and induce the deformation of mitochondria. In this approach, light-gated protein–protein heterodimerization recruits force-generating molecular motors to the outer mitochondrial membrane, enabling direct exertion of mechanical force on mitochondria. Details for the design, application, and experimental procedures are laid out in this chapter. This method presents a mitochondria-specific mechano-stimulator for studying the correlation between mitochondrial morphology and functions as well as mitochondrial mechanobiology.

Key words Mitochondria, Optogenetics, Mitochondrial morphology, Organelle mechanobiology, Molecular motor

1 Introduction

Mitochondria are double membrane-bound organelles that generate the energy needed in eukaryotic cells, while also regulating other essential cellular activities, including lipid synthesis, reactive oxygen species (ROS) production, calcium homeostasis, cell growth and differentiation, cell cycle control, and cell death [1]. The dynamic mitochondrial morphology has been found to be influenced and regulated by mechanical stimuli, which affects several mitochondrial functions such as cytochrome c release and ROS production [2–4]. However, the mechano-sensing and -responding properties of mitochondria and the correlation between mitochondrial morphologies and functions remain

Armin Baumschlager (ed.), *Optogenetics: Methods and Protocols*, Methods in Molecular Biology, vol. 2840,
https://doi.org/10.1007/978-1-0716-4047-0_14,

elusive. To reveal how mitochondria perceive and react to mechanical cues, a mechano-stimulating method that can apply force directly toward intracellular mitochondria with high specificity and precision is greatly needed [5].

Optogenetic technologies based on light-gated protein–protein interactions have been used to establish optical control of various biological processes [6–17]. The advantages of optogenetic approaches, such as noninvasiveness, reversibility, high throughput, and high spatiotemporal precision, have provided new opportunities to establish targeted manipulation of specific intracellular organelles. Previous studies have successfully used light to maneuver the intracellular movement of organelles by connecting organelle membranes to molecular motors, a class of natural force-generating protein molecules that mobilize on the cytoskeleton, with optical heterodimerizers [18–20]. Via a similar strategy, we developed an optogenetic method that directs the mechanical force generated by molecular motors toward mitochondria, achieving light-inducible intracellular mechano-stimulation and deformation of mitochondria [21]. This method can control mitochondrial deformation with spatial precision at the subcellular level and is capable of deforming both the outer mitochondrial membrane (OMM) and inner mitochondrial membrane (IMM). Also, this strategy is generalizable to different optical hetero-dimerizing pairs and different motors [22]. With spatiotemporal precision, mitochondria-specificity, and good modularity, this optical method can be a useful tool to interrogate the connection between mitochondrial morphology and functions as well as the mechano-regulation of various mitochondrial functions.

In this chapter, we first outline the design of our optogenetic method and explain how it induces the deformation of mitochondria upon optical triggers. In addition, we describe how to track the deformation of the inner mitochondrial membrane and how to achieve spatially precise optical control of mitochondrial deformation. Optical heterodimerizers and motors that have been shown applicable to this system are also listed. The experimental procedures and materials used are summarized in detail, including cell culture, plasmid construction, and live cell imaging.

2 Materials

2.1 Cell Culture

1. Cell lines: COS-7 (ATCC® CRL-1651™); HeLa (ATCC® CCL-2™) cells; U2OS (ATCC® HTB-96™); NIH/3T3 (ATCC® CRL-1658™).
2. Dulbecco's Modified Eagle Medium (DMEM) with high glucose and pyruvate.

3. Fetal Bovine Serum.
4. Penicillin-Streptomycin.
5. Phosphate Buffered Saline (PBS), pH 7.4.
6. 0.05% Trypsin-EDTA.

2.2 DNA Cloning

1. DNA plasmids: vector plasmid and template plasmid, purified and dissolved in autoclaved ultrapure water.
2. Invitrogen Anza buffer (Thermo Fisher Scientific).
3. Anza Restriction enzymes (Thermo Fisher Scientific).
4. Autoclaved ultrapure water.
5. CloneAmp HiFi PCR Premix (Takara).
6. Primers for PCR.
7. Agarose.
8. 1X Tris-acetate-EDTA buffer (TAE).
9. 10,000X SYBR Safe DNA Gel stain (Thermo Fisher Scientific).
10. GeneJET Gel Extraction Kit (Thermo Fisher Scientific).
11. 5X In-Fusion HD Enzyme Premix (Takara).
12. T4 DNA Ligase Buffer and T4 DNA Ligase (Thermo Fisher Scientific).
13. Competent cells: use XL10-Gold or DH5α.
14. Luria-Bertani (LB) Broth and LB Agar.
15. Antibiotics: Kanamycin and Ampicillin.
16. GeneJET Plasmid Miniprep Kit (Thermo Fisher Scientific).
17. Plasmids: Vector plasmids, CRY2-mCh-Miro1 and KIF5A-GFP-CIBN, are deposited in Addgene by Bianxiao Cui [18]. The following insert DNA segments, iLID, SspB (Micro), SspB(Nano), LOVpep, ePDZ, KIF1A, KIF5B, and Tau, can be cloned from plasmids deposited in Addgene that contain these sequences.

2.3 Transfection

1. Lipofectamine™ 2000 or Lipofectamine™ 3000 Transfection Reagent (Thermo Fisher).
2. Opti-MEM (Thermo Fisher).
3. DNA plasmids: purified and dissolved in autoclaved ultrapure water.
4. Confocal dishes.
5. 0.01% Poly-L-Lysine Solution (Sigma A-005-C).

3 Methods

3.1 Design of the Light-Inducible System for Mechanostimulation and Deformation of Mitochondria in Live Cells

1. In the primary design of this method, the optical heterodimerizer includes the photolyase homology region (PHR) of CRY2 (a.a. 1-498) and its binding partner CIBN (the N-terminal region of CIB1, a.a. 1-170), which dimerize in the presence of blue light and dissociate after removal of blue light [23–25]. The force generator is the kinesins, which are a type of molecular motors that can exert mechanical forces to their cargoes and move toward the plus end of microtubules [26].
2. To induce mitochondrial deformation with blue light, fusion proteins that make CRY2 localized on mitochondria or CIBN connected to kinesins need to be expressed in the cells (Fig. 1a). In order to achieve that, fuse CRY2 to the transmembrane domain of mitochondrial Rho GTPase, Miro1, which targets the outer mitochondrial membrane [28]. Fuse CIBN to the truncated kinesin-1 heavy chain isoform KIF5A, which moves on microtubules like full-length wildtype kinesin but lacks the light chains to bind to cargoes [29]. Place the fluorescence protein mCherry (mCh) into the CRY2 fusion protein to

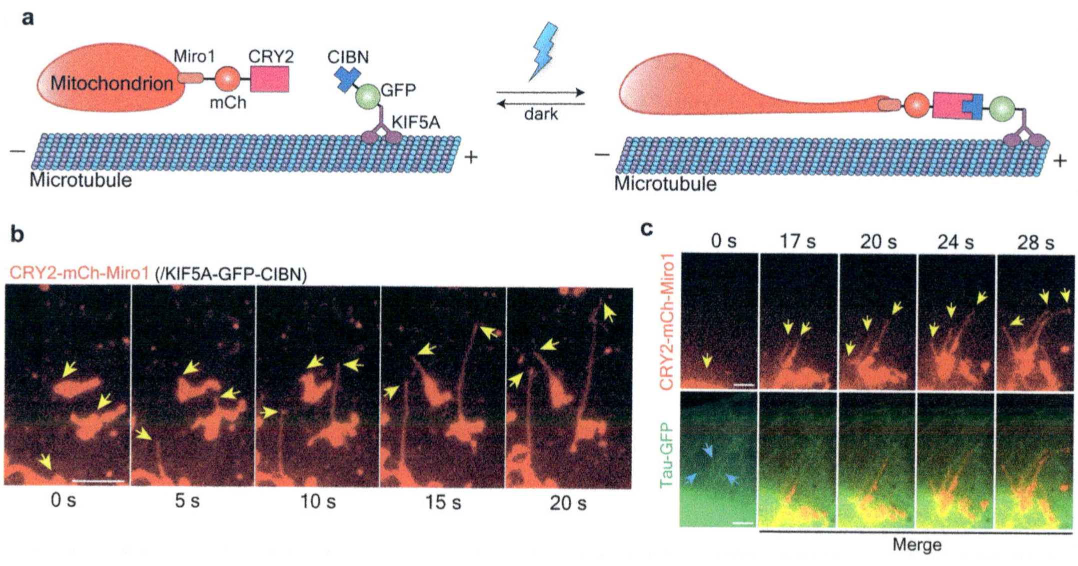

Fig. 1 Design of the light-inducible system for mechanostimulation and deformation of mitochondria. (**a**) Schematic representation of light-controlled mitochondrial deformation based on CRY2-CIBN heterodimerization. (**b**) Fluorescence images of mCh-tagged mitochondria showing deformation and elongation after blue light stimulation as indicated by arrows, in COS-7 cells transfected with CRY2-mCh-Miro1 and KIF5A-GFP-CIBN. Timepoints stand for time after blue light illumination. (**c**) Fluorescence images showing mitochondria deform along underlying microtubules. Mitochondrial tubules, indicated by yellow arrows, extend along the underlying microtubules marked by blue arrows after blue light stimulation in the COS-7 cell expressing CRY2-mCh-Miro1, KIF5A-CIBN, and Tau-GFP. Tau-GFP marks microtubules by the microtubule-associated protein Tau [27]. Scale bars, 5 μm. (Reproduced from (Song et al. 2022) with permission from Elsevier)

visualize mitochondria while avoiding blue light-inducible activation of the system when inspecting cells under a microscope. Insert GFP into the CIBN fusion protein to confirm the expression of the protein and visualize microtubule structures simultaneously. Construction methods for plasmids that encode these fusion proteins are in Subheading 3.2.

3. In cells expressing CRY2-mCh-Miro1 and KIF5A-GFP-CIBN, kinesins are recruited to mitochondria via CRY2-CIBN heterodimerization upon blue light stimulation, inducing mitochondrial tubule formation and elongation via mechanical forces provided by the motors (Fig. 1b) (*see* **Note 1**). Because kinesins mobilize along the underlying microtubules, the direction of the induced deformation is guided by the microtubule network (Fig. 1c).

4. To investigate the morphological response of IMM to light-induced mechanostimulation on OMM, first, visualize the boundary of IMM by tagging a yellow fluorescence protein (YFP) to the mitochondrial matrix-targeting sequence, Mito [30] (Fig. 2a), producing Mito-YFP.

5. In cells expressing CRY2-mCh-Miro1, KIF5A-CIBN, and Mito-YFP, OMM tubules indicated by mCh signals can be pulled out from mitochondrial bodies after blue light

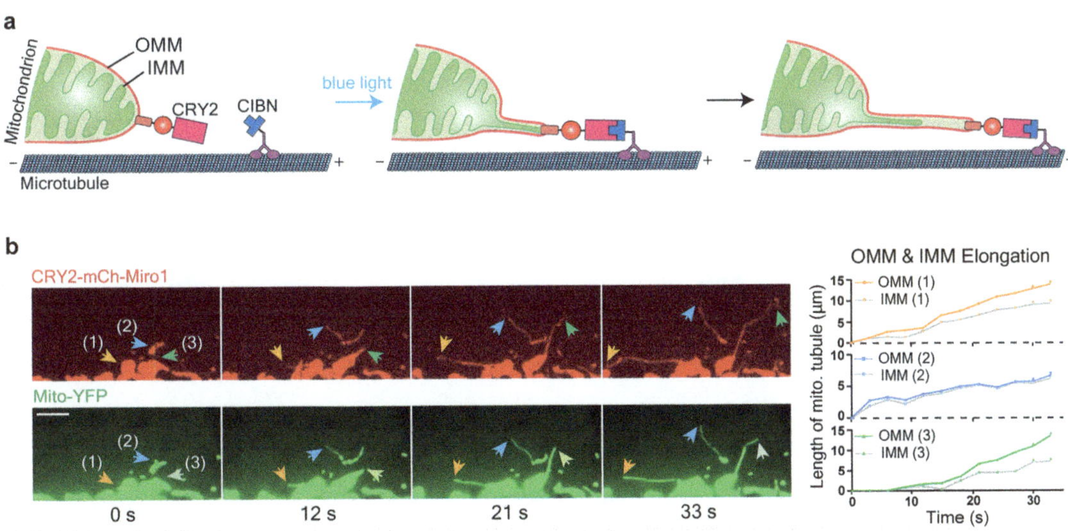

Fig. 2 Inducible deformation of inner mitochondrial membrane (IMM) during light-induced outer mitochondrial membrane (OMM) deformation. (**a**) Schematic representation of IMM deformation following the light-induced stretching of OMM. (**b**) The deformation and elongation of IMM tubules followed the same paths as the stretching OMM tubules induced by light, as indicated by arrows. The right graph quantifies the lengths of OMM and IMM tubules of the three indicated mitochondria with respect to time after blue light illumination. Number and color of the lines in the right graph correspond to that of the arrows shown in the left images. Scale bars, 5 μm. (Reproduced from (Song et al. 2022) with permission from Elsevier)

activation. Simultaneously, IMM inside of these mitochondria form extending tubules that stretch along with their respective leading OMM tubule (Fig. 2b). In some other cases, IMM tubules may lag behind the fast-stretching OMM tubules or not deform at all while OMM keeps extending.

6. Other optical hetero-dimerizing pairs can be incorporated into this method such as the improved light-induced dimer (iLID) derived from the *Avena sativa* light-oxygen-voltage 2 (AsLOV2) domain, and its binding partner SspB [31]. To integrate iLID-SspB into the optogenetic constructs, fuse iLID to Miro1, and link SspB(Micro), a version of SspB with medium binding affinity to iLID, to KIF5A (Fig. 3a). In cells expressing iLID-mCh-Miro1 and KIF5A-GFP-SspB(Micro), mitochondria can be deformed in response to blue light as well.
7. Swapping the position of the iLID-SspB, connecting SspB (Micro) to Miro1 and iLID to KIF5A can also result in a functional system that can induce deformation upon blue light exposure (Fig. 3b).
8. Another version of SspB with a high binding affinity to iLID, SspB(Nano), works as well in the system as a replacement for SspB(Micro).

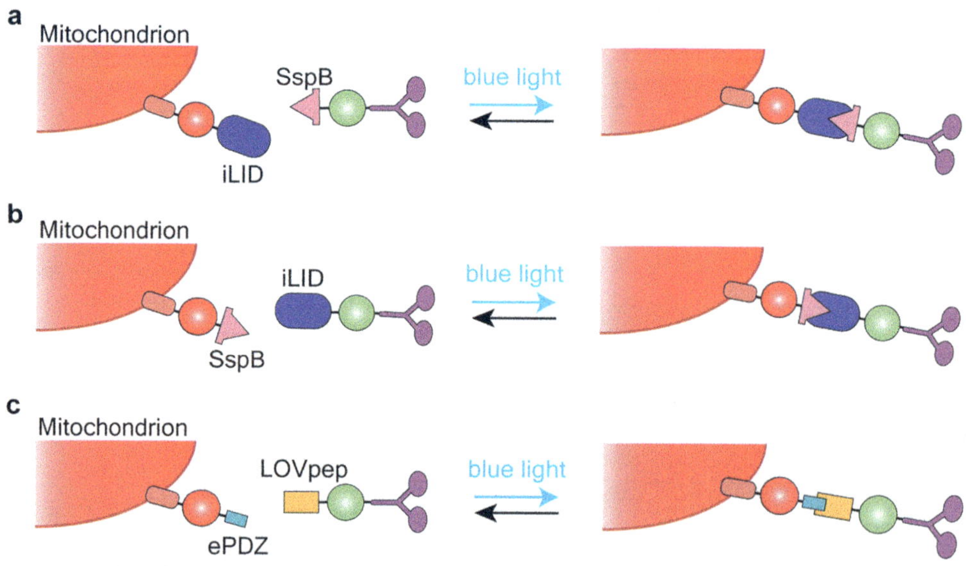

Fig. 3 The light-inducible mitochondrial membrane deformation system can incorporate other optical dimerizers. (**a**) Schematic representation of light-induced mitochondrial deformation using the iLID-SspB pair. (**b**) Scheme for another iLID-SspB system where positions of iLID and SspB are exchanged with regards to that in (**a**). (**c**) Scheme for light-induced mitochondrial deformation using the LOVpep-ePDZ pair. (Reproduced from (Song et al. 2022) with permission from Elsevier)

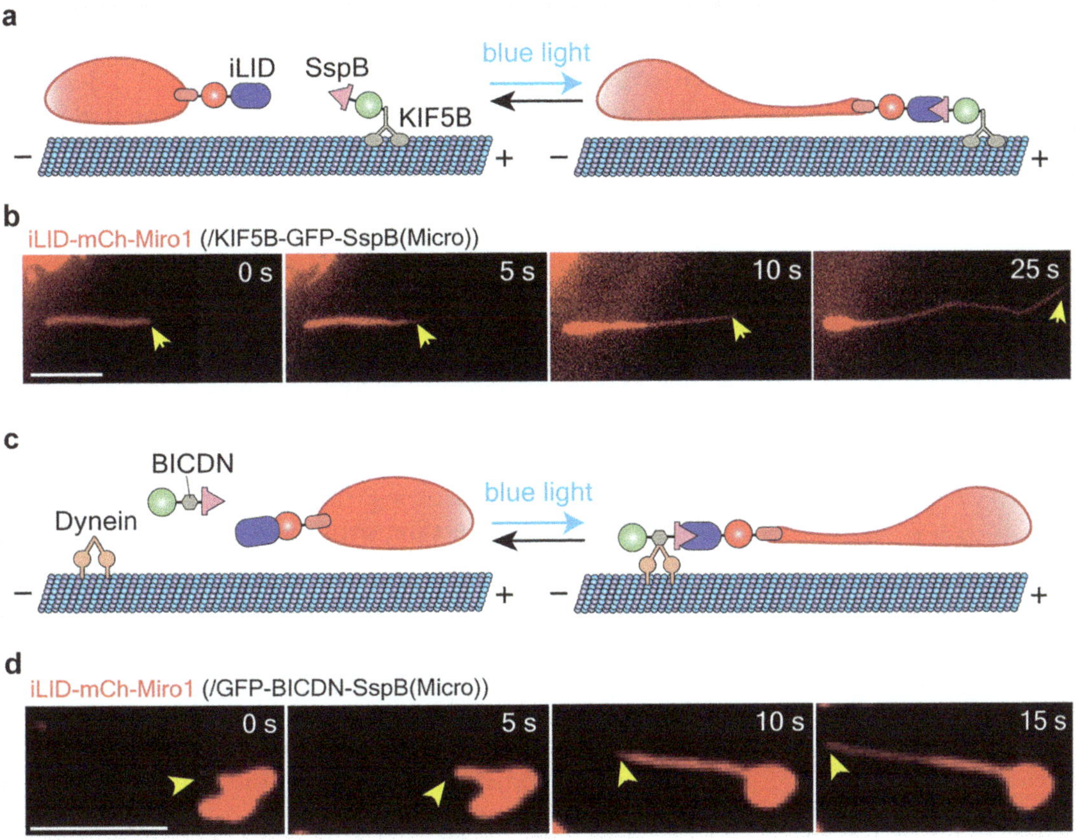

Fig. 4 The light-inducible membrane deformation system is generally applicable to other molecular motors. (**a**) Schematic representation of light-induced mitochondrial deformation implementing kinesin with the iLID-SspB pair. (**b**) Fluorescence images show that mitochondrial deformation can be driven by kinesin-1 heavy chain isoform KIF5B. (**c**) Schematic representation of light-induced mitochondrial deformation using dynein as force-generating motors. (**d**) Fluorescence images show that mitochondrial deformation can be driven by dynein. Scale bars, 5 μm. (Reproduced from (Song et al. 2022) with permission from Elsevier)

9. Another pair of optical dimerizers, LOVpep and ePDZ [32], is also capable of facilitating light-controlled deformation of mitochondria. Fuse LOVpep to KIF5A and its binding partner ePDZ to Miro1 (Fig. 3c). Formation of mitochondrial tubules can be triggered by blue light illumination in cells expressing ePDZ-mCh-Miro1 and KIF5A-GFP-LOVpep.
10. Different types of motors can also be utilized (Fig. 4a). Another kinesin-1 heavy chain isoform, KIF5B, is capable of driving formation of mitochondrial tubules in cells transfected with iLID-mCh-Miro1 and KIF5B-GFP-SspB(Micro) upon blue light exposure (Fig. 4b). The truncated isoform of kinesin-3, KIF1A, can also serve as the motor to provide sufficient mechanical forces.

11. Dynein, another type of molecular motor that moves toward the minus end of microtubules, can also generate mechanical force to elicit mitochondrial deformation. To employ dynein as the motor, fuse the dynein-interacting adapter protein BICDN to SspB(Micro) (Fig. 4c). In cells expressing iLID-mCh-Miro1 and GFP-BCIDN-SspB(Micro), mitochondrial deformation can be induced after blue light stimulation (Fig. 4d).

3.2 Plasmid Construction for the Optogenetic System

1. In general, the In-Fusion cloning method and DNA ligation method are used to construct plasmids.
2. To construct a new plasmid, first linearize the existing plasmid as the vector by cutting at specific sites with restriction enzymes.
3. Design and synthesize forward and reverse primers for PCR of the insert fragment following the guidelines for In-Fusion cloning. A 15-bp overlapping with the sticky ends of the linearized vector should be placed at the 5′ end of each primer.
4. Use restriction enzymes directed at desired cutting sites to linearize vector DNA. A typical reaction is composed of 2 μL 10X Invitrogen Anza buffer, 1 μL of each selected restriction enzyme, plasmid solution containing 1 μg of the vector, and ultrapure water to make up the final volume to 20 μL.
5. Put the final mixed solution in 37 °C incubation for 1 h for complete digestion.
6. Amplify the insert fragment from its original template plasmid by PCR using CloneAmp HiFi PCR Premix. A typical reaction is composed of 12.5 μL CloneAmp HiFi PCR Premix, around 20 ng template DNA, 0.75 μL of forward primer, and 0.75 μL of reverse primer both at 10 μM, and 10 μL ultrapure water to make up the final volume to 25 μL.
7. Put the mixed solution into a regular PCR machine and perform PCR. For cycling conditions, please refer to the CloneAmp HiFi PCR Premix Protocol provided by Takara on the Internet.
8. For gel electrophoresis, mix regular agarose with 1X Tris-acetate-EDTA (TAE) buffer to a concentration of 0.01 g/mL. Heat up the solution to completely dissolve agarose. Add 10,000X SYBR Safe DNA Gel stain into the solution and wait until the gel solidifies.
9. Run linearized vector and amplified insert fragments by electrophoresis to check if the sizes of the DNA fragments are correct.
10. Purify desired DNA fragments from the gel using the GeneJET Gel Extraction Kit.

11. For In-fusion cloning, mix the vector and the insert fragment with a molar ratio of 1:2 for one insert or 1:2:2 for two inserts, and make the total volume to 2 μL. Add 0.5 μL 5X In-Fusion HD Enzyme Premix to ligate the linearized vector and the fragment. Put the mixed solution into a regular PCR machine and run at 50 °C for 15 min.
12. For DNA ligation, mix the vector and the insert fragment with a molar ratio of 1:5 to 1:7. Mix the linearized vector and the insert fragment with T4 DNA Ligase and buffer. The recipe of ligation depends on the ligase product. Put the mixed solution at room temperature for at least 15 min.
13. Transform the plasmid into *E. coli* competent cells by heat shock at 42 °C for 45–60 s.
14. Prepare LB Agar and LB Broth for cultivating bacterial cells. Add antibiotics (Kanamycin or Ampicillin, based on what resistance the plasmid carries) into the agar and the broth to select cells that are successfully transformed by the plasmid.
15. Strike transformed bacteria on an LB agar plate that contains antibiotics. Incubate at 37 °C overnight.
16. On the next day, pick single colonies on agar plates and inoculate into LB broth containing antibiotics. Shake the cultures at 37 °C overnight.
17. On the next day, purify the amplified plasmids by Mini-prep using the GeneJET Plasmid Miniprep Kit. Keep the final concentration of the plasmids in a range of between 100 and 1000 ng/μL for easier transfection.
18. All plasmids used in this study are cloned in the mammalian expression vector pEGFP-N1 or pmCherry-C1. CRY2-mCh-Miro1 and KIF5A-GFP-CIBN are from previous work [18]. iLID-mCh-Miro1 is made by inserting iLID into and replacing CRY2 in CRY2-mCh-Miro1 using In-Fusion. SspB (Micro)-mCh-Miro1 and SspB(Nano)-mCh-Miro1 are made in a similar way, by replacing CRY2 in CRY2-mCh-Miro1 with SspB(Micro) or SspB(Nano), respectively, using In-Fusion. ePDZ-mCh-Miro1 is made by replacing iLID in iLID-mCh-Miro1 with ePDZ using In-Fusion. KIF5A-GFP-SspB(Micro) and KIF5A-GFP-iLID are constructed by inserting SspB (Micro) and iLID into KIF5A-GFP-CIBN, respectively, to replace CIBN using In-Fusion. KIF5A-GFP-LOVpep is made by replacing CIBN in KIF5A-GFP-CIBN with LOVpep using In-Fusion. Motor domain of KIF1A (a.a 1-383) and KIF5B (a. a 1-807) are inserted to KIF5A-GFP-SspB(Micro) and replaced KIF5A to make KIF1A-GFP-SspB(Micro) and KIF5B-GFP-SspB(Micro), respectively. GFP-BICDN-SspB (Micro) is constructed by inserting SspB(Micro) into GFP-BICDN-CIBN and replacing CIBN using In-Fusion.

Tau-GFP was constructed by replacing YFP in Tau-YFP with GFP using ligation. Construction details of the plasmids are listed in Table 1.

19. Please note that the construction strategies described above are strategies used in this study based on available plasmids at the time. They do not need to be strictly followed to generate the final DNA sequences. There are other ways to construct the exact same plasmid. Also, it is suggested to have at least a 3-a.a. flexible linker between two domains within fusion proteins.

3.3 Cell Culture and Transfection

1. Use COS-7 cells to express this optogenetic system and the other proteins. HeLa cells, U2OS cells, or 3T3 cells are also available options.
2. Culture cells in DMEM supplemented with 10% Fetal Bovine Serum (FBS) and 1% Penicillin-Streptomycin. Prepare cell culture for passaging in 6-well plates. Maintain all cultures at 37 ° C in a 5% CO_2 environment.
3. During cell passaging, use PBS buffer to wash off the remaining medium from the previous passage and 0.05% Trypsin-EDTA to detach cells from plate surfaces.
4. Use cells with passage numbers under 20 for the experiment.
5. Prepare the cell culture for imaging experiments in a confocal dish that is compatible with the microscope used.
6. The glass surface of a confocal dish can be coated with Poly-L-Lysine (PLL) solution for better cell adhesion. Cover the glass surface with PLL and incubate in a cell incubator at 37 °C for 15 min, then rinse the glass with autoclaved ultrapure water to wash off the excessive PLL.
7. To express the optical control system, transfect cells with designated groups of plasmids for desired light-induced deformation mentioned in the previous sections. To achieve optimal transfection efficiency, seed around 30,000 cells onto the PLL-coated confocal dish at around 40–50% confluency. Then allow cells to grow for 1 day and reach around 80% confluency before transfection.
8. Transfect cells by Lipofectamine™ 2000 Transfection Reagent following the manufacturer's protocol.
9. Here provided is a simplified version. Mix 0.5 μg total amounts of plasmid DNA in 50 μL Opti-MEM® I Reduced-Serum Medium in a 1.5 mL microcentrifuge tube. Meanwhile, prepare 1.5 μL Lipofectamine™ 2000 in 50 μL Opti-MEM in another tube.
10. Mix the two solutions together and incubate them at room temperature for 15 min before adding them to the cell culture. The amount of DNA, reagents, and Opti-MEM can be adjusted proportionally according to the cell numbers.

Table 1
Plasmids construction details

Plasmids	Method	Vector	Restriction sites	Forward primer	Reverse primer
Tau-GFP	Ligation	Tau-YFP	AgeI and NotI	–	–
iLID-mCh-Miro1	In-Fusion	CIBN-mCh-Miro1	NheI and AgeI	cgtcagatccgctagcgccaccggggagtttctggcaaccacac	catggtggcgaccggtccaaagtaattttcgtcgttcgc
KIF5A-GFP-SspB(Micro)	In-Fusion	KIF5A-GFP-CIBN	BSP1407I and AflII	atggacgagctgtacaagggcattgatctgagcggcctgac	caatttacgccttaagttaaccaatattcagctcgtcatag
KIF5A-GFP-LOVpep	In-Fusion	KIF5A-GFP-CIBN	BSP1407I and AflII	ggtggatctggaggttca	tataagctgcaataaacaagttctcgagaagcttacaccc
SspB(Micro)-mCh-Miro1	In-Fusion	CIBN-mCh-Miro1	NheI and AgeI	cgtcagatccgctagcgccaccatggaattcagctccccgaaacgcc	atggtggcgaccggtccactaccaccaatattcagctcgtcatagatttc
SspB(Nano)-mCh-Miro1	In-Fusion	CIBN-mCh-Miro1	NheI and AgeI	cgtcagatccgctagcgccaccatggaattcagctccccgaaacgcc	atggtggcgaccggtccactaccaccaatattcagctcgtcatagatttc
KIF5A-GFP-iLID	In-Fusion	KIF5A-GFP-CIBN	BSP1407I and AflII	atggacgagctgtacaagggtagtggtagtggggagtttc	caatttacgccttaagttaaaagtaattttcgtcgttcgctgc
KIF1A-GFP-SspB(Micro)	In-Fusion	KIF5A-GFP-SspB(Micro)	NheI and SalI	cgtcagatccgctagcgccaccatggctggggcctctgtgaag	gctcaccatagtcgacccggatcccagcagatctcgcagcctgg
KIF5B-GFP-SspB(Micro)	Ligation	KIF5A-GFP-SspB(Micro)	NheI and SalI	–	–
GFP-BICDN-SspB(Micro)	In-Fusion	GFP-CIBN	EcoRI and BamHI	ggtgggccgagaattcggcattgatctgagcggcctg	tagatccggtggatccttaaccaatattcagctcgtc

11. 4 h after the addition of the mix solution, replace the medium with fresh culture medium.
12. Allow cells to recover for 18–24 h after transfection in the complete culture medium overnight before imaging experiments. Proceed immediately to imaging once the cells are taken out of the incubator (*see* **Note 2**).

3.4 Optical Stimulation and Imaging

1. We examined transfected cells on a Leica DMi8 S microscope equipped with an on-stage temperature-controlled CO_2 incubation chamber and a motorized stage. The microscope was also equipped with the Infinity Scanner system that serves as the laser source for spatial control experiments. Images were captured by a Leica DFC9000 sCMOS Fluorescence Microscope Camera. Other microscopes with similar resolution and functions are all viable options for imaging.
2. Record time-lapse movies in two separate channels. Use 550 nm green light along with a 590/25 nm emission filter to check the mCh-labeled protein expression of the mitochondrial deformation system. For examining GFP-labeled or YFP-labeled proteins, as well as stimulating blue light-responsive photosensitive proteins, use 470 nm blue light as the excitation light source and combine it with a 510/25 nm emission filter.
3. Set the input light exposure time at 200 ms per pulse with 1-s or 5-s intervals in both channels (*see* **Note 3**). Set the light intensity cells receiving at the bottom of the confocal dishes at around 9.7 W/cm^2.
4. To deform mitochondria, select cells to be positioned in the field of view and adjust the focus plane to where most mitochondria can be clearly discerned in the mCh channel (*see* **Note 4**). Then set up the time-lapse imaging sequence with the parameters mentioned above.
5. Upon blue light activation of the optogenetic system, changes in mitochondrial morphology can be observed in the mCh channel (*see* **Note 5**).
6. For stimulation and observation of IMM deformation, transfect cells with plasmids indicated in Subheading 3.1 and follow the same procedure as described above. This time, since the mitochondrial matrix is tagged by YFP, IMM deformation can be visualized under blue light illumination. The periphery of the mitochondrial matrix indicates the shape of IMM. Changes in IMM morphology and phenomenon highlighted in the paper will be observed.
7. This optogenetic system has the exceptional advantage of subcellular spatial resolution so that it can precisely control the deformation of only the targeted mitochondria in a cell without

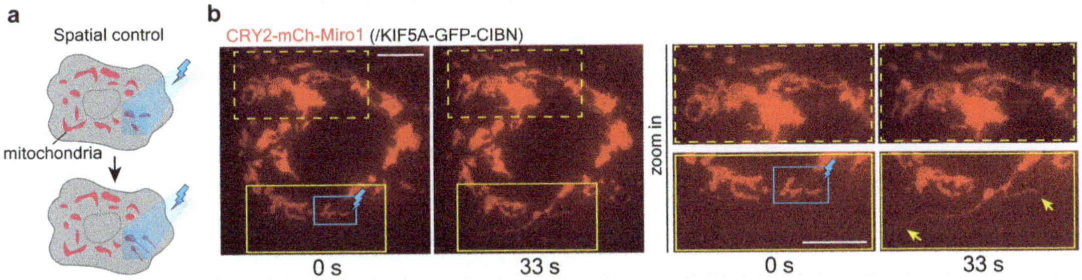

Fig. 5 Spatial control of mitochondrial deformation at subcellular locations. (**a**) Schematic representation of how light-induced mitochondrial deformation can be spatially controlled with subcellular precision. Restrict the blue light illumination to a subcellular region and only mitochondria inside will be activated and induced tubule stretching. (**b**) Fluorescence images show only mitochondria inside the illuminated area undergo deformation after blue light delivery. Zoomed-in images of subcellular areas indicated by dashed and solid yellow lines show that tubule formation can only be observed in mitochondria within the illuminated region indicated by the blue box, while mitochondria outside the region showed no obvious stretching deformation. Scale bars, 10 μm. (Reproduced from (Song et al. 2022) with permission from Elsevier)

perturbing the rest (Fig. 5a). To achieve spatial control of mitochondrial deformation, the blue light laser source (488 nm was used) must be capable of restricting the area of illumination at the subcellular level. Make sure blue light can only be delivered to a specific region of interest (ROI) within a cell.

8. After laser setup, select an ROI that includes several healthy individual mitochondria inside a cell.
9. Deliver blue light laser pulses to the ROI at 5–10 scanning iterations per pulse with a 5-s interval between pulses for a total of 1 min.
10. Use the same 550 nm green light to check the changes in mitochondrial morphology. Tubule formation in only the mitochondria within the illuminated ROI will be observed (Fig. 5b) (*see* **Note 6**).

4 Notes

1. Due to the difficulty in quantitatively assessing or precisely manipulating the number of molecular motors recruited to the mitochondrial membrane after blue light activation, the total force received by deforming mitochondria can not be quantitatively determined and controlled.
2. Although most transfected cells will not be stimulated by ambient light, it is still possible that some cells will have mitochondrial deformation before imaging due to exposure to ambient light or dark binding of the hetero-dimerizing pairs.

Avoidance of ambient light exposure during transfection and before imaging is thus advised. Keep the cells in a dark box during imaging. Choose cells with no obvious activation before blue light when selecting cells for imaging.

3. High-power pulses also cause photobleaching of fluorescence. Avoid consistent blue light illumination on cells that is over 10 min.
4. Ensure the selected cell has been successfully transfected by all optogenetic constructs by examining mCh and GFP signals. Make sure there is clear fluorescence in both channels in the cell during imaging. Cells missing either construct can not show light-induced deformation of mitochondria.
5. Notice that not all mitochondria in a transfected cell will be deformed upon blue light exposure. Deformation might not be successfully induced when cell conditions are not good, especially in cells showing abnormal mitochondrial shapes like hyperfused, overly fractured, or swelled mitochondrial bodies. Or, when the cell has disturbed cytoskeletal structure and does not possess normal microtubules, deformation of mitochondria can not be induced because this method depends on functioning motors mobilizing on microtubules. In healthy cells, some mitochondria are not in the vicinity of any microtubule, thus their membrane cannot be associated with motors moving on microtubules and forming stretching tubules.
6. Spatial control may not be successful in every transfected cell due to leakage of blue light from the designated region or activation of heterodimerization in areas outside of the ROI. Deviations between ROI and the actual laser scanning region also affect the precision. In such cases, adjust and calibrate the illumination settings to achieve more precise spatial control.

References

1. McBride HM, Neuspiel M, Wasiak S (2006) Mitochondria: more than just a powerhouse. Curr Biol 16:R551–R560
2. Bartolák-Suki E, Imsirovic J, Nishibori Y, Krishnan R, Suki B (2017) Regulation of mitochondrial structure and dynamics by the cytoskeleton and mechanical factors. Int J Mol Sci 18:1812
3. Chapman KE, Sinclair SE, Zhuang D, Hassid A, Desai LP, Waters CM (2005) Cyclic mechanical strain increases reactive oxygen species production in pulmonary epithelial cells. Am J Physiol Lung Cell Mol Physiol 289: L834–L841
4. Feng Q, Kornmann B (2018) Mechanical forces on cellular organelles. J Cell Sci 131: jcs218479
5. Feng Q, Lee SS, Kornmann B (2019) A toolbox for organelle mechanobiology research—current needs and challenges. Micromachines 10:538
6. Duan L, Hope JM, Guo S, Ong Q, François A, Kaplan L, Scherrer G, Cui B (2018) Optical activation of TrkA signaling. ACS Synth Biol 7: 1685–1693
7. Huang P, Liu A, Song Y, Hope JM, Cui B, Duan L (2020) Optical activation of TrkB signaling. J Mol Biol 432:3761–3770

8. Zhao EM, Suek N, Wilson MZ, Dine E, Pannucci NL, Gitai Z, Avalos JL, Toettcher JE (2019) Light-based control of metabolic flux through assembly of synthetic organelles. Nat Chem Biol 15:589–597
9. Hannanta-Anan P, Chow BY (2018) Optogenetic inhibition of Gα protein signaling reduces calcium oscillation stochasticity. ACS Synth Biol 7:1488–1495
10. Dowbaj AM, Jenkins RP, Williamson D, Heddleston JM, Ciccarelli A, Fallesen T, Hahn KM, O'Dea RD, King JR, Montagner M et al (2021) An optogenetic method for interrogating YAP1 and TAZ nuclear-cytoplasmic shuttling. J Cell Sci 134. https://doi.org/10.1242/jcs.253484
11. Inglés-Prieto Á, Reichhart E, Muellner MK, Nowak M, Nijman SMB, Grusch M, Janovjak H (2015) Light-assisted small-molecule screening against protein kinases. Nat Chem Biol 11:952–954
12. Kim NY, Lee S, Yu J, Kim N, Won SS, Park H, Heo WD (2020) Optogenetic control of mRNA localization and translation in live cells. Nat Cell Biol 22:341–352
13. Nguyen NT, Huang K, Zeng H, Jing J, Wang R, Fang S, Chen J, Liu X, Huang Z, You MJ et al (2021) Nano-optogenetic engineering of CAR T cells for precision immunotherapy with enhanced safety. Nat Nanotechnol 16:1424–1434
14. Niopek D, Wehler P, Roensch J, Eils R, Di Ventura B (2016) Optogenetic control of nuclear protein export. Nat Commun 7:10624
15. Wang X, Chen X, Yang Y (2012) Spatiotemporal control of gene expression by a light-switchable transgene system. Nat Methods 9: 266–269
16. Hörner M, Jerez-Longres C, Hudek A, Hook S, Yousefi OS, Schamel WWA, Hörner C, Zurbriggen MD, Ye H, Wagner HJ et al (2021) Spatiotemporally confined red light-controlled gene delivery at single-cell resolution using adeno-associated viral vectors. Sci Adv 7. https://doi.org/10.1126/sciadv.abf0797
17. Song Y, Zhao Z, Xu L, Huang P, Gao J, Li J, Wang X, Zhou Y, Wang J, Zhao W, Wang L, Zheng C, Gao B, Jiang L, Liu K, Guo Y, Yao X, Duan L (2024) Using an ER-specific optogenetic mechanostimulator to understand the mechanosensitivity of the endoplasmic reticulum. Dev Cell 59(11):1396–1409
18. Duan L, Che D, Zhang K, Ong Q, Guo S, Cui B (2015) Optogenetic control of molecular motors and organelle distributions in cells. Chem Biol 22:671–682
19. van Bergeijk P, Adrian M, Hoogenraad CC, Kapitein LC (2015) Optogenetic control of organelle transport and positioning. Nature 518:111–114
20. Harterink M, van Bergeijk P, Allier C, de Haan B, van den Heuvel S, Hoogenraad CC, Kapitein LC (2016) Light-controlled intracellular transport in Caenorhabditis elegans. Curr Biol 26:R153–R154
21. Song Y, Huang P, Liu X, Zhao Z, Wang Y, Cui B, Duan L (2022) Light-inducible deformation of mitochondria in live cells. Cell Chem Biol 29:109–119.e3
22. Liu X, Xu L, Song Y, Zhao Z, Li X, Wong CY, Chen R, Feng J, Gou Y, Qi Y, Chow HM, Yao S, Wang Y, Gao S, Liu X, Duan L (2024) Force-induced tail-autotomy mitochondrial fission and biogenesis of matrix-excluded mitochondrial-derived vesicles for quality control. Proc Natl Acad Sci USA 121(14): e2217019121
23. Kennedy MJ, Hughes RM, Peteya LA, Schwartz JW, Ehlers MD, Tucker CL (2010) Rapid blue-light-mediated induction of protein interactions in living cells. Nat Methods 7:973–975
24. Duan L, Hope J, Ong Q, Lou H-Y, Kim N, McCarthy C, Acero V, Lin MZ, Cui B (2017) Understanding CRY2 interactions for optical control of intracellular signaling. Nat Commun 8:547
25. Che DL, Duan L, Zhang K, Cui B (2015) The dual characteristics of light-induced cryptochrome 2, homo-oligomerization and heterodimerization, for optogenetic manipulation in mammalian cells. ACS Synth Biol 4:1124–1135
26. Mallik R, Gross SP (2004) Molecular motors: strategies to get along. Curr Biol 14:R971–R982
27. Hoover BR, Reed MN, Su J, Penrod RD, Kotilinek LA, Grant MK, Pitstick R, Carlson GA, Lanier LM, Yuan L-L et al (2010) Tau mislocalization to dendritic spines mediates synaptic dysfunction independently of neurodegeneration. Neuron 68:1067–1081
28. Fransson S, Ruusala A, Aspenström P (2006) The atypical Rho GTPases Miro-1 and Miro-2 have essential roles in mitochondrial trafficking. Biochem Biophys Res Commun 344:500–510
29. Cai D, Hoppe AD, Swanson JA, Verhey KJ (2007) Kinesin-1 structural organization and conformational changes revealed by FRET stoichiometry in live cells. J Cell Biol 176:51–63

30. Rizzuto R, Brini M, Pizzo P, Murgia M, Pozzan T (1995) Chimeric green fluorescent protein as a tool for visualizing subcellular organelles in living cells. Curr Biol 5:635–642
31. Guntas G, Hallett RA, Zimmerman SP, Williams T, Yumerefendi H, Bear JE, Kuhlman B (2015) Engineering an improved light-induced dimer (iLID) for controlling the localization and activity of signaling proteins. Proc Natl Acad Sci USA 112:112–117
32. Strickland D, Lin Y, Wagner E, Hope CM, Zayner J, Antoniou C, Sosnick TR, Weiss EL, Glotzer M (2012) TULIPs: tunable, light-controlled interacting protein tags for cell biology. Nat Methods 9:379–384

Chapter 15

Chlorophyll-Based Optogenetics to Control Membraneless Organelles

Manjia Li, Byung Min Park, Zhaoxia Li, Weiqi Huang, and Fei Sun

Abstract

Membraneless organelles (MLOs) formed via protein phase separation have garnered significant attention recently due to their relevance to cellular physiology and pathology. However, there is a lack of tools available to study their behavior and control their bioactivity in complex biological systems. This chapter describes a new optogenetic tool based on water-soluble chlorophyll protein (WSCP), a red light-induced singlet oxygen-generating protein, to control synthetic MLOs. Upon exposure to red light, WSCP generates singlet oxygen, which triggers the crosslinking of the proteins in the MLOs, resulting in their liquid-to-solid phase transition. The effective delivery of chlorophylls enables the successful reconstitution of WSCP in living cells, thus offering a potential approach to biological regulation at the subcellular level.

Key words Optogenetics, Photo-responsive protein, Chlorophyll, Membraneless organelle, Protein phase transition

1 Introduction

In recent years, membraneless organelles (MLOs) have gained significant attention due to their diverse physiological functions and implications in disease pathologies. MLOs are typically formed by liquid-liquid phase separation (LLPS) of biomacromolecules and can be found in various subcellular compartments, such as P bodies, postsynaptic density (PSD), and stress granules (SGs) [1–3]. Aberrant phase separation, particularly of intrinsically disordered proteins like Tau, FUS, and TDP-43, can result in pathological aggregation, which is associated with neurodegenerative diseases [4–7]. During pathological aggregation, the phase-separated condensates transform from dynamic liquid-like droplets

Authors Manjia Li and Byung Min Park have equally contributed to this chapter.

Armin Baumschlager (ed.), *Optogenetics: Methods and Protocols*, Methods in Molecular Biology, vol. 2840,
https://doi.org/10.1007/978-1-0716-4047-0_15,

to static solid-like particles, similar to material aging [8]. This phase transition can reduce molecular exchange and limit molecular dynamics, leading to loss of normal function. Although significant progress has been made in studying the solid state of protein condensates and their pathological relevance, there is a lack of tools to control the phase transition of MLOs in complex living systems.

Optogenetic tools based on photo-responsive proteins, such as cryptochrome 2 (Cry2) and Dronpa, have been developed to control protein behaviors in living cells [9, 10]. However, most optogenetic tools for phase separation rely on blue light illumination, which penetrates poorly into deep tissues. Here, we describe a red light-dependent optogenetic tool based on a water soluble chlorophyll protein (WSCP) for controlling MLOs. The WSCP derived from the *Brassicaceae* family is a photosensitizer that can generate singlet oxygen upon red light illumination [11, 12]. WSCPs exist as tetramers, harboring 2 to 4 chlorophylls, and are soluble in aqueous solutions, unlike many membrane-bound chlorophyll-binding proteins involved in photosynthesis [11]. WSCP tetramers are also super stable and remain intact under harsh conditions such as extreme pH or high temperature, making them an excellent candidate to be a robust optogenetic tool.

The synthetic MLO system is created via the LLPS of a recombinant protein RGG-Mfp-3-RGG (RMR). The RGG domain, an intrinsically disordered region (IDR), serves as the scaffolding domain to drive phase separation, while Mfp-3, a tyrosine-rich protein from *Mytilus edulis*, is amenable to singlet oxygen-mediated crosslinking [13, 14]. The recombinant RMR protein can undergo liquid–liquid phase separation in a concentration/temperature-dependent manner in vitro, and the resulting protein condensates can be decorated with WSCP via hydrophobic interactions. Red light-induced generation of singlet oxygen by WSCP facilitates crosslinking within RMR condensates, leading to the liquid-to-solid phase transition of protein condensates. This red light-triggered phase transition system can be introduced into living cells, providing a new strategy to study biological aging and pathology at the subcellular level (Fig. 1).

2 Materials

Milli-Q ultrapure Type 1 water is used for all buffer solutions, except when working with DNA, where we used nuclease-free water. All chemicals are purchased from Sigma Aldrich unless otherwise noted.

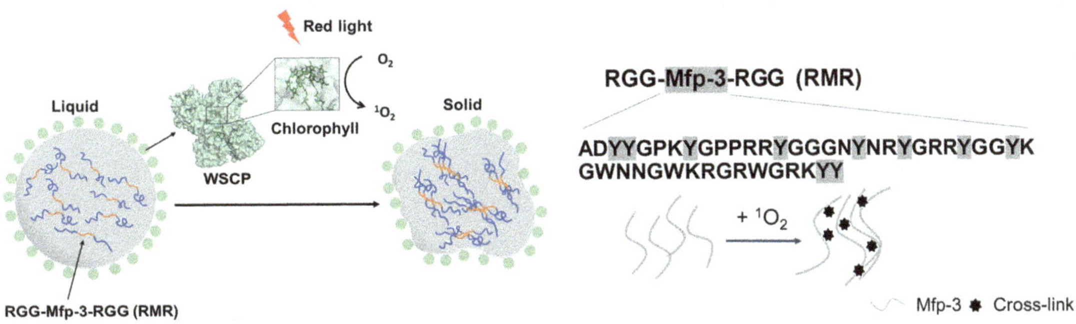

Fig. 1 *Schematic representation of a synthetic protein condensate's liquid-to-solid phase transition induced by red light.* RGG-Mfp-3-RGG (RMR) acts as the scaffold protein and undergoes LLPS to form a synthetic protein condensate. The protein condensate is decorated with WSCP via hydrophobic interactions. WSCP acts as a photosensitizer that generates singlet oxygen upon red light exposure. The resulting singlet oxygen triggers the crosslinking of RMR, resulting in the liquid-to-solid phase transition of the protein condensate. (Adapted from Ref. [14] with permission from Nature Publishing Group)

2.1 Plasmid Construction

1. pQE80l::*His6-RGG-mfp-3-RGG*: The plasmid, previously described by Li and coworkers [14], was used to express the RGG-Mfp-3-RGG (RMR) protein in *Escherichia coli*. The *mfp-3* gene (gBlocks) and the *RGG* gene (amplified from the construct pET::*RGG*, Addgene, Plasmid #124929) were inserted into pQE80l::*AAA* [15] using the restriction sites, NheI/SalI, SacI/SpeI, and XhoI/KpnI. The corresponding amino acid sequence is shown in the Appendix.
2. pCAGGS::*RGG-mfp-3-RGG-mCherry*: The plasmid, previously described by Li and coworkers [14], was used for the RMR expression in mammalian cells. The gene of *RGG-mfp-3-RGG* was amplified from pQE80l::*His6-RGG-mfp-3-RGG* and then inserted into pCAGGS:: *hPhyBY276H-mCherry-HRasCT* (Addgene, Plasmid #100540).
3. pET22b::*WSCP* and pEGFP-N1::*WSCP-egfp*: The plasmids, previously described by Li and coworkers [14], were used to express WSCP in *E. coli* and mammalian cells, respectively. The WSCP domain was derived from *Lepidium virginicum*, with the N-terminal signaling and C-terminal extension sequences removed [16, 17]. The corresponding amino acid sequence is shown in the Appendix. The *WSCP* gene was synthesized by GENEWIZ and cloned into pET22b and pEGFP-N1.
4. *E. coli* DH5α and BL21(DE3) strains (New England Biolabs) were used for plasmid amplification and protein expression, respectively.

2.2 Protein Expression and Purification

1. Luria Broth (LB): Mix 5 g/L yeast extract, 5 g/L NaCl, 10 g/L tryptone, and 15 g/L bacteriological agar (for solid media only) in water. Autoclave the mixture for sterilization.

2. Isopropyl β-D-1-thiogalactopyranoside (IPTG) Solution: Dissolve IPTG in water to prepare a 1 M stock solution. Filter the solution through a 0.22 μm filter for sterilization.
3. Liquid nitrogen.
4. RMR Lysis Buffer: 500 mM NaCl, 20 mM Tris-HCl, 0.1 mM PMSF, pH 7.5. Filter it into a sterile bottle with a 0.22 μm pore bottle cap filter.
5. RMR Wash Buffer: 500 mM NaCl, 20 mM Tris-HCl, 25 mM imidazole, 8 M urea, pH 7.5. Filter it into a sterile bottle with a 0.22 μm pore bottle cap filter.
6. RMR Elution Buffer: 500 mM NaCl, 20 mM Tris-HCl, 500 mM imidazole, 8 M urea, pH 7.5. Filter it into a sterile bottle with a 0.22 μm pore bottle cap filter.
7. RMR Storage Buffer: 500 mM NaCl, 20 mM Tris-HCl, pH 7.5. Filter it into a sterile bottle with a 0.22 μm pore bottle cap filter.
8. WSCP Lysis Buffer: 300 mM NaCl, 20 mM Tris-HCl, 0.1 mM PMSF, pH 7.5. Filter it into a sterile bottle with a 0.22 μm pore bottle cap filter.
9. WSCP Elution Buffer: 300 mM NaCl, 20 mM Tris-HCl, 500 mM imidazole, pH 7.5. Filter it into a sterile bottle with a 0.22 μm pore bottle cap filter.
10. Phosphate Buffered Saline (PBS): 100 mM Na_2HPO_4, 18 mM KH_2PO_4, 1.37 M NaCl, and 27 mM KCl, pH 7.4.
11. Fresh Spinach Leaves (from local market).
12. Acetone (Sigma-Aldrich, 179973).
13. HisTrap HF 5 mL column (Cytiva).
14. Fast Protein Liquid Chromatography (FPLC) system (ÄKTA pure, GE Healthcare).
15. Branson Sonifier 450 (Marshall Scientific).
16. Size-exclusion Chromatography (SEC) column (Cytiva, Superose 6 Increase 10/300 prepacked column & Superdex 200 pg 26/600 prepacked column).
17. Amicon® Ultra-15 Centrifugal Filter Unit.
18. Dialysis tubing cellulose membrane (Sigma-Aldrich, D9652).
19. Whatman Grade 2 Qualitative Filter Papers (Cytiva).

2.3 Mammalian Cell Culture and Transfection

1. HEK293 cell [American Type Culture Collection (ATCC)].
2. Hela cell (ATCC).
3. Decco's Modified Eagle Medium (DMEM) (Gibco™, catalog number:12100046): with additives: high glucose, glutamine, and phenol red; without additives: HEPES, sodium pyruvate, and phenol red.

4. Fetal Bovine Serum (FBS) (Gibco™, catalog number: 26140079).
5. Penicillin-streptomycin solution (Sangon Biotech): 10,000 units/mL penicillin and 10 mg/mL streptomycin.
6. Trypsin-EDTA solution (Sangon Biotech): 2.5 mg/mL trypsin and 0.53 mM ethylenediaminetetraacetic acid (EDTA).
7. Hydrophobic PTFE Membrane Filters (Sartorius, Type 11807).
8. Glass bottom confocal dish (SPL Life Science, 200350).
9. Dimethyl sulfoxide (DMSO) (Sigma-Aldrich, D8418).
10. Lipofectamine 3000 reagent (ThermoFisher).
11. Confocal microscope system (Nikon® C2+) with a cell incubator (Chamlide TC).
12. Photometer: Power meter (PM100D-THORLABS) with a standard photodiode power sensor (S120VC-THORLABS).

2.4 Imaging Processing and Data Analysis

1. NLS-Elements viewer (Nikon®).
2. ImageJ.

3 Methods

3.1 RMR Protein Expression

1. Transform pQE80l::*His6-RGG-mfp-3-RGG* into competent *E. coli* BL21(DE3) cells using the INOUE method [18]. Plate the transformed cells on LB agar supplemented with ampicillin (100 mg/L). Incubate the plate at 37 °C overnight for colony formation.
2. Inoculate 10 mL of LB supplemented with ampicillin (200 mg/L) with a single colony. Incubate the culture overnight at 37 °C and 220 RPM.
3. Inoculate 1 L of LB supplemented with ampicillin (100 mg/L) in a baffled conical flask using the overnight culture. Grow the culture in a shaking incubator at 37 °C and 220 RPM until the OD600 reaches 0.8–1.0.
4. Induce the culture with IPTG at a final concentration of 400 μM. Continue to culture overnight at 16 °C and 220 RPM in a shaking incubator.
5. Harvest the culture by centrifugation at 4200 *g* and 4 °C for 20 min. Discard the supernatant and flash-freeze the pellet with liquid nitrogen. Store the frozen pellet at −80 °C before purification.

3.2 RMR Protein Purification

1. Resuspend the pelleted *E. coli* cells from Subheading 3.1 in the RMR Lysis Buffer. Sonicate the sample using a sonifier with the following settings: 20% duty cycle, 30% output power, 5-min sonication, and 5-min cooling. Repeat the sonication process four times. Keep the sample in an ice bath during sonication.
2. Centrifuge the homogenized solution at 4 °C and 18,000 *g* for 1 h. Discard the supernatant and wash the cell pellets 2–3 times with the RMR Lysis Buffer supplemented with 1 M urea.
3. Resuspend the resulting pellets in the RMR Lysis Buffer with 8 M urea. Incubate the solution at 30 °C for 1 h. Centrifuge the solution again at 4 °C and 18,000 *g* for 1 h to separate the supernatant from the insoluble fraction.
4. Load the resulting supernatant onto a Ni-NTA gravity flow column with a 5-mL column volume (CV). Wash the column with five CVs of the RMR Wash Buffer. Elute the protein with two CVs of the RMR Elution Buffer. Assess the purity of the RMR protein using the sodium dodecyl sulfate-polyacrylamide gel electrophoresis (SDS-PAGE).
5. Dialyze the purified RMR protein against the RMR Storage Buffer at 4 °C to prevent phase separation. After dialysis, aliquot the protein and flash-freeze each aliquot with liquid nitrogen. Store the aliquots of the protein at −80 °C for later use.

3.3 WSCP Expression

1. Transform the pET22b::*WSCP* plasmid into *E. coli* BL21 (DE3) cells. Plate the transformed cells on LB agar supplemented with ampicillin (100 mg/L). Incubate the plates at 37 °C overnight for colony formation.
2. Pick a single colony and inoculate 10 mL of LB with ampicillin (200 mg/L). Grow the culture overnight at 37 °C and 220 RPM.
3. Inoculate 1 L of LB supplemented with ampicillin (100 mg/L) in a baffled conical flask using the overnight culture. Grow the culture in a shaking incubator at 37 °C and 220 RPM until the OD600 reaches 0.8–1.0.
4. Induce the culture with IPTG at a final concentration of 400 μM. Continue to incubate the culture at 37 °C and 220 RPM for 4 h.
5. Harvest the bacterial cells by centrifugation at 5000 *g* and 4 °C for 30 min. Weigh the wet pellet. Discard the supernatant. Flash-freeze the pellet with liquid nitrogen. Store the frozen pellet at −80 °C before further purification.

3.4 Reconstitution of WSCP and Purification of Reconstituted WSCP

1. Combine the *E. coli* pellet from Subheading 3.3, fresh/frozen spinach leaves, and WSCP lysis buffer at a mass ratio of 1:8:8. Blend the mixture using a kitchen blender.
2. Sonicate the blended material using a sonifier in an ice water bath in the dark. Use the following settings: 20% duty cycle, 70% output power, 2-min sonication, and 3-min cooling. Repeat the sonication process 10 times.
3. Centrifuge the homogenized solution at 35,000 *g* at 4 °C for 1 h. Discard the pellet and carefully decant the supernatant. Sterile-filter the supernatant using a 0.45 μm syringe filter. Keep the filtered supernatant on ice under dark conditions.
4. Load a HisTrap HP 5 mL column onto an FPLC system. Set the flow rate at 5 mL/min. Turn off the flow restrictor in the loop. Wash the column with 50 mL of the WSCP Elution Buffer. Equilibrate the column with the WSCP Lysis Buffer (~50 mL).
5. Inject the filtered supernatant into the column at a flow rate of 4 mL/min. Keep the column away from light during this process.
6. Run the preprogrammed FPLC program (*see* **Note 1**). Collect the eluent in the dark and immediately store it at 4 °C before subsequent analysis.
7. Analyze the eluent using a Superose 6 Increase 10/300 prepacked SEC column (Fig. 2).
8. Further, purify the protein using a Superdex 200 pg 26/600 prepacked SEC column. Assess the protein purity using SDS-PAGE.

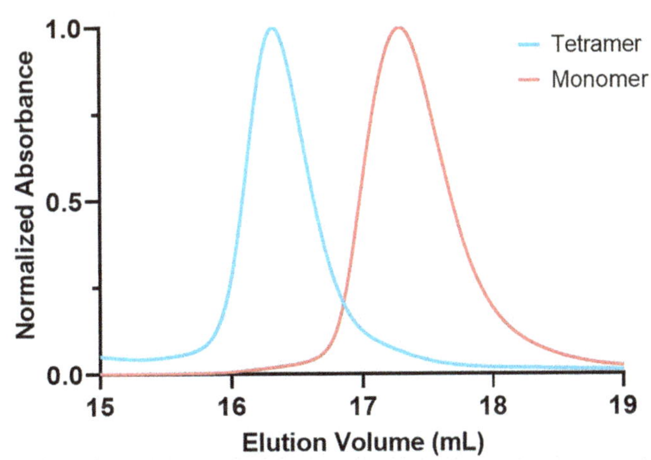

Fig. 2 *Size-exclusion chromatography (SEC) analysis comparing reconstituted WSCP in its tetrameric form and apoWSCP in its monomeric form.* (Reproduced from Ref. [14] with permission from Nature Publishing Group)

9. Remove salts and imidazole from the protein solution via dialysis. For each round of dialysis, dialyze the protein solution (10 mL) against 1 L of MilliQ water at 4 °C for 2 h. Perform four rounds of dialysis. The final concentration of imidazole should be below 1 nM.
10. Concentrate the protein to ~250 μM using a Centrifugal Filter. Perform buffer exchange with PBS using the same centrifugal filter unit.
11. Filter-sterilize the resulting protein solution using a 0.22-μm pore syringe filter. Store the protein solution in the dark at 4 °C before use. The protein solution can be stored for up to 2 weeks.

3.5 Phase Transition of RMR In Vitro

1. Thaw the frozen RMR protein (~25 μM) obtained from Subheading 3.2 with a 52 °C water bath. Initiate the liquid-to-liquid phase separation of RMR by cooling it at room temperature (23 °C) (Fig. 3) (*see* **Note 2**).
2. Add WSCP (obtained from Subheading 3.4) to the phase-separated RMR solution (Fig. 4).
3. Trigger the liquid-to-solid phase transition of RMR condensates by illuminating red LED light (630 nm) (Figs. 5 and 6) (*see* **Notes 2** and **3**).

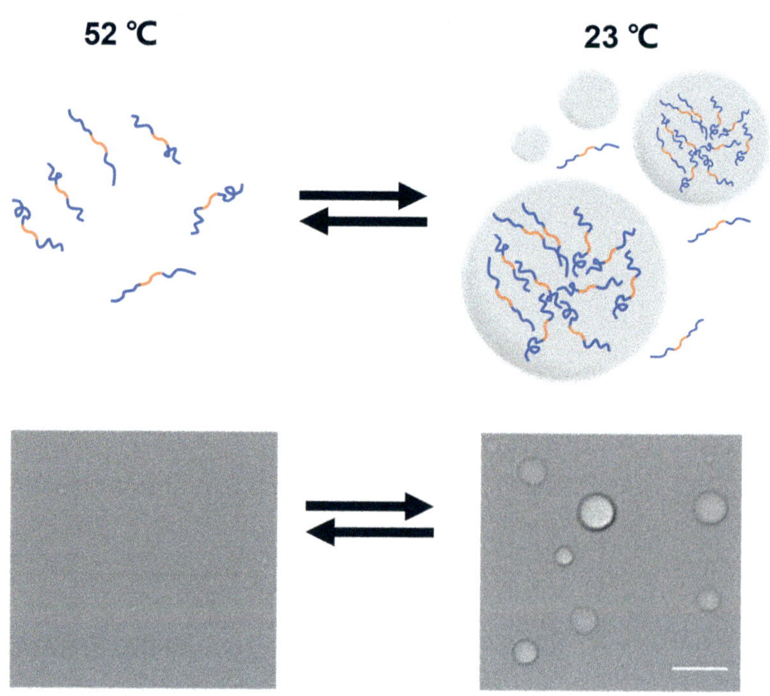

Fig. 3 *Temperature-dependent phase separation of RMR.* Scale bar: 5 μm. (Reproduced from Ref. [14] with permission from Nature Publishing Group)

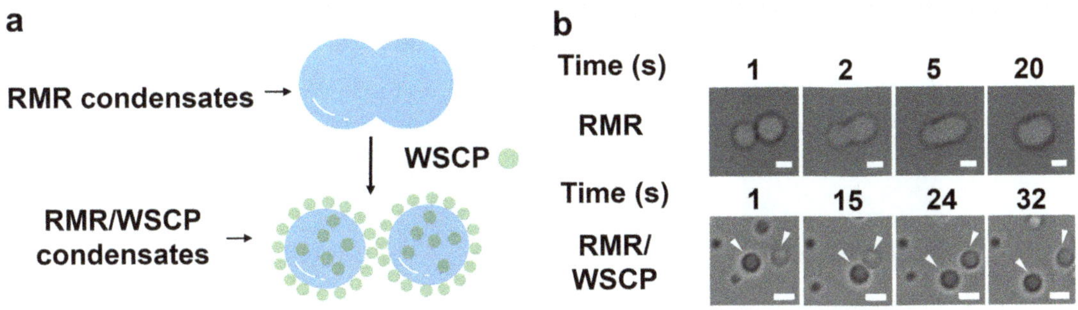

Fig. 4 *Modulation of RMR condensates with WSCP.* (**a**) Schematic illustration of the recruitment of WSCP into the condensates. (**b**) Representative images showing RMR condensates in the absence and presence of WSCP at 23 °C over time. WSCP decreases the surface tension of the RMR condensates, preventing the coalescence of two condensates. Scale bars: 1 μm. (Reproduced from Ref. [14] with permission from Nature Publishing Group)

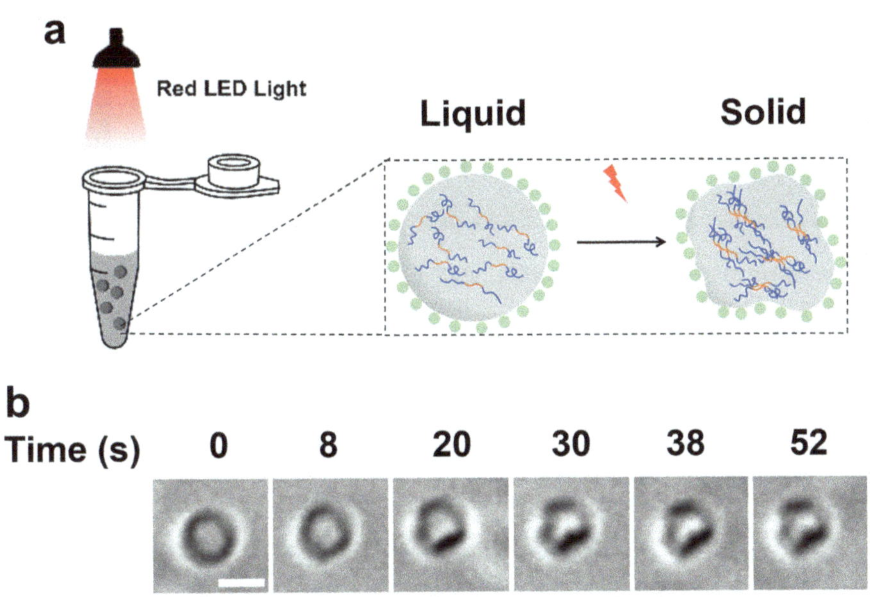

Fig. 5 *Red-light-induced liquid-to-solid phase transition of RMR/WSCP condensates.* (**a**) Schematic illustration of the red light-induced liquid-to-solid phase transition. (**b**) Representative images of a protein RMR/WSCP condensate at 23 °C under red light illumination, showing the condensate transitioning from liquid to solid over time. Scale bar: 1 μm. (Reproduced from Ref. [14] with permission from Nature Publishing Group)

3.6 Preparation of Spinach Extracts

1. Wash 500 g of fresh spinach leaves. Briefly heat the spinach leaves in boiling water for 1 min. Dry the spinach leaves using paper towels, followed by the addition of 800 mL of acetone. Stir the mixture at room temperature in the dark for 4 h.
2. Filter the resulting dark green supernatant using filter papers. Remove acetone using a rotary evaporator. Lyophilize the concentrated solution to obtain a black, hygroscopic solid.

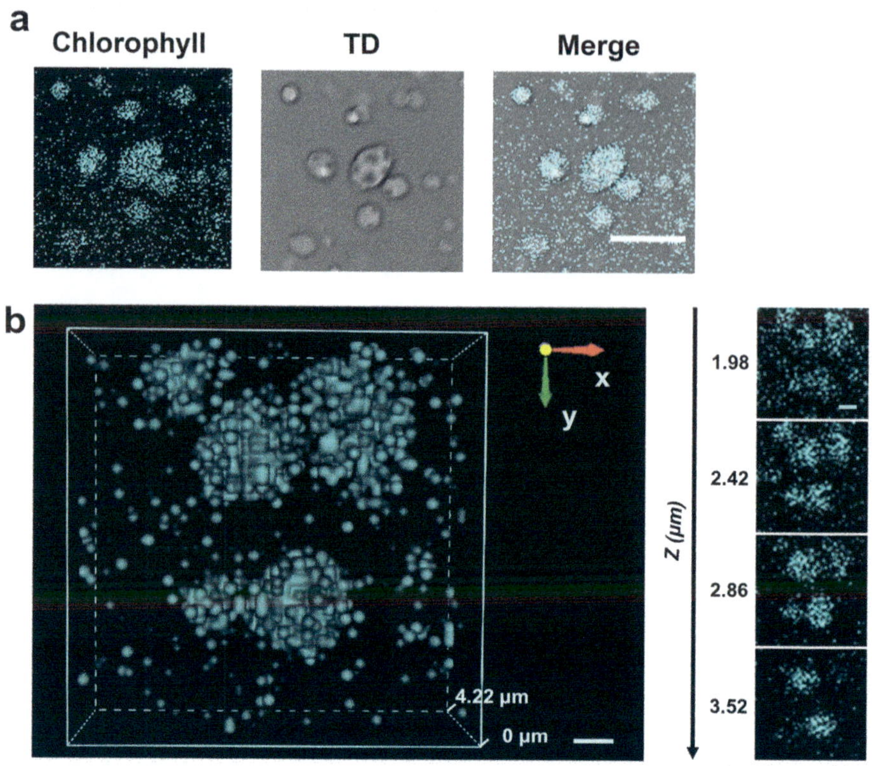

Fig. 6 *Representative confocal images of RMR/WSCP condensates.* (**a**) WSCP enrichment in RMR condensates. For fluorescence imaging, λ_{ex}, 405 nm. λ_{em}, 640 nm. *TD* transmitted detector image. Scale bar: 10 μm. (**b**) Representative 3D rendering and z-slice images of RMR/WSCP condensates. Scale bar: 1 μm. WSCPs are present on both the surface and interior of the RMR condensates. (Reproduced from Ref. [14] with permission from Nature Publishing Group)

3.7 Delivery of Chlorophyll into Mammalian Cells

1. Dissolve the freeze-dried spinach extract in DMSO to a concentration of 10 mg/mL. Filter the solution with a PTFE filter before use.
2. Add 50 μL spinach extract in DMSO to 2 mL DMEM supplemented with 10% (v/v) FBS and 1% (v/v) penicillin-streptomycin solution. Wash HEK293 cells with PBS in a 35-mm glass bottom dish. Add the prewarmed DMEM (37 °C) supplemented with the spinach extract to the cells (approximately 70% confluency). Incubate the cells under standard culturing conditions for 6 h.
3. After the 6-h incubation, wash the cells twice with PBS to remove any remaining spinach extract. Add fresh prewarmed DMEM to the cells. The cells are now ready for imaging or transfection (Fig. 7).

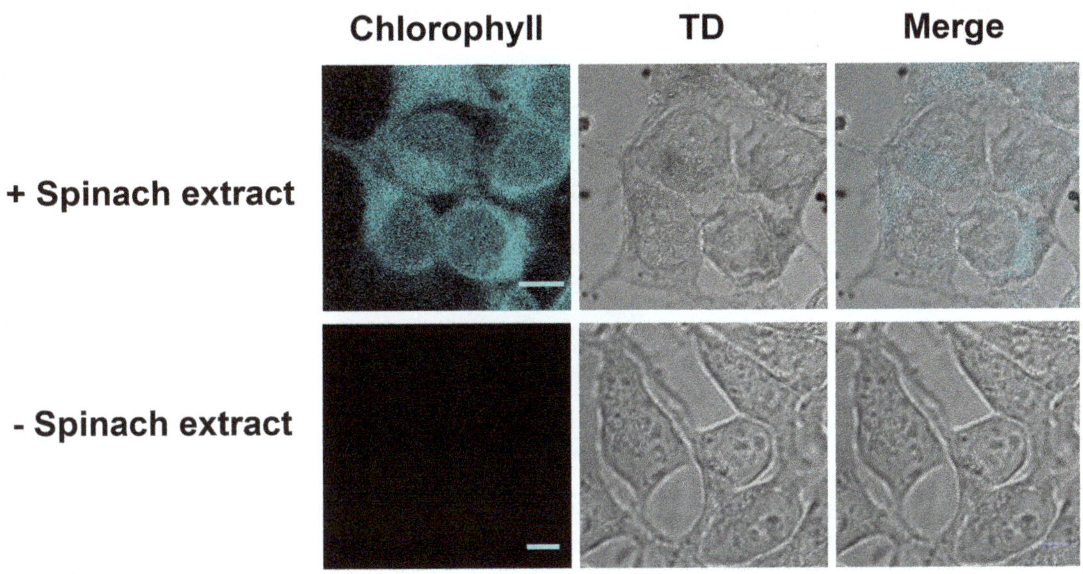

Fig. 7 *Delivery of chlorophyll into HEK293 cells.* The cells were treated with spinach extract (0.25 mg/mL) in DMEM for 6 h and imaged under standard cell culture conditions. Scale bar: 10 μm. (Adapted from Ref. [14] with permission from Nature Publishing Group)

3.8 Expression of Chlorophyll-Bound WSCP in Mammalian Cells

1. Dissolve the spinach extract in DMSO to a concentration of 1 mg/mL. Filter the spinach extract solution using a PTFE filter before use.
2. Add 40 μL of the spinach extract solution to 2 mL of DMEM supplemented with 10% (v/v) FBS and 1% (v/v) penicillin-streptomycin solution. Wash HEK293 cells with PBS in a glass-bottom dish. Add the prewarmed DMEM (37 °C) supplemented with the spinach extract to the cells (approximately 70% confluency). Incubate the cells under standard culturing conditions for 6 h.
3. Use the Lipofectamine 3000 reagent to transfect 1 μg of pEGFP::*WSCP-NLS-egfp* into the cells (*see* **Note 4**).
4. Culture the transfected HEK293 cells in the DMEM supplemented with the spinach extract for 14–16 h before imaging.
5. Image and determine the distribution of chlorophylls and WSCP inside the cells via fluorescence confocal microscopy (Fig. 8).

3.9 Production of RMR Condensates in Cellulo

1. Culture HEK293 cells in the DMEM supplemented with 10% (v/v) FBS and 1% (v/v) penicillin-streptomycin solution. Maintain the cells under a 5% CO_2 atmosphere at 37 °C [14]. Culture approximately 20,000 cells for transfection.
2. Add ~1 μg of pCAGGS::*RGG-mfp-3-RGG-mCherry* into the HEK293 cells in a glass-bottom dish. Use the Lipofectamine

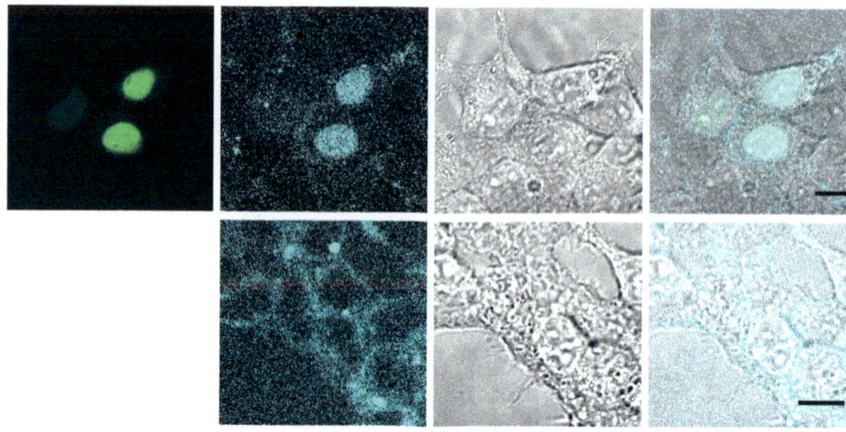

Fig. 8 *Colocalization of chlorophyll and WSCP in HEK293 cells.* Chlorophyll was predominantly co-localized with WSCP-NLS-EGFP in the nucleus, suggesting successful reconstitution of the WSCP/chlorophyll complex in living cells. In cells lacking WSCP-NLS expression, chlorophyll was mainly localized in the cytoplasm. *NLS* nuclear localization sequence. Scale bar: 10 μm

3000 reagent according to the manufacturer's protocol for transfection.

3. Culture the transfected HEK293 cells for 14–16 h before imaging.
4. Incubate the cells in an incubator during imaging. Image RMR-mCherry condensates using a fluorescence confocal microscope (Fig. 9).
5. Perform the Fluorescence Recovery After Photobleaching (FRAP) assay with the 561-nm laser from the confocal microscope (Fig. 10).

3.10 Photo-Induced Liquid-to-Solid Phase Transition in Cellulo

1. Transfect HEK293 cells with ~1 μg of pCAGGS::*RGG-mfp-3-RGG-mCherry* and ~1 μg of pEGFP::*WSCP-egfp*. Use the Lipofectamine 3000 reagent according to the manufacturer's instructions for transfection.
2. Culture the transfected HEK293 cells harboring chlorophylls for 14–16 h before imaging.
3. Use a confocal microscope to visualize and determine the localization of RMR-mCherry, WSCP-EGFP, and chlorophylls.
4. Irradiate the transfected HEK293 cells with the 405-nm laser beam from the confocal microscope for 5 min to induce the phase transition of RMR condensates in the cells (*see* **Note 5**).
5. Perform the FRAP assay using the 561-nm laser from the confocal microscope (Fig. 11) (*see* **Note 6**).

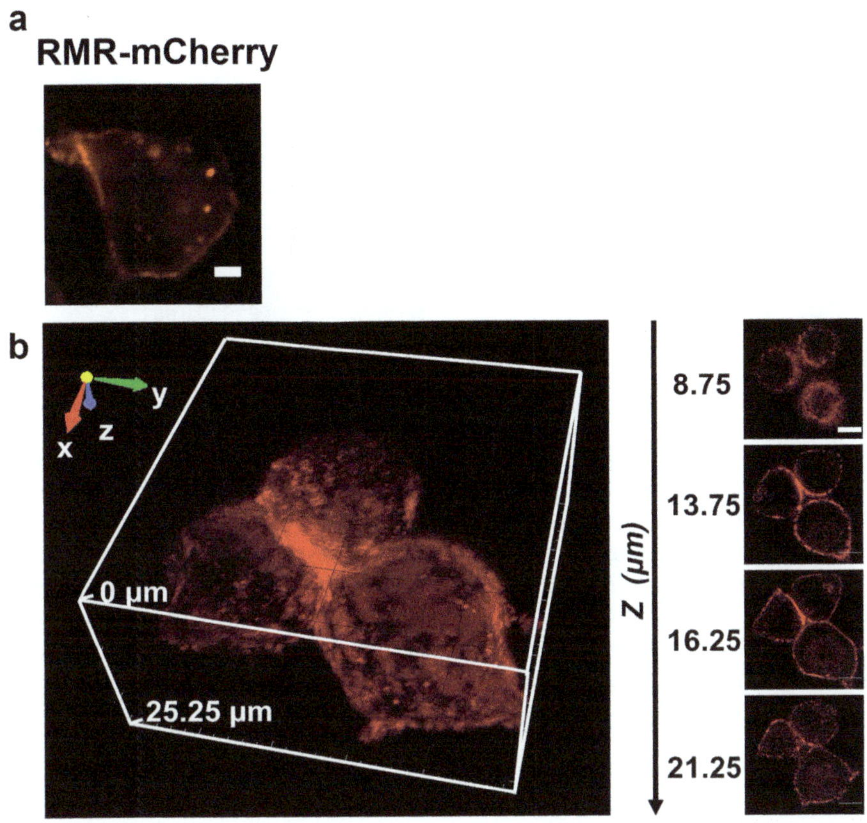

Fig. 9 *RMR phase separation in HEK293 cells.* (**a**) Fluorescent image of an HEK293 cell transfected with the gene encoding RGG-Mfp-3-RGG-mCherry (RMR-mCherry). Scale bar: 5 μm. (**b**) Representative 3D rendering and z-slice images of HEK293 cells producing RMR-mCherry. RMR is condensed not only in the cytoplasm but also along the cell membrane. Scale bar: 10 μm. (Reproduced from Ref. [14] with permission from Nature Publishing Group)

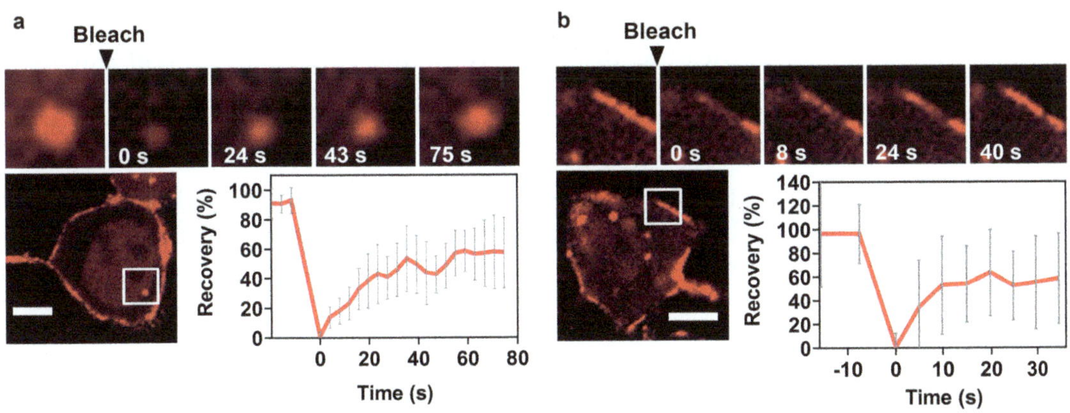

Fig. 10 *Fluorescence recovery after photobleaching (FRAP) assay of RMR condensates in the absence of WSCP in the cytoplasm* (**a**) *and along the cell membrane of HEK293 cells* (**b**). The images show the fluorescence of RMR condensates before and after photobleaching. The plots show the normalized fluorescence recovery after bleaching. Scale bar: 10 μm. (Reproduced from Ref. [14] with permission from Nature Publishing Group)

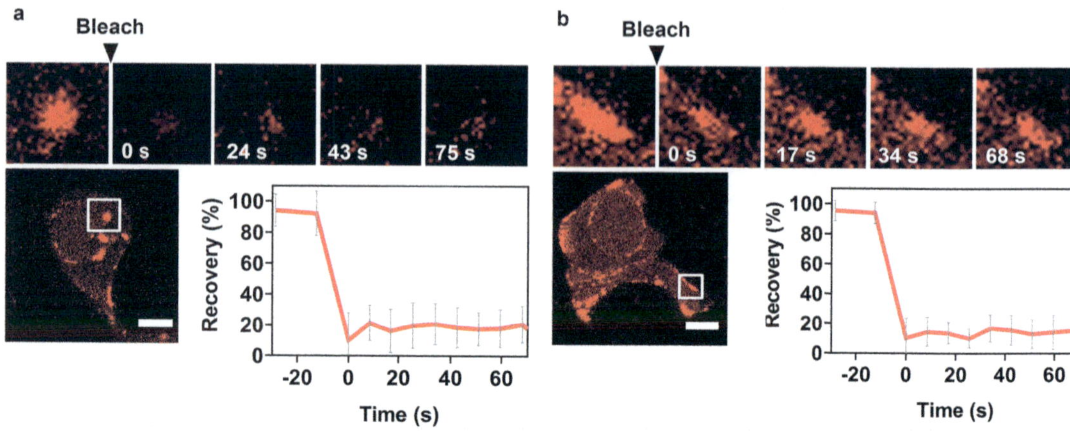

Fig. 11 *FRAP assays of RMR/WSCP condensates in the cytoplasm* (**a**) *and along the cell membrane of HEK293 cells* (**b**). The images show the fluorescence of RMR/WSCP condensates before and after photobleaching. The plots show the normalized fluorescence recovery after bleaching. After irradiation, compared to those in the absence of chlorophyll and WSCP (with ~60% fluorescence recovery) (Fig. 10), the condensates in the presence of WSCP-EGFP and chlorophyll barely recovered their fluorescence (<20%) after 1 min. Scale bar: 10 μm. (Reproduced from Ref. [14] with permission from Nature Publishing Group)

4 Notes

1. Set the FPLC at a mode of manual load and a flow rate of 5 mL/min. Keep A280 and pH monitoring "ON" and flow restrictor "OFF." Preprogram the FPLC by including an equilibration step with one column volume (CV) of the Lysis Buffer and 10 CVs of the Wash Buffer, and an elution step with 3 CVs of the Elution Buffer, linearly increased from 5% to 100%, and 2 CVs of the 100% Elution Buffer.
2. Measure the intensity of the LED light at 630 nm using the THORLABS Photometer. Ensure that the measured light intensity is less than the recommended threshold, 10 mW/cm^2.
3. Load the purified RMR protein onto a glass slide within a dual-beam optical tweezer system, the latter of which is equipped with an inverted microscope and an additional 532-nm laser source. Utilize the 532-nm laser to trigger the liquid-to-solid phase transition of protein condensates.
4. Nuclear localization sequence (NLS).
5. Use the THORLABS Photometer to determine the intensity of the laser beam before the photo-induced phase transition of RMR condensates in living cells.
6. Perform the FRAP assay to evaluate the material properties of RMR condensates in living cells [14]. Use the 561-nm laser

beam (100% power and 10-s irradiation) to quench the fluorescence. Quantify the fluorescence intensity using the NIS-Elements software.

Appendix

1. Amino acid sequence of RGG-Mfp-3-RGG.

MKGSSHHHHHHVEASESNQSNNGGSGNAALNRGGRY
VPPHLRGGDGGAAAAASAGGDDRRGGAGGGGYRRGGGN
SGGGGGGGYDRGYNDNRDDRDNRGGSGGYGRDRNYEDR
GYNGGGGGGGNRGYNNNRGGGGGGYNRQDRGDGGSSN
FSRGGYNNRDEGSDNRGSGRSYNNDRRDNGGDGVDGGG
GSELADYYGPKYGPPRRYGGGNYNRYGRRYGGYKGWNNG
WKRGRWGRKYYTSGGGGSLEESNQSNNGGSGNAALNRGG
RYVPPHLRGGDGGAAAAASAGGDDRRGGAGGGGYRRGG
NSGGGGGGGYDRGYNDNRDDRDNRGGSGGYGRDRNYE
DRGYNGGGGGGGNRGYNNNRGGGGGGYNRQDRGDGG
SSNFSRGGYNNRDEGSDNRGSGRSYNNDRRDNGGDGLE.

2. Amino acid sequence of LvWSCP.

INDEEPVKDTNGNPLKIETRYFIQPASDNNGGGLVPAN
VDLSHLCPLGIVRTSLPYQPGLPVTISTPSSSEGNDVLTNTN
IAITFDAPIWPCPSSKTWTVDSSSEEKYIITGGDPKSGESFFRI
EKYGNGKNTYKLVRYDNGEGKSVGSTKSLWGPALVLNDD
DDSDENAFPIKFREVDT.

Acknowledgments

This work received funding support from the Natural Science Foundation of China Excellent Young Scientists Fund to F.S. (22122707), from the Ministry of Science and Technology to F.S. (2020YFA0908100), from the Science, Technology, and Innovation Commission of Shenzhen Municipality to F.S. (Shenzhen-Hong Kong-Macau S&T Program (Category C) #SGDX2020110309460101), from the Natural Science Foundation of Fujian Province (2023J011395), from the Research Grants Council of Hong Kong SAR Government to F.S. (General Research Fund #16103421; Theme-based Research Scheme #T13-602/21-N), from the Hong Kong Innovation and Technology Commission—Midstream Research Program for Universities—Standalone Project (MRP/012/20) are acknowledged.

References

1. Jang G-J, Jang J-C, Wu S-H (2020) Dynamics and functions of stress granules and processing bodies in plants. Plan Theory 9(9):1122
2. Zeng M, Shang Y, Araki Y et al (2016) Phase transition in postsynaptic densities underlies formation of synaptic complexes and synaptic plasticity. Cell 166(5):1163–75 e12. https://doi.org/10.1016/j.cell.2016.07.008
3. Correction for Jang et al (2019) Processing bodies control the selective translation for optimal development of Arabidopsis young seedlings. Proc Natl Acad Sci USA 116(25):12575. https://doi.org/10.1073/pnas.1908132116
4. Arendt T, Stieler JT, Holzer M (2016) Tau and tauopathies. Brain Res Bull 126:238–292. https://doi.org/10.1016/j.brainresbull.2016.08.018
5. Patel A, Lee HO, Jawerth L et al (2015) A liquid-to-solid phase transition of the ALS protein FUS accelerated by disease mutation. Cell 162(5):1066–1077. https://doi.org/10.1016/j.cell.2015.07.047
6. Hofweber M, Hutten S, Bourgeois B et al (2018) Phase separation of FUS is suppressed by its nuclear import receptor and arginine methylation. Cell 173(3):706–19.e13. https://doi.org/10.1016/j.cell.2018.03.004
7. Zbinden A, Perez-Berlanga M, De Rossi P et al (2020) Phase separation and neurodegenerative diseases: a disturbance in the force. Dev Cell 55(1):45–68. https://doi.org/10.1016/j.devcel.2020.09.014
8. Jawerth L, Fischer-Friedrich E, Saha S et al (2020) Protein condensates as aging Maxwell fluids. Science 370(6522):1317–1323. https://doi.org/10.1126/science.aaw4951
9. Shin Y, Berry J, Pannucci N et al (2017) Spatiotemporal control of intracellular phase transitions using light-activated optoDroplets. Cell 168(1–2):159–71 e14. https://doi.org/10.1016/j.cell.2016.11.054
10. Lyu S, Fang J, Duan T et al (2017) Optically controlled reversible protein hydrogels based on photoswitchable fluorescent protein Dronpa. Chem Commun (Camb) 53(100): 13375–13378. https://doi.org/10.1039/c7cc06991j
11. Palm DM, Agostini A, Tenzer S et al (2017) Water-soluble chlorophyll protein (WSCP) stably binds two or four chlorophylls. Biochemistry 56(12):1726–1736. https://doi.org/10.1021/acs.biochem.7b00075
12. Bouargalne Y, Raguénès-Nicol C, Guilbaud F et al (2022) New insights into chlorophyll-WSCP (water-soluble chlorophyll proteins) interactions: the case study of BnD22 (*Brassica napus* drought-induced 22 kDa). Plant Physiol Biochem 181:71–80. https://doi.org/10.1016/j.plaphy.2022.03.023
13. Schuster BS, Reed EH, Parthasarathy R et al (2018) Controllable protein phase separation and modular recruitment to form responsive membraneless organelles. Nat Commun 9(1): 2985. https://doi.org/10.1038/s41467-018-05403-1
14. Li M, Park BM, Dai X et al (2022) Controlling synthetic membraneless organelles by a red-light-dependent singlet oxygen-generating protein. Nat Commun 13(1):3197. https://doi.org/10.1038/s41467-022-30933-0
15. Sun F, Zhang W-B, Mahdavi A et al (2014) Synthesis of bioactive protein hydrogels by genetically encoded SpyTag-SpyCatcher chemistry. Proc Natl Acad Sci 111(31): 11269–11274
16. Takahashi S, Uchida A, Nakayama K et al (2014) The C-terminal extension peptide of non-photoconvertible water-soluble chlorophyll-binding proteins (class II WSCPs) affects their solubility and stability: comparative analyses of the biochemical and chlorophyll-binding properties of recombinant Brassica, Raphanus and Lepidium WSCPs with or without their C-terminal extension peptides. Protein J 33(1):75–84. https://doi.org/10.1007/s10930-013-9539-5
17. Takahashi S, Yanai H, Oka-Takayama Y et al (2013) Molecular cloning, characterization and analysis of the intracellular localization of a water-soluble chlorophyll-binding protein (WSCP) from Virginia pepperweed (*Lepidium virginicum*), a unique WSCP that preferentially binds chlorophyll b in vitro. Planta 238(6): 1065–1080. https://doi.org/10.1007/s00425-013-1952-7
18. Green MR, Sambrook J (2020) The Inoue method for preparation and transformation of competent *Escherichia coli*: "ultracompetent" cells. Cold Spring Harb Protoc 2020(6): 101196

Chapter 16

Automated Plate Reader-Based Assays of Light-Activated GPCRs

Elliot J. Gerrard, Alexandra-Madelaine Tichy, and Harald Janovjak

Abstract

In the emerging field of optogenetics, light-sensitive G-protein coupled receptors (GPCRs) allow for the temporally precise control of canonical cell signaling pathways. Expressing, stimulating, and measuring the activity of light-sensitive GPCRs (e.g., opsins or chimeric OptoXRs) in mammalian cells is a nontrivial task as many standard assay practices are not compatible with light-sensitive molecular tools. In this chapter, we present a method for quantifying opsin activity in automated plate reader-based assays without the need for additional optical hardware (i.e., light sources). The protocol is applied to assess cAMP levels downstream of a chimeric OptoXR but can be expanded to other opsins and second messengers, such as Ca^{2+} mobilization. We describe how the internal optical components in commonly available plate readers can be utilized to both activate and detect kinetic and dose–response relationships, as well as provide general guidance for optimizing assays with light-sensitive molecular tools.

Key words Opsin, OptoXR, Optogenetics, Cell signaling, Automation, Synthetic biology

1 Introduction

The emerging field of optogenetics encompasses a broad range of methodologies that aim at rendering physiological processes controllable by light [1]. This is typically achieved through expression of light-sensitive proteins that convert photons into bioelectrical or biochemical signals that influence the behavior of cells. One of the largest families of receptors that control cell state are GPCRs, which sense a diverse range of stimuli including ions, hormones, and neurotransmitters [2]. Given their ubiquitous nature, GPCRs are a common target for established and emerging drugs [3]. Rendering GPCR signaling controllable by light is achieved either through (i) naturally occurring light-sensitive GPCRs, that is, invertebrate and vertebrate opsins, or (ii) fusions of the light-sensitive domain of an opsin with the signaling domain of a GPCR of interest (OptoXR chimeras [4–6]). Collectively, these technologies provide control

Armin Baumschlager (ed.), *Optogenetics: Methods and Protocols*, Methods in Molecular Biology, vol. 2840,
https://doi.org/10.1007/978-1-0716-4047-0_16,

over the main canonical signaling pathways, which are typically the rise or fall of cAMP levels (associated with $G\alpha_s$ or $G\alpha_{i/o/t}$ activation) and release of intracellular Ca_{2+} (associated with $G\alpha_q$ activation).

The real-time measurement of GPCR activity in cells is well established with a myriad of methods capable of reporting proximal and amplified signaling activity through luminescent, fluorescent, or FRET/BRET-based reporters [7, 8]. These are typically assayed in the 96-well plate format using commonly available plate readers allowing for relatively high throughput. However, quantifying opsin activity in plate reader assays is nontrivial, owing to specific considerations relevant to working with light-sensitive receptors. First, animal opsins, unlike some of their distant relatives from microbes/bacteria (i.e., channelrhodopsins), are extremely light sensitive. Consequently, assays must be carried out in complete darkness or under red light. Second, the delivery of the light stimulus must not only be temporally precise (as activation is near instant) but spatially precise to avoid unintentional activation of nearby wells. If this light stimulus is delivered separately to the plate reader optical path, precise synchronization is necessary to avoid inter-replicate errors or missing the peak of fast kinetic response [9, 10]. Third, a further consideration is the selection of the reporter system, with luminescent tools typically preferred over fluorescent reporters as excitation light may activate opsins in undesired ways [9, 11, 12]. Here we present a set of methods that utilize the optical components of a plate reader to both activate and measure OptoXRs in mammalian cells in an automated manner (Fig. 1). The fluorescence filter sets or, alternatively, monochromators in plate readers allow for spectrally defined and programmable light stimulation that is sufficiently bright to activate opsins. We utilize custom scripts within the plate reader software for automated reading of a luminescent reporter before and after light stimulation. These protocols can then be adapted to measure signaling cascades that are very fast (e.g., intracellular Ca^{2+} release, *see* **Note 1**) or slow (e.g., changes in cAMP levels) in high throughput formats. An additional forskolin injection step can also be incorporated prior to light stimulation to use this assay to measure $G\alpha_{i/o/t}$ activation, again using built-in plate reader hardware capabilities. Finally, we package this protocol with a series of recommendations to optimize the expression and measurement of OptoXRs in mammalian cells.

2 Materials

2.1 Cell Culture

1. HEK293 cells (or other mammalian cell lines suitable for heterologous expression of mammalian receptors).
2. Temperature- and CO_2- controlled cell incubator, humidified.
3. Fetal bovine serum (FBS).

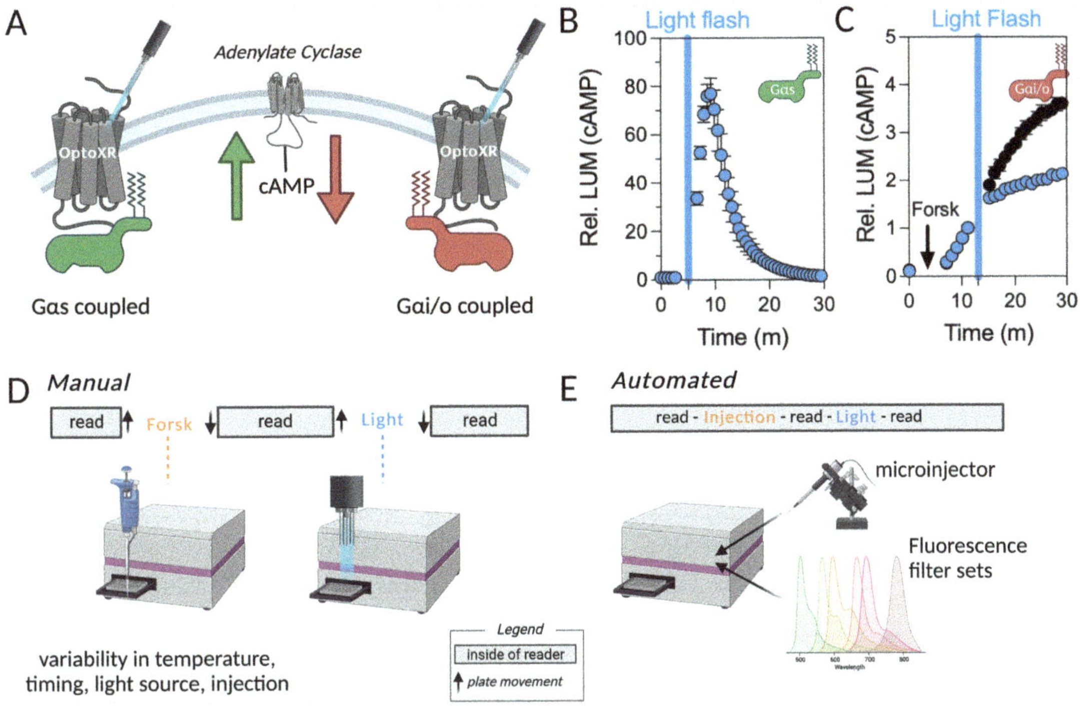

Fig. 1 *Advantages of automated plate reader-based assay.* (**a**) Overview of the two main G-protein coupling pathways ($G\alpha_{i/o/t}$ and $G\alpha_s$) regulating adenylate cyclase to influence intracellular cAMP levels. (**b**) Example data of cAMP induction in HEK293 cells transfected with a cAMP reporter, GloSensor, and the $G\alpha_s$ coupled opsin JellyOp. Cells were exposed to 200 flashes of light with a maximum wavelength of 470 nm. (**c**) Example data of cAMP suppression in HEK293 cells transfected with the same cAMP sensor but also with rhodopsin ($G\alpha_{i/o/t}$ coupled). In this experiment, cells were first treated with 1 μM forskolin before light stimulation. Black dots indicate cells not stimulated with light. Data are mean +/− SEM of three independent experiments. (**d**, **e**) Schematic of automated compared to manual assay formats

4. Dulbecco's modified Eagle medium (DMEM).
5. Trypsin (0.05% solution, with EDTA).
6. Phosphate buffered saline, sterile (PBS).
7. Leibovitz's L-15 medium (L-15).
8. 96-well, white, opaque bottom plates with lid.

Head torch with red-light setting (commonly available from hiking, biking, or home improvement stores, *see* **Note 2**).

2.2 DNA Constructs and Transfection Reagents

1. OptoXR (typically subcloned into a mammalian expression vector, for example, pcDNA3.1, *See* **Note 3**).
2. GloSensor cAMP reporter (Promega, or a synthetic gene subcloned into a mammalian expression vector, for example, pcDNA3.1, *See* **Note 4**).

3. Polyethyleneimine HCl Max (PEI Max, Polysciences) (or PEI of a different provider, or other suitable transfection reagent).
4. 150 mM Sodium Chloride, sterile filtered.
5. 1x Poly-*D*-lysine (PDL, optional, *see* **Note 5**).

2.3 Assay Reagents

1. D-luciferin (Cayman Chemical).
2. Forskolin, at 1 mM (1000x) dissolved in ethanol (Sigma-Aldrich Merck).
3. 9-*cis* retinal, at 10 mM (1000x) dissolved in ethanol in single-use aliquots, protected from light (Sigma-Aldrich Merck, *see* **Note 6**).

2.4 Plate Reader

1. A commonly used plate reader (e.g., FLUOstar Omega, BMG Labtech) capable of luminescence (broad emission) and fluorescence (spectrally defined excitation and emission) measurements. In addition, software must allow some form of scripting (i.e., the definition of a sequence of protocol events), either in a "wizard" or through a text editor.
2. For spectral sensitivity measurements, the plate reader requires a series of spectrally distinct wavelength filters that do not overlap with each other; or, alternatively, a monochromator-based optics system.
3. For automated forskolin/digitonin/ionomycin injection, microinjectors are capable of delivering up to 10 μL/well at a speed <100 μL/s (*see* **Note** 7).
4. External data analysis software (e.g., GraphPad Prism Version 9).

3 Methods

3.1 Cell Culture and Transfection

This subheading describes a typical transfection protocol using PEI (Max) aimed at inducing a high level of expression of an OptoXR and expression of complimentary second messenger sensor. Transfection efficiency should be tested and optimized in each laboratory for the individual cell lines; the following is a general protocol for seeding and transfecting a full 96-well plate with a single OptoXR and the cAMP sensor GloSensor (*see* **Notes 8** and **9** regarding transfection optimization).

1. In a sterile 15 mL falcon tube, mix 2500 ng of OptoXR expression plasmid and 2500 ng of GloSensor plasmid with 1 mL of 150 mM NaCl. In a separate tube, add 30 μL of PEI to 1 mL of 150 mM NaCl. Incubate at room temperature for 5 min.

2. Add the PEI/NaCl mix to the DNA/NaCl mix and invert 5 times to mix. Incubate at room temperature for 20 min.
3. During this incubation, detach, resuspend, and dilute HEK293 cells to a density of 1.25 million cells/mL in DMEM containing 10% FBS.
4. Add 2 mL of PEI/DNA mix to 8 mL of diluted cells and invert to mix thoroughly.
5. Seed 100 μL/well, regularly remixing through inversion. Avoid seeding wells directly next to each other on the plate (*see* **Note 10**).
6. Incubate for 24 h in DMEM containing a minimum of 5% FBS at 37 °C in the humidified incubator supplied with 5% CO_2.

   The following steps are to be performed under dim red light.

   This can be achieved using bike light/headtorches with an RGB LED (*see* **Note 2**).
7. After 24 h, under dim red light, aspirate media from the plate and replace with 100 μl DMEM containing 5% FBS and 10 μM 9-*cis* retinal.
8. Wrap the plate in aluminum foil and incubate overnight at 37 °C in a humidified incubator supplied with 5% CO_2.

3.2 Substrate Addition

1. For cAMP/GloSensor assays, replace media on cells with 100 μL L-15 medium containing *D*-luciferin diluted to a final concentration of 2 mM.
2. Incubate the cells at room temperature for 1 h, protected from light using aluminum foil.

3.3 Plate Reader Set Up

3.3.1 General Format

The general format of the protocol is to have the plate reader perform three back-to-back segments using a custom script in order to read, stimulate, and then read from a number of wells without the need for user interference. Light flashes are delivered to the cells via a fluorescence intensity protocol, which can be tuned both in spectral and intensity dimensions and is flanked by two luminescence protocols.

On most devices, the three segments must be created as new separate protocols and are as follows:

Segment 1: Baseline measurement—Luminescence (Plate Mode). This protocol measures the baseline luminescence for a minimum of five cycles. This protocol can include an optional fluid injection step of forskolin (see Subheading 3.3.3).

Segment 2: Light stimulation—Fluorescence Intensity (Well Mode). This protocol is used to stimulate relevant wells with light using a fluorescence intensity setting. Light wavelength can be specified using the available filter sets. The "dose" of

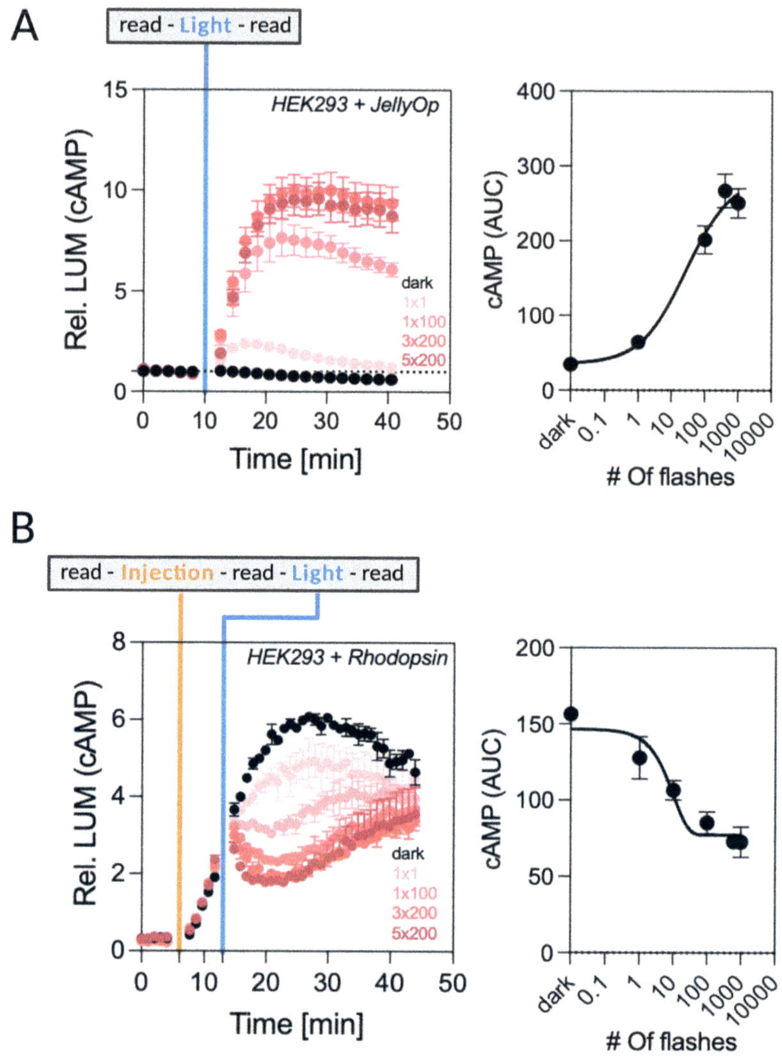

Fig. 2 *Example data for constructing dose–response curves with automated plate reader assays.* (**a**) Dose–response curve generated from stimulating HEK293 cells expressing JellyOp and GloSensor with increasing flashes of light (λ_{max}: 485 nm). (**b**) Dose–response curve generated from stimulating HEK293 cells expressing rhodopsin (a G$\alpha_{i/o/t}$ coupled opsin) and GloSensor with increasing flashes of light (λ_{max}: 485 nm). Data are mean +/– SEM of a minimum of three independent experiments

light can be adjusted by the number of flashes (1–200) and number of intervals (1–1000), reaching a maximal possible dose of 200,000 flashes of light per well and per protocol. For producing dose–response curves, multiple protocols will need to be executed in sequence specifying the amount of light to be applied in each relevant well (see Subheading 3.3.5, *See* **Note 11**, Fig. 2). The data acquired during this fluorescence segment is irrelevant and not included in the analysis.

Protocol 3: Post-stimulation measurement—Luminescence (Plate Mode). This protocol then measures the luminescence after light stimulation.

To array these protocols into a single loop, select "create new protocol from script wizard" on the BMG LabTech reader control software and specify the three protocols to be executed as "<st1>," "<st2>," and "<st3>." Specify the number of repeats of the loop and then confirm. This protocol will now be selectable as a button on the main screen.

Below are specific protocols to measure the main signaling pathways that govern cAMP increase/decrease and how to adjust them to measure different features (kinetics or dose–responses) (*see* **Note 12**). This protocol can be adapted to measure Gα_q coupled receptors but requires alternate cell transfection, light stimulation parameters, and positive controls (*see* **Notes 13** and **14**).

3.3.2 Gα_s Measurement (cAMP Increase)

Protocol 1—Baseline (Luminescence, Plate Mode).

Number of cycles: 5.

Interval: 120 s.

Number of wells: up to 48.

Gain: Reader dependent (*see* **Note 15**).

Protocol 2—Light stimulation (Fluorescence, Well Mode).

Protocol 3—Post-light stimulation (Luminescence, Plate Mode).

Number of cycles: 20.

Interval: 120 s.

Number of wells: up to 48.

3.3.3 G$\alpha_{i/o/t}$ Measurement (cAMP Decrease)

Protocol 1—Baseline and forskolin injection (Luminescence, Plate Mode).

Number of cycles: 8.

Interval: 120 s.

Number of wells: up to 48.

Injection: If the plate reader has microinjectors, preload the injectors with 10x forskolin diluted in buffer (final concentration 1 μM) and inject at a speed between 50 and 100 μL/s. Set the protocol to inject forskolin after 5 cycles of baseline (*see* **Notes 7** and **16**).

Protocol 2—Light stimulation (Fluorescence, Well Mode).

Protocol 3—Post-light stimulation (Luminescence, Plate Mode).

Number of cycles: 20.

Interval: 120 s.

Number of wells: up to 48.

3.3.4 Setting Light Stimulation Parameters

1. Determine the fluorescent filter set to use, if broad spectrum light/maximal activation is desired consider using a luminescence optic module (*see* **Note 11**). This will be dependent on the optic modules installed in the plate reader, or if the plate reader is equipped with a monochromator (*see* **Notes 17** and **18**). If spectrally defined activation is desired, consider using a narrow-bandwidth fluorescence filter set (*see* **Notes 19** and **20**). Take care to ensure that if a fluorescence filter set is used, the desired wavelength of stimulation is the excitation wavelength and not the emission wavelength. External light delivery is possible to include at this step instead of using plate reader built-in stimulation (*see* **Notes 21** and **22**).
2. Determine the number or intervals (typically 1–1000), this can be treated as a multiplier for the number of flashes (typically 1–200). A larger range of light stimulation intensities can be constructed using a combination of interval and flash numbers.

3.3.5 Adjusted for Dose–Response Curves

The previous protocols are set up to deliver a single dose of light stimulation to all wells other than those defined as "dark" (i.e., not illuminated) control wells. To acquire a dose–response curve using different intensities of light stimulation, the script for a standard assay as outlined below will need to be opened and edited in a text editor. Multiple fluorescence intensity protocols will need to be inserted between the pre-light flash baseline and post-light flash luminescence reads in order to deliver multiple doses of light to different wells. A standard experiment could use the following gradient of flashes/intervals to achieve a dose–response: 1 × 1, 5 × 1, 10 × 1, 50 × 1, 100 × 1, 200 × 1, 200 × 10, and 200 × 100. Given the slow response kinetics of cAMP assays, light flashes can be delivered sequentially to many wells before resumption of the luminescence measurement. Typically up to 48 wells can be measured in this way in a single experiment. Alternatively, the prestimulation, light stimulation, and post-stimulation measurements can be conducted in sequence for each well or a subset of wells, reducing the cycle time to read all wells and therefore giving greater kinetic resolution for faster responses.

3.4 Data Analysis

This subheading contains some general guidance for data analysis. The output data files from separate protocols can be combined within the BMG LabTech MARS analysis software into a single file for exporting.

3.4.1 Normalization

The following provides a two-step normalization process that can be performed within GraphPad Prism.

1. Normalize all values post-light stimulation to the average baseline luminescence of each individual well. This can be done through selecting "Analyze," "Remove Baseline and Column

Math." Define the baseline as selected rows and mean of the first x cycles (with x being the number of baseline cycles). Define the calculation as "Ratio:Value/Baseline" and to occur as "Repeat Measures" (*see* **Note 23**).

2. Normalize the baseline-normalized data to the average dark treated well using the same process as above but now define the baseline as selected columns: the column containing the dark control wells. Perform a subtraction calculation and use replicates to average the dark control wells.

3.4.2 Measuring Responses

Calculating a response from a light stimulation experiment can be performed in two main ways, either area under the curve (AUC) or peak response.

3.4.3 Area under the Curve (AUC)

1. Plot double normalized data in a new X-Y scatter graph with a single y value for each x value.
2. Select "Analyze," "Area under the Curve," and specify that peaks may go above or below the baseline.

3.4.4 Peak Response

1. Select "Analyze", "Column analyses", "Descriptive statistics". Ensure that "minimum and maximum, range" box is ticked.

4 Notes

1. This general assay protocol can also be used for the measurement of intracellular Ca^{2+} release induced by $G\alpha_q$ coupled OptoXRs/opsins (e.g., melanopsin). In this case, a suitable Ca^{2+} sensor should be transfected (e.g., aequorin [13] with coelentrazine-h substrate, Nanolight Technologies) or a Ca^{2+} sensitive dye added (e.g., Fluo-8-AM) prior to experimentation.
2. When using a bike light or a camping headtorch, red light for light-sensitive experiments should be provided by a "pure" red or RGB LED rather than a white LED filtered through a red plastic casing. RGB LED strips may also be mounted in cell culture rooms to produce red light. It is advisable to turn all other lights off and lower blinds covering windows.
3. To express an OptoXR in a mammalian cell, the coding sequence for the receptor is typically subcloned into a mammalian expression vector (such as pcDNA3.1) preceded (in frame) by a signal peptide and followed by an optional epitope tag (e.g., FLAG- or HA-tags). The start codon is preceded by a Kozak sequence, such as commonly GCCACC.
4. GloSensor is available to purchase from Promega but has also been cloned from synthetic gene fragments into a pcDNA3.1 vector [11].

5. For increased cell adhesion, plates may be precoated with PDL for 1 h at 37 °C or overnight at 4 °C. Wash the plates twice with sterile PBS prior to cell plating.
6. 9-*cis* retinal is a less expensive and more widely available vitamin-A chromophore isomer than 11-*cis* retinal. For best results, store in opaque tubes at −20 °C in ethanol. Be aware whenever using HEK293 cells and retinaldehyde: These cells possess intrinsic enzymes capable of converting retinaldehyde isomers, so no experiment can be considered "purely" the isomer introduced [14].
7. If using microinjectors, it is advisable to optimize the injection speed/volume using PBS to decrease the likelihood of injection artifacts. We find using speeds in excess of 100 μL/s and volumes greater than 10 μL increase the likelihood of artifacts significantly. This is particularly relevant to measuring intracellular Ca^{2+} release.
8. For optimal transfection efficiency, perform trials varying the cell number, PEI:DNA ratio, and total amount of DNA transfected. Inspect cell health after transfection by seeding some cells into a clear bottom plate, so that they can be viewed under a microscope.
9. This protocol is compatible with cell lines stably expressing OptoXR of interest. In this instance, a recommended 25 ng/well of GloSensor should be transfected into these cells and halving of the amount of PEI used. Viruses to induce constitutive expression of genes of interest are commercially available.
10. To avoid plate or kinetic artifacts, place light-stimulated and dark control wells in different areas of the plate between replicates, avoiding directly neighboring wells (i.e., A1 and B2, not A1 and A2 or A1 and B1). In a similar way, when performing a dose–response curve, ensure that multiple technical replicates and dark control wells are placed in different parts of the plate.
11. Most plate readers employ flash lamps as the light source. Thus, to acquire dose–response curves, only the number of flashes can be adjusted rather than intensity of the flash itself. Among various opsins and OptoXRs that we have studied, we did not find any that were near-maximally activated by a single flash, allowing for a dose–response curve to be constructed by varying flash number.
12. This protocol can be used to detect light-sensitive OptoXRs and opsins that couple to all G-protein pathways and second messenger signaling (e.g., JellyOp—$G\alpha_s$, melanopsin—$G\alpha_q$, or rhodopsin—$G\alpha_{i/o/t}$ [10, 15].

13. For measuring intracellular Ca^{2+} release, the kinetics of the response are too fast to be able to stimulate and record multiple wells at a time. In this case, the script can be edited to repeat a cycle of three segments for each well, using a command to edit the plate layout in between in order to measure each well in turn. This can be achieved with the following command syntax:

    R_EditLayout "<st1>" "A1=Empty A2=X1"

14. Possible positive controls for intracellular Ca^{2+} mobilization include ionomycin or digitonin. For cAMP, forskolin or isoproterenol can be used.
15. In luminescence assays, typically a high gain is used of around 80% of the maximum. Keep the gain consistent between independent replicates to reduce variability. Take care to avoid reader saturation, which may not always lead to an error message but can be identified by seeing values that match or exceed a small red arrow on the y-axis of the live view of each well during the experiment.
16. To measure the response of a $G\alpha_{i/o/t}$ coupled receptor, where a forskolin injection has preceded the light flash(es), consider adding both dark forskolin- and dark vehicle-treated wells to assess the signal window. Additionally, to measure the response of the receptor, adjust the baseline measurement to include the forskolin treatment window to normalize for inter-well variability in forskolin responses.
17. Multichromatic filters are built into common plate reader hardware in order to be compatible with commonly used fluorescence/BRET/FRET assays, and thus separate sourcing of LEDs/light sources is not required.
18. Monochromator-based plate readers should work equally well with this protocol as they have similar bandwidths as common excitation filters and the same light sources, resulting in comparable light intensities.
19. If the maximum sensitivity wavelength is not known, or wavelength-dependent information is not required, consider using a bandwidth filter for light stimulation. The peak sensitivity for most animal opsin-based receptors/OptoXRs lies between 450 and 550 nm.
20. To create an action spectrum (x-axis = wavelength, y-axis = response), use spatially segregated wavelength filters to execute dose–response curves at multiple wavelengths and plot the relative potency at each wavelength. These can then be fit with Govardovskii's nomogram equation to estimate the wavelength of maximal sensitivity (λ_{max}) [13].

21. Plate readers can also be modified to allow external light sources to directly illuminate plates whilst the plate reader is recording. In a previous example, the reader bottom optic was coupled to a liquid light guide from a pE-4000 light source (CoolLED). To avoid bleaching of the photomultiplier tube during light stimulation a motorized shutter was built and synchronized to the light source using an Arduino microcontroller [16]. Another example included an LED mounted in the solution injector port of a plate reader [17].
22. Measuring the direct light intensity emitted by the fluorescence optic module is challenging due to an inability to place large detection equipment in the plate reader and loss of light along the optic path. Some plate reader models of some manufacturers allow intensity measurement (e.g., Agilent/BioTek Synergy HTX or BMG LabTech Omega series).
23. For the vehicle subtraction in the normalization process, consider selecting "Repeat Measures" if there are dark-treated wells interspersed between light-treated wells.

References

1. Fenno L, Yizhar O, Deisseroth K (2011) The development and application of optogenetics. Annu Rev Neurosci 34:389–412. https://doi.org/10.1146/annurev-neuro-061010-113817
2. Rosenbaum DM, Rasmussen SG, Kobilka BK (2009) The structure and function of G-protein-coupled receptors. Nature 459(7245):356–363. https://doi.org/10.1038/nature08144
3. Hauser AS et al (2017) Trends in GPCR drug discovery: new agents, targets and indications. Nat Rev Drug Discov 16(12):829–842. https://doi.org/10.1038/nrd.2017.178
4. Airan RD et al (2009) Temporally precise in vivo control of intracellular signalling. Nature 458(7241):1025–1029. https://doi.org/10.1038/nature07926
5. Tichy AM, Gerrard EJ, Sexton PM, Janovjak H (2019) Light-activated chimeric GPCRs: limitations and opportunities. Curr Opin Struct Biol 57:196–203. https://doi.org/10.1016/j.sbi.2019.05.006
6. Spangler SM, Bruchas MR (2017) Optogenetic approaches for dissecting neuromodulation and GPCR signaling in neural circuits. Curr Opin Pharmacol 32:56–70. https://doi.org/10.1016/j.coph.2016.11.001
7. Zhou Y, Meng J, Xu C, Liu J (2021) Multiple GPCR functional assays based on resonance energy transfer sensors. Front Cell Dev Biol 9:611443. https://doi.org/10.3389/fcell.2021.611443
8. Zhang R, Xie X (2012) Tools for GPCR drug discovery. Acta Pharmacol Sin 33(3):372–384. https://doi.org/10.1038/aps.2011.173
9. Tichy AM, So WL, Gerrard EJ, Janovjak H (2022) Structure-guided optimization of light-activated chimeric G-protein-coupled receptors. Structure 30(8):1075–1087 e1074. https://doi.org/10.1016/j.str.2022.04.012
10. Moglich A, Moffat K (2010) Engineered photoreceptors as novel optogenetic tools. Photochem Photobiol Sci 9(10):1286–1300. https://doi.org/10.1039/c0pp00167h
11. Morri M et al (2018) Optical functionalization of human Class A orphan G-protein-coupled receptors. Nat Commun 9(1):1950. https://doi.org/10.1038/s41467-018-04342-1
12. Capek D et al (2019) Light-activated Frizzled7 reveals a permissive role of non-canonical wnt signaling in mesendoderm cell migration. elife 8. https://doi.org/10.7554/eLife.42093
13. Bailes HJ, Lucas RJ (2013) Human melanopsin forms a pigment maximally sensitive to blue light (lambdamax approximately 479 nm) supporting activation of G(q/11) and G(i/o) signalling cascades. Proc Biol Sci 280(1759):20122987. https://doi.org/10.1098/rspb.2012.2987
14. Brueggemann LI, Sullivan JM (2002) HEK293S cells have functional retinoid

processing machinery. J Gen Physiol 119(6): 593–612. https://doi.org/10.1085/jgp.20018495

15. Bailes HJ, Zhuang LY, Lucas RJ (2012) Reproducible and sustained regulation of Galphas signalling using a metazoan opsin as an optogenetic tool. PLoS One 7(1):e30774. https://doi.org/10.1371/journal.pone.0030774
16. Rodgers J et al (2021) Using a bistable animal opsin for switchable and scalable optogenetic inhibition of neurons. EMBO Rep 22(5): e51866. https://doi.org/10.15252/embr.202051866
17. Richter F et al (2015) Upgrading a microplate reader for photobiology and all-optical experiments. Photochem Photobiol Sci 14(2): 270–279. https://doi.org/10.1039/c4pp00361f

Chapter 17

Modular Light-Emitting Diode Shelving Systems for Scalable Optogenetics

Christina G. Gangemi and Harald Janovjak

Abstract

Optogenetic experiments rely on the controlled delivery of light to diverse biological systems. Impressive devices have been recently developed to stimulate cells and small animals with multiple wavelengths and intensities. However, existing hardware solutions are often limited to a single sample holder, and their design and cost can further limit scalability. This chapter describes an illumination system that is modular (through the use of accessible components) and scalable (through a shelving structure and low cost). Assembly and operation require no or minimal electrical engineering or programming expertise. Multi-intensity wavelength and temporal experiments can be performed in dozens of small and large samples. This chapter also introduces methods for temperature and light intensity measurements toward appropriate illumination conditions. This work aims to provide a greater level of accessibility and complementary opportunities for large-scale optogenetics in a broad range of biological samples.

Key words Optogenetics, Optical control, Optics, Illumination hardware, LED, Synthetic biology

1 Introduction

In the emerging field of optogenetics, light-sensitive proteins are delivered into genetically targeted cells ex vivo or in vivo to optically regulate their function and behavior with unprecedented spatio-temporal precision. The success of optogenetic experiments is linked to the availability of tailored light stimulation hardware. For instance, optogenetics in rodents is commonly achieved using implanted light fibers or wireless devices [1, 2], whereas in vivo experiments in zebrafish or *Drosophila*, or ex vivo experiments in cells, employ external light sources, often in microscopes or using light-emitting diodes (LEDs) in a wide variety of configurations (e.g., as arrays or "boxes") [3–9].

Armin Baumschlager (ed.), *Optogenetics: Methods and Protocols*, Methods in Molecular Biology, vol. 2840,
https://doi.org/10.1007/978-1-0716-4047-0_17,

General requirements in external (i.e., non-implantable) illumination hardware are flexibility and compatibility with the experimental system. For instance, in order to determine the optimal light protocol for a new optogenetic method, evaluation of multiple parameters, in particular light intensities and temporal patterns, is required. Furthermore, illumination devices need to be functional in the appropriate environments, such as in an incubator housing a specific type of tissue culture plate. General challenges in device design and operation are undesired heating of samples and the need to precisely control the amount of delivered light. Some of these challenges are met by LED arrays that are often elaborate. The optoPlate is appropriate for samples in 96- or 384-well plates and particularly useful if more than one wavelength is applied to individual wells [3]. The LED illumination tool for optogenetic stimulation (LITOS) and a light activation at variable amplitudes (LAVA) device likewise employs a single LED array and can be used for a large variety of culture plates, dishes, and flasks [5, 9]. Complex illumination patterns can be designed without extensive coding knowledge. Commercial solutions include the optoWELL (Opto Biolabs GmbH) [10] and Lumos (Axion Biosystems, Inc.) [11], which also offer multi-color illumination, customizable pulsing protocols, and single well control.

Complementary to existing technologies, this chapter describes an easy-to-use LED array shelving system (Fig. 1). A key property of this system is modularity as its electrical components, such as dimmers, timers, and LED arrays, can be easily reconfigured with minimal or no requirements for soldering or programming. In particular, LED arrays and sample holders in shelves can be quickly exchanged between experiments aimed at any plate or dish type (Fig. 1a, b). A further key property of this system is scalability. Changing the number of shelves and/or combining multiple shelving systems increases the total number of LED arrays in an experiment (Fig. 1c). For instance, on a laboratory bench of 1 m length, nine stacked shelving systems would allow for the illumination of up to 216 wells (in the 12-well plate format) containing small animals or cells, ideal for experiments where large sample numbers are required. The total parts cost per system is also very low (~250 Australian Dollar or 160 Euro). While we here demonstrate application in a tissue culture incubator, the system is also suited for samples that require other growth conditions (e.g., organoids, flies, or zebrafish), and it can be used to validate new optogenetic tools in long-term experiments. This chapter describes procedures for assembly of the modular system, including sample holders and LED arrays, as well as their operation, including characterization of light and environmental conditions.

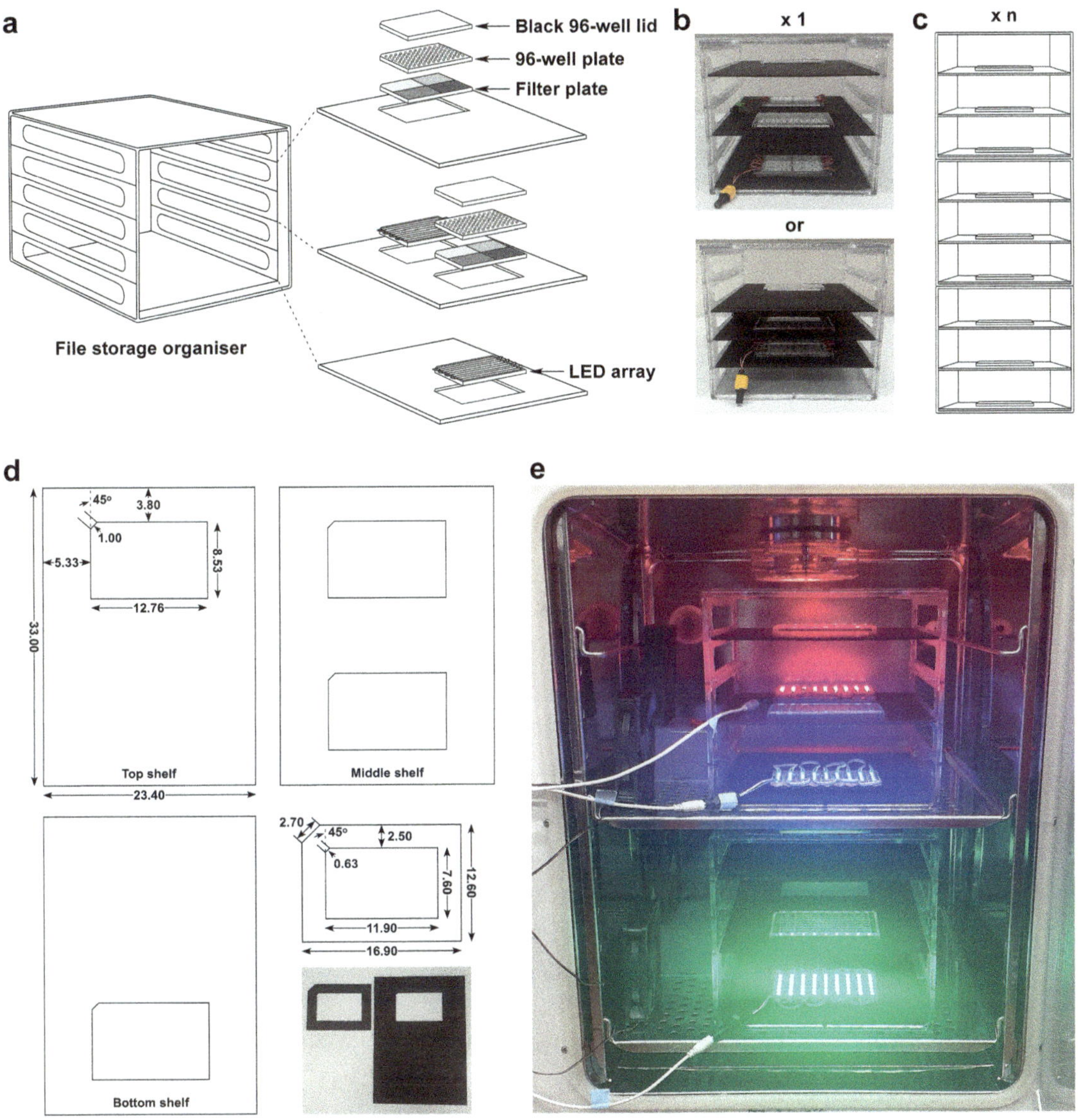

Fig. 1 *Overview of LED shelving system.* (**a**) Schematic of all main components and their assembly. (**b**) Front view of a single shelving unit containing three custom acrylic shelves inserted in different locations to alter the distance between sample and arrays. (**c**) Schematic front view of multiple shelving units vertically stacked for large-scale applications. (**d**) Dimensions (in cm) for custom shelving to hold 96-well plates. Three different shelves were made to hold two plates and two LED arrays (top and bottom left). A smaller piece (bottom right) was designed to be glued underneath the shelves to hold arrays and plates in place. (**e**) Example of light hardware in a standard tissue culture incubator installed with three different modular LED arrays. Two shelving units can be placed on separate incubator shelves (as shown here) or directly stacked on top of one another

2 Materials

All materials can readily be sourced from local hardware and stationary stores, or common online platforms such as eBay, Amazon, or AliExpress (unless otherwise indicated). Most components are shown in the figures for clarity.

2.1 Shelving Organizer and Custom Shelving

1. Desktop file storage organizer 5 drawer (clear) (e.g., SEDD105P, J. Burrows Officeworks) (*see* **Note 1**).
2. SOLIDWORKS software (*see* **Note 2**).
3. 3 mm thick acrylic sheets.
4. Laser cutting equipment or facility (*see* **Note 2**).
5. Black spray paint with primer suited for plastic (*see* **Note 3**).
6. 5 min epoxy adhesive.

2.2 LED Arrays

2.2.1 Module Version

This version uses array-style LEDs that come fabricated in 20 connected modules, each containing 6 LEDs, and fit perfectly onto 96-well plate lids.

1. 20-piece 6 LED module waterproof fairy strip lights (e.g., SMD5050, waterproof, DC 12 V).
2. Black electrical tape.
3. Manual wire cutter/stripper.
4. Female DC barrel jack adapters (*see* **Note 4**).
5. Small screwdriver set with Phillips and flat head drive tips.
6. Power supplies (AC 220 V to DC 12 V, 1 A, 12 W).
7. 96-well plate lids (*see* **Note 5**).

2.2.2 Custom Version

This version uses the same components listed for the module version, except for the type of LED and additional materials required for soldering steps.

1. LED strip lights (e.g., SMD2835 or SMD5050, 5 m length, waterproof, DC 12 V).
2. Single-use surgical blades.
3. Soldering iron or station.
4. 0.5 mm diameter lead-free solder.
5. 22 AWG 0.3 mm^2 2 pin extension cable wire.

2.3 Light Control and Characterization

1. LED single-color light strip manual dimmers with DC adapters.
2. TC420 programmable LED time controller (Spectral Designs).

3. Power supply (AC 220 V to DC 12 V, 10 A, 120 W).
4. 1 m extension cables (male to female) with DC power adaptor connectors (5.5 mm × 2.1 mm).
5. LI-180 Spectrometer (LI-COR Biosciences) (*see* **Note 6**).
6. Laser Power Meter LP-1 (Sanwa Electric Instrument).

2.4 Temperature Monitoring and Control

1. Multi-port USB hub (with at least 4 ports and powered through a single power supply).
2. Computer cooling fans (5 V USB powered, 120 × 120 × 25 mm).
3. Small waterproof digital thermometers (with wired probes, such as those designed for reptile enclosures or aquariums).

2.5 Filter Plates

1. Neutral density filters (LEE Filters) (*see* **Note 7**).
2. Clear 96-well plate lids.
3. Black adhesive tape.

2.6 Tissue Culture Plates

1. Black polystyrene 96-well plates with clear bottoms.
2. Black 96-well plate lids.
3. Clear 96-well plate lids.
4. Clear 35 mm round dishes.

3 Methods

3.1 Custom Shelving

1. Generate SOLIDWORKS files that contain the dimensions for custom shelving (e.g., for 96-well plates and supporting lids the required dimensions are outlined in Fig. 1d) (*see* **Notes 2** and **8**).
2. Laser cut acrylic sheets according to the SOLIDWORKS design. One of each shelf designs is required, along with a smaller shelf part (four in total) that allows for LED arrays, 96-well plates, and supporting lids to slot into place.
3. Spray paint the shelving black until opaque or use black acrylic sheets.
4. Secure smaller shelf parts to main shelves using epoxy glue so that they are ready for mounting of LEDs, lids, and plates (Fig. 1d, e).

3.2 LED Arrays

This subheading describes two different types of arrays that can be assembled. The first type termed "module version" uses waterproof LED modules that do not require soldering and can be assembled in minutes (Fig. 2a) (*see* **Note 9**). The second type termed "custom

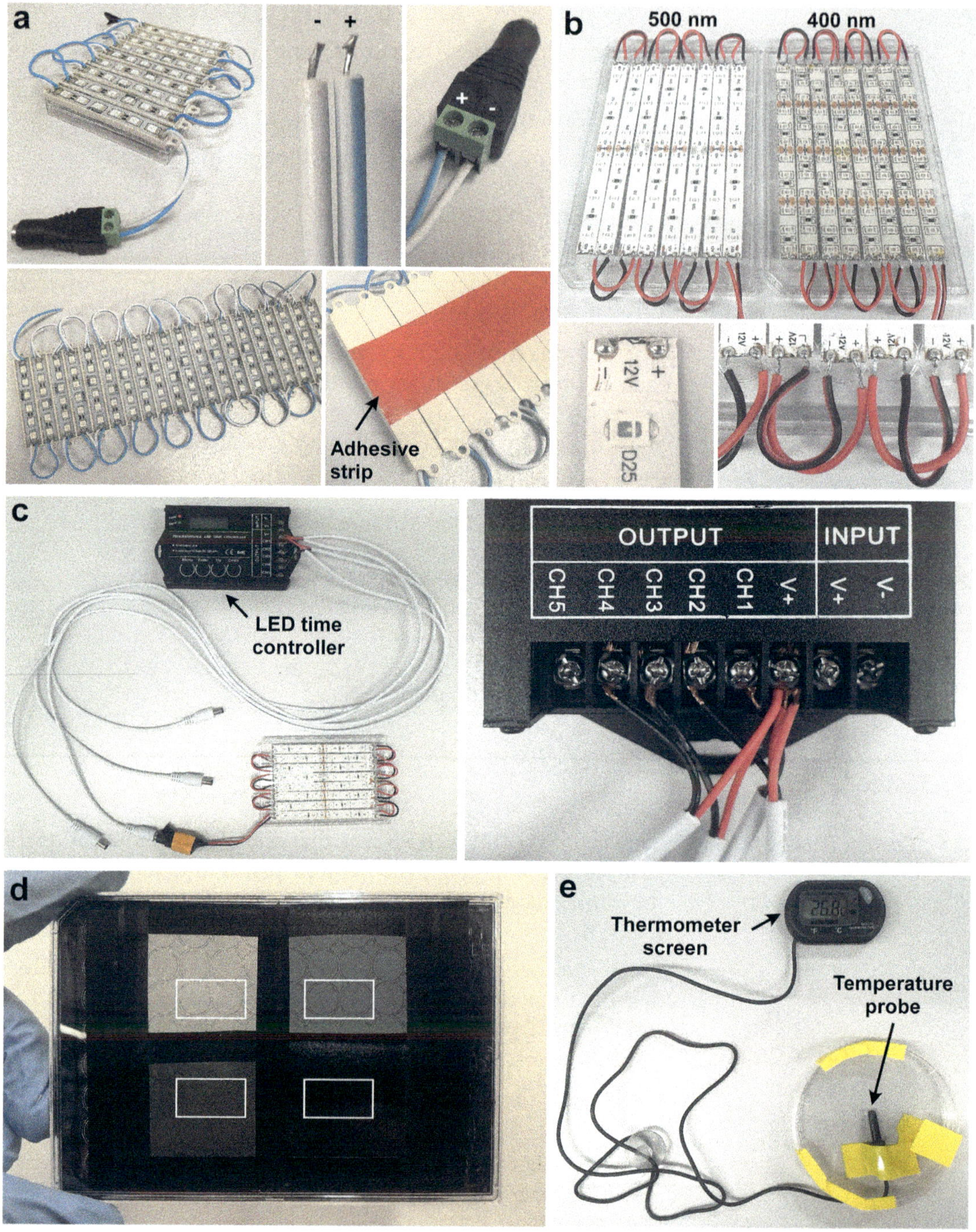

Fig. 2 *Assembly of LED array components*. (**a**) Example of a fully assembled 450 nm module version LED array (top left). LEDs are pre-fabricated in panels containing 20 modules, each consisting of six LEDs per module (bottom left). Anode (−) and cathode (+) wires from LEDs can be connected to a DC barrel jack adapter (top middle and right). Adhesive strip on the back of panels can be used to secure LEDs to a surface (bottom right) and a desired number of modules can be cut to suit culture plate size. (**b**) Examples of two custom version LED arrays (400 and 500 nm) fully assembled (top). LED strips can be cut into 10 cm pieces from a 5 m reel. The waterproof seal just after the last LED on the end of each strip is removed to expose

version" uses LED strips that come in a roll format and can be cut to a desired length (Fig. 2b). Assembly into an LED array follows a similar principle to that of the module LEDs, except that strips are connected by wire that is soldered between strips. This requires basic soldering skills and arrays can still be assembled in a few hours. The custom method is particularly useful when the desired wavelength is not available in the module LED form (e.g., cyan light of 500 nm wavelength). Both types are attached to well plate lids (or holders of comparable size) to illuminate samples on the shelf above them from below (as detailed in Subheading 3.5).

3.2.1 Module Version

1. Cut module LEDs such that nine LED modules can completely cover the top of a single 96-well plate lid (Fig. 2a). The adhesive cover on the back of each module can be removed to allow the modules to be secured to a desired surface.
2. Use a wire stripper to expose ~2–3 mm of wire from one end of the array and connect to a DC barrel jack adapter using a small screwdriver.
3. Power LED arrays and control their intensity by connecting an extension cable to the adapter along with the manual LED dimmer and a 12 V power supply (*see* **Note 10**).

3.2.2 Custom Version

1. Cut the LED strip into shorter ~11 cm pieces at the indicated cutting lines.
2. Scrape the waterproof seal on either ends of the strips using a surgical blade to expose the end copper anode (−) and cathode (+) points (*see* **Note 11**).
3. Apply solder to completely coat both points (*see* **Note 12**).
4. Cut 22 AWG 0.3 mm^2 extension wires into 5 cm long pieces and expose ~2–3 mm of wire on either end using wire strippers.
5. Coat exposed wire ends with solder before adhering to a cathode (+) point on the LED strip. This can be achieved by first heating the solder on the LED strip with the soldering iron tip and placing the wire end onto the melted solder and allowing it to cool and set (*see* **Note 13**).

Fig. 2 (continued) anode (−) and cathode (+) points before being coated with solder (bottom left). Ends of extension cable wires are also coated with solder before being used to connect multiple LED strips together (bottom right). (**c**) Pulsed light hardware assembly. LED arrays can be connected to a single LED time controller via modified 1 m extension cables (left). Close-up view of four extension cable connections to the LED time controller (right). (**d**) Example of a custom filter lid for multi-intensity experiments. The underside of a 96-well plate lid is lined with layers of neutral-density filter sheets to achieve four different intensity conditions per plate. Samples can be seeded in wells equivalent to positions outlined as white squares. This ensures equal distance from the edge of the plate, resulting in equal input intensity of light for sample wells. (**e**) Temperature measurements can be performed by securing the probe of digital thermometers to the center of 100 mm round dishes

6. Solder the opposite end of the wires onto another LED strip and repeat this process until a total of nine LED strips are connected.
7. Connect one of the end strips to an ~8 cm long extension wire and connect the wire to a DC barrel jack adapter using a small screwdriver.
8. Adhere the connected LED strips onto a 96-well plate lid in a similar fashion to the module version LEDs by removing the adhesive cover on the back of each strip (Fig. 2b) (*see* **Note 14**).
9. Power LEDs by connecting a manual LED dimmer, extension cable, and power supply.

3.3 Pulsed Light Hardware

This subheading describes the use of an LED time controller to achieve pulsed light at different intensities (Fig. 2c). The device does not need any coding or programming expertise and is provided with its own software. A range of different pulsing protocols can be applied by simple input of intensity and time values.

1. Cut and discard the female connector of each 1 m extension cable.
2. Use a wire stripper to remove external installation to expose internal anode (−) (red) and cathode (+) (black) wires. Then remove a further ~2–3 mm of insulation from the internal wires.
3. Attach anode wires of all extension cables onto the single V+ port on the "OUTPUT" section of the time controller using a screwdriver.
4. On the same side, secure each cathode wire to an individual channel (e.g., CH1) using a screwdriver (Fig. 2c).
5. Connect the LED time controller to a 12 V power supply via its DC socket.
6. Design a custom pulse program/mode using the PLed software interface that is provided with the device (*see* **Note 15**).
7. Load the custom program onto the time controller according to the manufacturer's instructions.
8. Install LED arrays into the shelving system and connect the time controller to each array via the extension cables (as detailed in Subheading 3.5).
9. Set the clock time on the time controller and run the appropriate custom mode/pulsing program.

3.4 Custom Filter Lids

This subheading describes custom filter lids that can provide different intensity conditions across a single 96-well sample plate (Fig. 2d). A similar approach can be applied to other plate formats. In this method, neutral density filters are applied to reduce the photon dose that reaches parts of the sample (*see* **Note 16**).

1. Cut neutral-density filter sheets into 4.5 × 3.6 cm pieces.
2. Line the underside of a clear 96-well plate lid with filters and secure with black tape (*see* **Note 17**).
3. Place filter lids on the shelf in the position above an LED array. 96-well plates containing samples are then placed on top of the filter lid. Ensure samples are placed at equal distance from the edges of the plate so that the input intensity is the same between conditions (Fig. 2d) (*see* **Note 18**).

3.5 Assembly of Shelving System

This subheading describes the assembly of two shelving systems in a standard tissue culture incubator, with the capacity to hold up to four LED arrays and four sample plates/dishes. Up to four manual dimmers can be installed or alternatively, a single LED time controller (*see* Subheading 3.3) that can apply not only different pulsing protocols per array but also different intensity settings.

1. Remove the drawers of two identical file storage organizers and replace with custom shelving (Fig. 1) (*see* **Note 19**).
2. For cell culture research, install both storage organizers in a tissue culture incubator with one incubator shelf between the organizers (*see* **Note 20**).
3. Place LED arrays on shelves directly below where samples will be located (Fig. 1a).
4. Place either a custom filter lid or clear 96-well plate lid in the shelf directly above each array.
5. Place tissue culture plate containing samples on top of the lid (Fig. 1a).
6. Connect each LED array to a 1 m extension cable followed by a manual dimmer and power supply. For pulsed light experiments, manual dimmers can be replaced with a single LED time controller.
7. Secure dimmers or the LED time controller to the outside of the incubator.
8. If temperature control is required, install fans directly up against the left of the organizers and connect each to a multiport USB hub (that can be secured to the outside of the incubator) (*see* **Note 21**).

3.6 Illumination of Samples

This subheading describes two examples to achieve ex vivo illumination of primary tissue (in 96-well plates) or immortalized cell lines (in either 96-well plates or 35 mm round dishes). A range of other plate types can be used and the hardware adapted depending on sample and illumination conditions. Note the 35 mm dish format is only applicable when all installed LED arrays are of the same wavelength.

3.6.1 96-well Plate Format

1. Culture cells in 96-well black clear bottom plates with opaque lids (*see* **Note 22**) as per your standard protocol. This type of plate allows for light to only enter from the bottom to directly illuminate samples in each well.
2. Place an LED array on a shelf below each 96-well plate.
3. Place either a filter lid or a clear 96-well plate lid into the holder.
4. Position the 96-well plate containing samples. A non-illuminated "dark control" plate can be placed on the top shelf away from LED arrays and wrapped in aluminum foil. In addition, a dark condition plate that is placed in the light (e.g., to test for any positional and temperature effects) can be included; in this case, utilize a black lid in place of a filter lid.

3.6.2 35 mm Dish Format

1. Culture cells in clear 35 mm dishes as per your standard protocol.
2. Place two dishes on top of clear 96-well plate lids in the holder.
3. Place an LED array on a shelf below the dishes.
4. Wrap a dark control dish in aluminum foil. One dish is placed in the light, and the other is placed in light but does not receive any light during illumination.

3.7 Temperature Measurements

This subheading describes a method to assess if a specific illumination condition produces temperature increases. This can determine if installation of fans and/or light pulsing is required (*see* **Note 23**).

1. Secure the probes of a digital thermometer inside the center of a 100 mm tissue culture dish (Fig. 2e).
2. Place dish in the same position as the samples (i.e., on top of a clear 96-well plate lid).
3. Secure the digital screen outside of the incubator.
4. Repeat **Steps 1–3** to install thermometers for each sample position in the shelving system.
5. Allow incubator to return to and stabilize at 37 °C before commencing temperature measurements.
6. Record a baseline measurement before illumination.
7. Turn on LED arrays and commence temperature measurements at timed intervals over the course of 3 h (*see* **Note 24**).

3.8 Measuring and Modifying Light Intensity

This subheading describes a method that enables intensity measurements across each LED array installed in the shelving system. It is recommended that measurements be performed prior to each experiment to maintain consistent illumination conditions (*see* **Note 25**).

1. Place the sensor probe of an optical power meter (e.g., Sanwa LP-1) onto a clear 96-well plate lid in an area corresponding to where samples would be positioned. If needed, secure the probe with tape; this will allow the probe and lid to be easily moved between arrays when obtaining readings.
2. Position the lid and probe on the shelf directly above an LED array in the incubator.
3. Adjust intensity of array via the manual LED dimmers until a desired reading is achieved.
4. Obtain final intensity by applying correction factors if required (according to manufacturer's instructions of the optical power meter) to obtain light intensity (in units of W/m^2) (*see* **Note 26**).

3.9 Wavelength Measurements

1. Wavelengths of the LEDs can be validated and characterized by using a portable spectrophotometer (e.g., LI-180) according to manufacturer's instructions.
2. Place the device directly above powered LED arrays connected to a power supply.
3. The spectrophotometer will report both wavelength characteristics and photon dose at the maximum wavelength (in units of W/m^2).

4 Notes

1. The desktop file storage organizer is also available in a black version (OWDD105PBK) which can be combined with the black shelving.
2. Engineering workshops in universities and research institutes, and student-focused workshops, may provide local support with shelf design as that is routine work for them. Workshops in engineering departments often offer laser cutting services and have ample experience with SOLIDWORKS. Laser cutting can also be performed by a local commercial provider.
3. If acrylic sheets are not completely opaque, black spray paint can be applied to absorb excess light and minimize light scattering between shelves.
4. The exact type of adapter can easily be changed if the power supplies and dimmer sourced are different from what is described here (e.g., slight changes in connector diameter are not uncommon).
5. Lids can be either clear or black. Black lids are recommended for multi-wavelength and intensity experiments as they prevent light from entering through the top of the plate.

6. Light characterization devices (light intensity and spectrum) are often available from common university or research institute fee-for-service facilities, such as advanced microscopy centers. These are not a part of the permanent setup but only occasionally accessed for calibration purposes.
7. A range of filters, also to change color or create "safe light" can be found at http://www.leefilters.com/lighting/technical-list.html. These can be sourced in a roll format that can be cut to the desired size. For neutral density filters, a range of different types are available with different transmission percentages (ODs).
8. Shelves can easily be customized to hold different plate formats. Shelving also can be readily exchanged depending on experimental requirements.
9. A range of other module lights (that are of different sizes and number of LEDs per module) can be sourced depending on plate type and size of array required.
10. In the designs described here, a total of four LED arrays can be used across two shelving units. Each array is powered by its own power supply and dimmer.
11. Perform this step with caution to prevent damage to LEDs or removal of copper. Ensure local guidelines are followed for work with and dispose of blades.
12. Soldering should be performed in a fume cupboard. This is to prevent exposure to fumes generated from the heating of the solder and the sash of the cupboard offers face and eye protection. Alternatively, soldering fans can be used together with safety goggles. Consult your local safety officer for approval of a safe setup before commencing.
13. Free online videos and tutorials are available that demonstrate this step if required.
14. LEDs can be adhered to either clear or black 96-well plate lids. Black is recommended when performing multi-wavelength and intensity experiments as this can further reduce light scattering.
15. The LED time controller runs on 24 h time and the clock can be set to the time of the day as required. Different operating modes containing a light protocol can be created in the PLed software. Transitions between different light conditions occur between each "step" that is created. Up to 50 steps can be created per mode. Pulses can be programmed to range between 1 and 59 mins per step. Intensity of each channel (i.e., LED array) can be adjusted by setting the percentage of light brightness. There is also the option for light to fade in and out or immediately jump to the next step that is programmed.

16. Light intensities can either be measured on the other side of the filter plate (in the sample position) or calculated using the known parameters of the filter (e.g., an OD = 1 filter corresponds to a transmission rate of 10%, with a ten-fold intensity loss for every further OD unit).
17. Neutral-density filter pieces can be layered to achieve the desired light intensity. In the shown example, four filter combinations were used to expose selected wells to four defined intensities. Along with securing the filters, apply black tape to the edges of the lid to minimize light scattering.
18. With any LED array intensity can be higher in the center compared to its edge. Thus, strategic placement of samples can be important to ensure each condition is receiving the same dose of light. Samples can be seeded in symmetrical wells across the plate to ensure exposure at equal intensities. Symmetrical wells also permit easy measurement and adjustment of intensity settings, as only one well per plate needs to be measured each time.
19. Shelving location can be modified to alter the distance between LED array and sample. This is a further method, in addition to the use of dimmers, to alter light intensity.
20. Organizers can also be scaled up, stacked, and/or installed in any other temperature-controlled environment.
21. Fans can be installed to minimize LED-induced heat build-up between shelves (as in Subheading 3.5). Two fans are installed per shelving unit inside the incubator, each to the left of an LED array. Install the fans such that they are each in line with each LED array. The fan closest to the back of the incubator may need to be elevated. This ensures the fan is at the same level as the array on the higher shelf.
22. Black clear bottom 96-well plates with opaque black lids are suitable for multi-wavelength and intensity experiments to prevent light scattering from other LED arrays in the shelving system (e.g., because of light reflected on the internal metal incubator surface). These also allow for illumination solely from the bottom of each well as the LED array is placed on the shelf below the sample plate. Illumination from the top of the plate would result in the need for light to travel past the black well walls to the sample resulting in a significant amount of light loss. Illumination from the bottom of the plate overcomes this and allows for a wider range of intensities to be tested and accurately quantified.
23. With any type of illumination hardware, light sources can generate some heat which has the potential to negatively impact sample function and viability. Inclusion of fans on the side of the shelving provides a method to effectively dissipate heat. Our temperature measurements demonstrated that

intensities at or below 500 μW/cm^2 are compatible with long-term continuous illumination experiments when fans are included. For higher intensities, reduction of the incubator temperature settings can be employed to offset the temperature increases caused by illumination devices. Alternatively, a pulsing/timer device can be used for contexts where continuous illumination is not required. The ON/OFF kinetics of the optogenetic tool (if being used) will also inform pulsing duration. In the case of short-term illumination in the scale of minutes, a wider range of intensities can be used. However, with any illumination protocol, temperature measurements should first be performed, and appropriate dark controls included (*see* Subheading 3.5).

24. In our cell culture incubators, 3 h of closed-door operation of the incubator during measurements are sufficient to allow temperature stability. Further temperature changes (if any) should not occur after this point.
25. In addition to setting different light intensities for each LED array via dimmers and/or using custom filter plate lids, the slot location of the shelving can be modified to change the distance between the LED arrays and sample plates.
26. Some optical power meters are provided with a conversion table that lists sensitivity coefficients for a range of different wavelengths (e.g., between 400 and 1100 nm). The correction factor is determined by the wavelength of the LED array and this coefficient.

References

1. Gutruf P, Rogers JA (2018) Implantable, wireless device platforms for neuroscience research. Curr Opin Neurobiol 50:42–49
2. Zhang F, Aravanis AM, Adamantidis A, de Lecea L, Deisseroth K (2007) Circuit-breakers: optical technologies for probing neural signals and systems. Nat Rev Neurosci 8(8):577–581
3. Bugaj LJ, Lim WA (2019) High-throughput multicolor optogenetics in microwell plates. Nat Protoc 14(7):2205–2228
4. Capek D, Smutny M, Tichy AM, Morri M, Janovjak H, Heisenberg CP (2019) Light-activated Frizzled7 reveals a permissive role of non-canonical wnt signaling in mesendoderm cell migration. elife 8(e42093):e42093
5. Hohener TC, Landolt AE, Dessauges C, Hinderling L, Gagliardi PA, Pertz O (2022) LITOS: a versatile LED illumination tool for optogenetic stimulation. Sci Rep 12(1):13139
6. Johnson HE, Goyal Y, Pannucci NL, Schupbach T, Shvartsman SY, Toettcher JE (2017) The spatiotemporal limits of developmental erk signaling. Dev Cell 40(2):185–192
7. Morri M, Sanchez-Romero I, Tichy AM, Kainrath S, Gerrard EJ, Hirschfeld PP, Schwarz J, Janovjak H (2018) Optical functionalization of human Class A orphan G-protein-coupled receptors. Nat Commun 9(1):1950
8. Agus V, Janovjak H (2017) Optogenetic methods in drug screening: technologies and applications. Curr Opin Biotechnol 48:8–14
9. Repina NA, McClave T, Johnson HJ, Bao X, Kane RS, Schaffer DV (2020) Engineered illumination devices for optogenetic control of cellular signaling dynamics. Cell Rep 31(10): 107737
10. GmbH OB (2023) optoWELL. https://optobiolabs.com/products/optowell/. Accessed 25 Feb 2023
11. Axion Biosystems I (2023) Lumos. https://www.axionbiosystems.com/products/mea/lumos. Accessed 25 Feb 2023

Chapter 18

Analysis of Light-Controlled Artificial Cell–Cell Adhesions

Seraphine V. Wegner and Christopher A. Raab

Abstract

The precise spatial and temporal regulation of cell–cell adhesions is crucial for understanding the underlying biological processes and for assembling multicellular structures in tissue engineering. Traditional approaches have relied on chemical membrane functionalization and regulated gene expression of native cell adhesion molecules (CAMs), but these methods lack the necessary control and can be detrimental to cells. In contrast, engineered photoswitchable cell–cell adhesions offer a reversible and dynamic regulation at a single-cell resolution. This is achieved by expressing different photodimerizers as artificial CAMs on the cell surfaces. Here, we describe a straightforward method for the functional analysis of these photoswitchable cell–cell adhesions in a 3D suspension culture.

Key words Photoswitchable, Cell–cell adhesions, Artificial cell adhesions, Reversibility, Extracellular optogenetics, 3D clustering, Bottom-up tissue engineering

1 Introduction

Cell–cell adhesions are fundamental in many biological processes, including tissue homeostasis, development, and collective cell behavior, and they are dynamically altered as part of wound healing, cell sorting, and cancer metastasis. In addition, in the context of bottom-up tissue engineering, the assembly of cell-rich multicellular structures that replicate the cellular organization in vivo from single cells as the building blocks is a promising approach [1]. For these purposes, various approaches have been developed to control cell–cell adhesions with artificial cell adhesion molecules (CAM) through the cell surface medication with chemicals and expression of CAMs. Chemical modifications include but are not limited to biotin-streptavidin [2], complementary DNA strands [3, 4], and clickable groups [5]. However, such chemical modifications are not sustainable over long periods of time and do not provide dynamic exchange between cells as observed in native CAMs [6]. The expression of native and artificial CAMs overcomes these problems but still provides no spatial and temporal control over them, which

Armin Baumschlager (ed.), *Optogenetics: Methods and Protocols*, Methods in Molecular Biology, vol. 2840,
https://doi.org/10.1007/978-1-0716-4047-0_18,

a

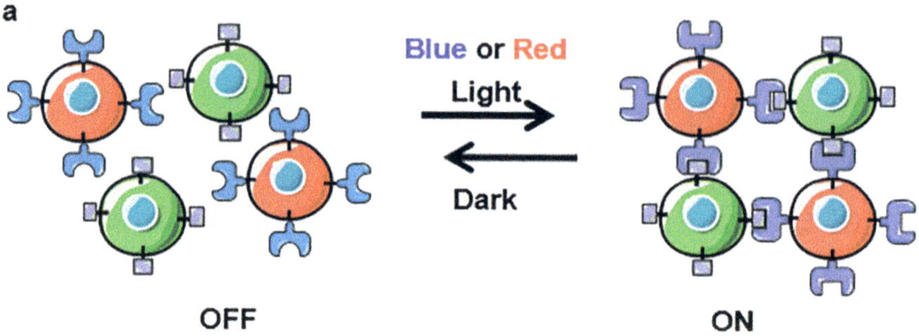

b

Cell A	Cell B	Interaction mode	Light spectrum
VVD	VVD	Homodimerization	Blue
Cph1	Cph1	Homodimerization	Red
CarH	CarH	Homotetramerization	Green
CRY2	CIBN	Heterodimerization	Blue
iLID	Nano	Heterodimerization	Blue
pMag	nMag	Heterodimerization	Blue

Fig. 1 Designing photoswitchable cell–cell adhesions. Panel (**a**) shows the principal of how cell adhesions between two populations are facilitated by light. In this example, two cell populations express one interaction partner of a heterodimer on their cell surface. Light activation leads to formation of multicellular assemblies between the two cell populations, while light removal will revert the process. Panel (**b**) shows a selection of photoswitchable proteins, their mode of interaction and the light spectrum, which induces change in formation that have been expressed on the cell surface to create artificial adhesions between cells

is required for investigating the underlying biology and producing multicellular architectures with precise microstructures. To address this issue, light-responsive small molecules [7] and proteins have been introduced on the cell membrane as artificial CAMs. In particular, CAMs based on photoswitchable proteins provide reversible and noninvasive control with visible light with high precision (Fig. 1a) [8–12]. Major advantages of the photoswitchable cell–cell adhesion include the fast activation or deactivation of adhesions with spatial control at single-cell resolutions, reversibility in the dark, the possibility of modulating adhesions through light intensity and pulsed light illumination, and lastly, the orthogonal regulation of different cell types using blue and red light, which were not achievable with previous approaches.

When considering the design of a bottom-up approach to tissue engineering, where individual cells or cell clusters are utilized as modular building blocks to create complex structures, it is crucial to understand how these building blocks interact with each other. At a basic level, Steinberg's Differential Adhesion Hypothesis proposes that cell sorting occurs due to variations in cell–cell adhesivity, assuming that the connections are dynamic [13, 14]. The assembly and self-sorting of single cells into multicellular structures rely on the effective binding strength between CAMs of adjacent cells, which is determined by the type and expression level of the CAMs [15]. Similarly, differences in CAMs at the cell surface also are one of the major driving forces for morphogenesis during organogenesis and the sorting out of cells [16–18].

In the design of photoswitchable cell-cell adhesions, we have developed a general approach where various proteins are expressed on the extracellular side of plasma membrane using the Display plasmid system (Fig. 1a). In particular, the plasmid contains an N-terminal secretion signal and a C-terminal transmembrane anchoring domain of platelet-derived growth factor receptor. This approach has been successful with a variety of photoswitchable protein–protein interactions, which form or disassemble depending on the light illumination. Moreover, the photoswitchable protein-protein interactions have different properties in terms of the wavelength of light they respond to (blue, green, or red), the type of cell–cell interaction they mediate (homophilic vs. heterophilic) and the dark reversion rates (Fig. 1b) [19–22]. Together the identity of the photoswitchable protein and the expression levels at the cell surface determine the strength of the cell–cell adhesions and the fold-change between activated and inactivated states.

Depending on the research question at hand, the preferred choice of photoswitchable protein–protein pair may vary. Most of the established photoswitchable cell–cell adhesions form upon illumination and are reversible in the dark except for CarH-mediated adhesions, which form in the dark and can irreversibly be broken upon green light illumination [12]. Other options exist to bring cells of the same type together; with blue light by expressing VVD and with red light expressing Cph1 as artificial adhesion receptors on the cell surface [11]. These blue and red light-triggered cell–cell adhesions can even be employed in the same system, which allows to orthogonally address different cell populations. In the case where different cell types should be brought together, different blue light-triggered cell–cell adhesion with proteins that have varying dark reversion rates and different strengths are available (e.g., iLID/Nano pair mediates cell–cell adhesion that reverses faster than nMagHigh/pMagHigh mediated adhesions). In addition, the use of pulsed light illumination alters the adhesion dynamics systematically, rendering the formed multicellular aggregates rounder and more compact.

Lastly, regardless of the specific light-switchable system employed, varying expression levels of the artificial photoswitchable adhesion receptors will yield different outcomes regarding self-assembly kinetics, cluster size and morphology, self-sorting, and background level cell–cell adhesion. Therefore, it is crucial to establish an easily implementable experimental setup for the functional analysis of cells expressing light-switchable receptors. 3D cell

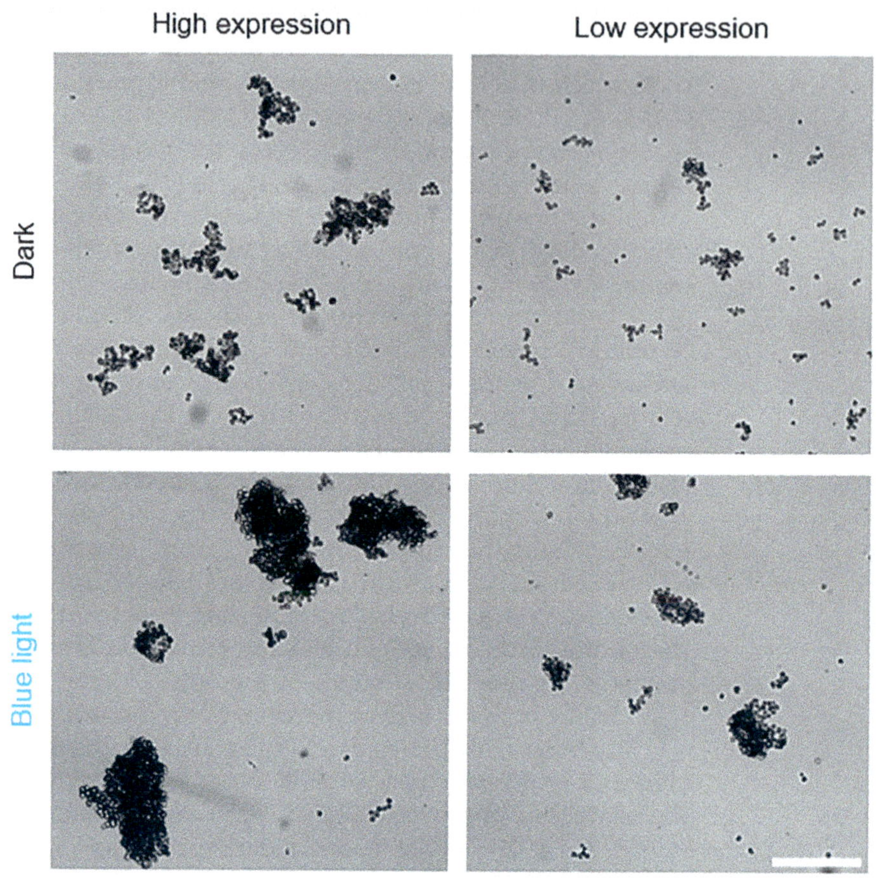

Fig. 2 3D cell clustering. Cell clones that express the same light-mediated adhesion system but have different expression levels of the protein of interest. Low expressing clones will have minimal background activity in the dark, while high expressing clones form large and compact clusters when activated with light. Here, MDA-MB-231 cells express the photoswitchable homodimer VVD. Under blue light, VVD undergoes conformational changes, which increase binding affinity. Relative expression levels were determined by detection of a tagged fluorescent protein by flow cytometry. The scalebar is 300 μm

clustering provides a rapid and cost-efficient method to screen multiple monoclonal cell lines expressing various photoswitchable proteins that yield results within hours. We have employed this approach to evaluate the fold-change between ON/OFF states (Fig. 2), bi-directionality, cycling conditions, light protocols to maximize cell–cell adhesion, and the sorting of mixed heterogeneous populations [10, 11].

Here, we describe how to perform 3D clustering experiments to validate the functionality of artificial CAMs with high reproducibility as well as the subsequent analysis and interpretation of the results (*see* **Notes 1** and **10**). Furthermore, we shed light on the caveats and pitfalls regarding this protocol, which makes this a fast and easy-to-use tool.

2 Materials

MDA-MB-231 cells were chosen due to low levels of native cell–cell adhesion molecules. Other cell types with low native cell–cell adhesions may be used.

2.1 Cell Culture Media

1. Dulbecco's Modified Eagle Medium (DMEM); 10% fetal bovine serum (FBS), 1% penicillin-streptomycin (1000 U/mL), 12.5 mM N-2-hydroxyethylpiperazine-N-2-ethane sulfonic acid (HEPES).

2.2 3D Clustering of Cells

1. DMEM.
2. Accutase.
3. mL LoBind tubes.
4. 3D Orbital Shaker.
5. 4% Paraformaldehyde.
6. 12-well standard culture plate.
7. Brightfield microscope.
8. ImageJ.

3 Methods

3.1 3D Clustering

1. Remove cell culture medium from adherent cells in a T25 flask and wash twice with 1 mL PBS (*see* **Notes 2** and **3**).
2. Add 500 μL of Accutase and place the flask into the incubator at 37 °C and 5% CO_2 to lift the cells.
3. Add 5 mL of supplemented DMEM media into the flask, suspend cells, and transfer into a 15 mL Falcon tube.
4. Centrifuge the cells at 600 *g* for 5 min.
5. Remove the supernatant and resuspend the cell pellet in 3 mL supplemented DMEM media gently with a 1 mL Pipette.
6. Determine the cell count and verify if cells are dispersed as single cells in suspension. Otherwise, repeat **Step 5**.
7. Adjust the cell density in suspension to 5×10^4 cells/mL and add the appropriate cofactor for the photoswitchable proteins (*see* **Note 4**) (0.5 μM FAD for VVD, Cry2/CIBN, pMag/nMag; 0.5 μM FMN for iLID/Nano; 10 μM AdoB12 for CarH; 5 μM PCB for Cph1).
8. Aliquot 1 mL cell suspension into 1.5 mL LoBind microfuge tubes (Eppendorf). Prepare duplicates for each condition, for example, four aliquots if testing dark and light responses. In heterodimerization systems, use 0.5 mL of each population keeping the total number of cells the same.

9. Place the tubes laterally onto a 3D orbital shaker with a 5° angle at 30 rpm for 1–6 h depending on the experiment—typically 2 h. Either keep the cells under approximately 270 μW/cm^2 light (light sample) or protect the sample from light with aluminum foil (dark sample) (*see* **Notes 5–7**).
10. Carefully transfer the cell suspension from the LoBind tube into 12-well cell culture dishes preloaded with 500 μL 4% PFA. Use a 1000 μL pipette with a cut tip to avoid disrupting cell clusters (*see* **Note 8**).
11. Acquire bright field images of the entire well using a microscope with a scanning platform (*see* **Note 9**). Stitch the images together to create a merged image of the entire well. For this purpose, low magnification objectives (4x objective) with large fields of view are recommended to decrease the number of imaged tiles while still retaining the sharp edges of the cell clusters.

3.2 Image J Analysis

ImageJ is an open-source Java-based image-processing program used by a large portion of the scientific community, which provides a wide array of plugin options that can be easily installed

1. Analyse the cell clusters in ImageJ with the particle analysis tool in which objects with a projected area of >5000 μm^2 (>ca. 20 cells) are considered to be a cluster (Fig. 3). The area for a single cell described here is 200–250 μm^2. Depending on the cell type used in the experiment, the analysis parameters may need adjustment. Below are useful ImageJ commands that can be used to write a macro for batch processing of many samples at once. The command language is IJ1Macro.

3.2.1 ImageJ Commands

1. run("8-bit"); If the acquired bright field images are in 16-bit, this command will help reduce the computational workload. As merged images are typically big in size, this is especially helpful during batch processing.
2. run("Sharpen"); This increases the contrast between background and objects. It may accentuate noise.
3. run("Find Edges"); This command will highlight sharp changes in intensity found in the image (Foreground vs. Background),
4. run("Threshold..."); Thresholding is required for segmentation. Adjust the threshold such that the entirety of a cluster is within range and keep the settings identical in-between experimental replicates.
5. run("Set Scale...", "distance=... known=... unit=μm"); This is necessary if the raw images do not contain the information on pixel size or, if it is in the wrong unit and it needs to be set manually.

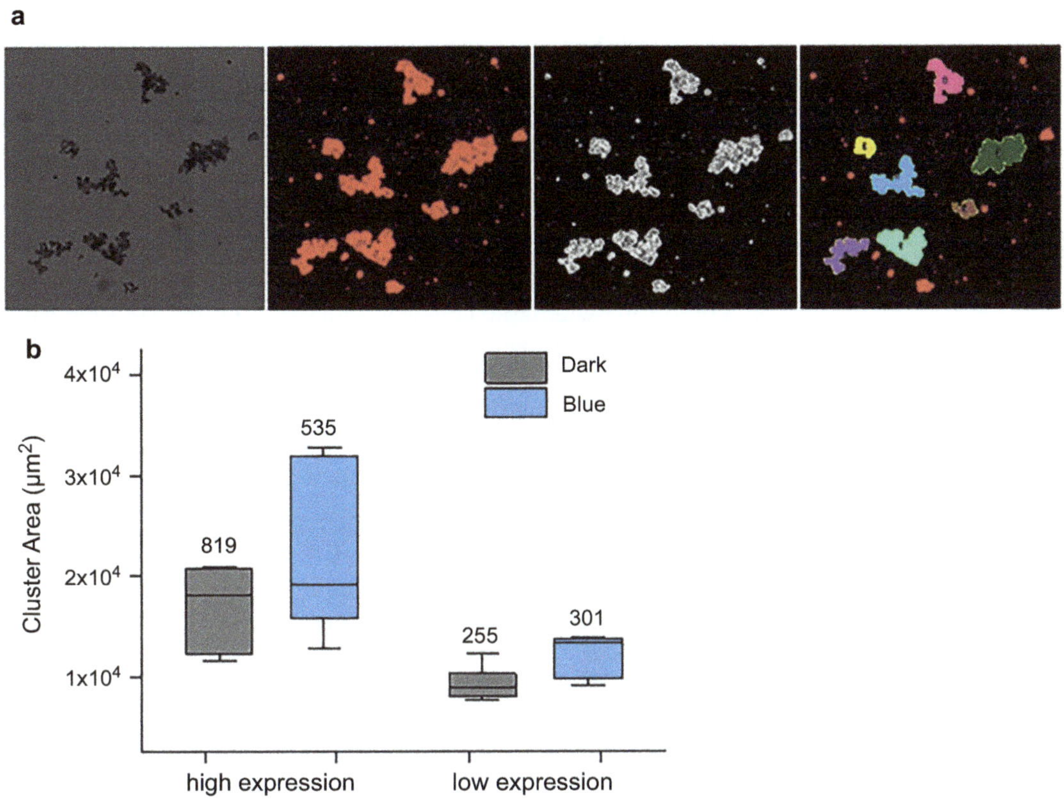

Fig. 3 Analysis of cell clusters formed in suspension. Panel (**a** from l-r) shows a bright field image of cell clusters of varying sizes, after edge detection with, after threshold adjust and after particle analysis detection with ImageJ. Note that only cell clusters with a projected area of at least 5000 > μm^2 (20 cells) are considered for quantification as indicated by the coloured regions of interest. Panel (**b**) shows the quantification of cell clusters of a high and low expressing clone (2 technical- and 3 biological replicates). The numbers above the box plots indicate the number of clusters detected in average per technical replicate

6. run("Analyze Particles...", "size=5000-Infinity display exclude clear add"); With the particle analysis tool, every object in the image that is within the range of the thresholded intensity value and >5000 μm^2 will be identified as a unique object and added to the ROI Manager. The results on morphology (area size, roundness, circularity, mean gray value) will appear in another window, which can be either exported or copy-pasted into the preferred data analysis program.
7. waitForUser("[Enter text here] and press OK."); This command will stop any macro and open a dialog box with a pre-written note (e.g., Check Cluster Quality) and an OK button. Only upon confirmation will the macro resume its function. This is helpful when manual handling of the image is still necessary during batch processing to ensure correct cluster analysis. It is possible that the particle analysis tool recognizes

artifacts from incorrect tile stitching or dirt as cell clusters, which will have an impact on cluster size analysis. As a quality control step stop the macro with this command prior to particle analysis and use the paintbrush tool to remove such obstacles.

4 Notes

1. The photoswitchable cell–cell adhesions, clone-to-clone variations, and optimized light protocols can also be established in 2D cell culture clustering but the analysis of reversion is more straightforward with the 3D clustering analysis. Moreover, photoswitchable cell–cell adhesions have also been used in wound healing assays, invasion assays, and spheroids.
2. We find that prior to the experiment, samples should be seeded at low confluence, typically 20–30%, in a T25 flask and grown until 70–80% on the day of the experiment. Confluent cell layers required longer detachment times with accutase, with a higher risk of CAM degradation.
3. On the day of the experiment handle the cells and conduct the experiment under light that doesn't lead to premature activation of the photoswitchable proteins. If the cell adhesions are activated by blue light, the cell culture work should be performed under red light and vice versa.
4. When working with a heterodimerization system, it is possible to employ live cell plasma membrane stains of different color prior to mixing (25,000 cells of both cell populations are mixed per tube).
5. It is important to use media with 12.5–25 mM HEPES final concentration to provide extra buffering capacity for prolonged experiments outside of a CO_2 incubator.
6. The samples within the LoBind tubes are incubated on top of the orbital shaker, ensuring that the cells in suspension move and don't settle down.
7. Depending on the light-switchable system lower or higher rpm on the orbital shaker may provide better results.
8. It is not recommended to increase the cell number or decrease the well size for imaging. Both will lead to crowding of the clusters and make image analysis more laborious or impossible.
9. Taking note of the shape and the number of clusters analyzed of each sample is just as important as the average cluster size because analysis of the average cluster size only considers the projected area from a top-down view. A sample with strong adhesions will lead to compact clusters, while loose clusters may cover a larger area and thus, falsely seem to be better.

10. Membrane-bound light-switchable proteins have decreased changes in light-dependent affinity and slower versions in the dark [23, 24] when compared to studies performed in vitro. Optimization of the membrane tether has been shown to improve spatial control and dimerization affinities and should be considered when troubleshooting [25, 26].

References

1. Schmidt T, Xiang Y, Bao X et al (2021) A paradigm shift in tissue engineering: from a top–down to a bottom–up strategy. PRO 9: 935. https://doi.org/10.3390/pr9060935
2. Sarkar D, Spencer JA, Phillips JA et al (2011) Engineered cell homing. Blood 118:e184–e191. https://doi.org/10.1182/blood-2010-10-311464
3. Selden NS, Todhunter ME, Jee NY et al (2012) Chemically programmed cell adhesion with membrane-anchored oligonucleotides. J Am Chem Soc 134:765–768. https://doi.org/10.1021/ja2080949
4. Shi P, Zhao N, Lai J et al (2018) Polyvalent display of biomolecules on the live cells. Angew Chem Int Ed Engl 57:6800–6804. https://doi.org/10.1002/anie.201712596
5. Koo H, Choi M, Kim E et al (2015) Bioorthogonal click chemistry-based synthetic cell glue. Small 11:6458–6466. https://doi.org/10.1002/smll.201502972
6. Vicente-Manzanares M, Horwitz AR (2011) Adhesion dynamics at a glance. J Cell Sci 124: 3923–3927. https://doi.org/10.1242/jcs.095653
7. Luo W, Pulsipher A, Dutta D et al (2014) Remote control of tissue interactions via engineered photo-switchable cell surfaces. Sci Rep 4:6313. https://doi.org/10.1038/srep06313
8. Yüz SG, Ricken J, Wegner SV (2018) Independent control over multiple cell types in space and time using orthogonal blue and red light switchable cell interactions. Adv Sci (Weinh) 5: 1800446. https://doi.org/10.1002/advs.201800446
9. Yüz SG, Rasoulinejad S, Mueller M et al (2019) Blue light switchable cell-cell interactions provide reversible and spatiotemporal control towards bottom-up tissue engineering. Adv Biosyst 3:e1800310. https://doi.org/10.1002/adbi.201800310
10. Mueller M, Rasoulinejad S, Garg S et al (2020) The importance of cell-cell interaction dynamics in bottom-up tissue engineering: concepts of colloidal self-assembly in the fabrication of multicellular architectures. Nano Lett 20: 2257–2263. https://doi.org/10.1021/acs.nanolett.9b04160
11. Rasoulinejad S, Mueller M, Nzigou Mombo B et al (2020) Orthogonal blue and red light controlled cell-cell adhesions enable sorting-out in multicellular structures. ACS Synth Biol 9:2076–2086. https://doi.org/10.1021/acssynbio.0c00150
12. Nzigou Mombo B, Bijonowski BM, Rasoulinejad S et al (2021) Spatiotemporal control over multicellular migration using green light reversible cell-cell interactions. Adv Biol (Weinh) 5:e2000199. https://doi.org/10.1002/adbi.202000199
13. Steinberg MS (1963) Reconstruction of tissues by dissociated cells. Some morphogenetic tissue movements and the sorting out of embryonic cells may have a common explanation. Science 141:401–408. https://doi.org/10.1126/science.141.3579.401
14. Foty RA, Steinberg MS (2005) The differential adhesion hypothesis: a direct evaluation. Dev Biol 278:255–263. https://doi.org/10.1016/j.ydbio.2004.11.012
15. Friedlander DR, Mège RM, Cunningham BA et al (1989) Cell sorting-out is modulated by both the specificity and amount of different cell adhesion molecules (CAMs) expressed on cell surfaces. Proc Natl Acad Sci USA 86:7043–7047
16. Laukaitis CM, Webb DJ, Donais K et al (2001) Differential dynamics of alpha 5 integrin, paxillin, and alpha-actinin during formation and disassembly of adhesions in migrating cells. J Cell Biol 153:1427–1440. https://doi.org/10.1083/jcb.153.7.1427
17. Levayer R, Lecuit T (2013) Oscillation and polarity of E-cadherin asymmetries control actomyosin flow patterns during morphogenesis. Dev Cell 26:162–175. https://doi.org/10.1016/j.devcel.2013.06.020
18. Fagotto F, Rohani N, Touret A-S et al (2013) A molecular base for cell sorting at embryonic boundaries: contact inhibition of cadherin adhesion by ephrin/ Eph-dependent

contractility. Dev Cell 27:72–87. https://doi.org/10.1016/j.devcel.2013.09.004

19. Kawano F, Suzuki H, Furuya A et al (2015) Engineered pairs of distinct photoswitches for optogenetic control of cellular proteins. Nat Commun 6:6256. https://doi.org/10.1038/ncomms7256
20. Schwerdtfeger C, Linden H (2003) VIVID is a flavoprotein and serves as a fungal blue light photoreceptor for photoadaptation. EMBO J 22:4846–4855. https://doi.org/10.1093/emboj/cdg451
21. Ortiz-Guerrero JM, Polanco MC, Murillo FJ et al (2011) Light-dependent gene regulation by a coenzyme B12-based photoreceptor. PNAS 108:7565–7570. https://doi.org/10.1073/pnas.1018972108
22. Levskaya A, Chevalier AA, Tabor JJ et al (2005) Synthetic biology: engineering *Escherichia coli* to see light. Nature 438:441–442. https://doi.org/10.1038/nature04405
23. Bartelt SM, Chervyachkova E, Steinkühler J et al (2018) Dynamic blue light-switchable protein patterns on giant unilamellar vesicles. Chem Commun (Camb) 54:948–951. https://doi.org/10.1039/c7cc08758f
24. Hallett RA, Zimmerman SP, Yumerefendi H et al (2016) Correlating in vitro and in vivo activities of light-inducible dimers: a cellular optogenetics guide. ACS Synth Biol 5:53–64. https://doi.org/10.1021/acssynbio.5b00119
25. Natwick DE, Collins SR (2021) Optimized iLID membrane anchors for local optogenetic protein recruitment. ACS Synth Biol 10:1009–1023. https://doi.org/10.1021/acssynbio.0c00511
26. Di Iorio D, Bergmann J, Higashi SL et al (2023) A disordered tether to iLID improves photoswitchable protein patterning on model membranes. Chem Commun (Camb) 59:4380–4383. https://doi.org/10.1039/d3cc00709j

Index

Armin Baumschlager (ed.), *Optogenetics: Methods and Protocols*, Methods in Molecular Biology, vol. 2840,
https://doi.org/10.1007/978-1-0716-4047-0,

GPSR Compliance
The European Union's (EU) General Product Safety Regulation (GPSR) is a set of rules that requires consumer products to be safe and our obligations to ensure this.

If you have any concerns about our products, you can contact us on

ProductSafety@springernature.com

In case Publisher is established outside the EU, the EU authorized representative is:

Springer Nature Customer Service Center GmbH
Europaplatz 3
69115 Heidelberg, Germany

www.ingramcontent.com/pod-product-compliance
Ingram Content Group UK Ltd.
Pitfield, Milton Keynes, MK11 3LW, UK
UKHW052004051026
15624UKWH00003B/563
9781071640463